Studies in Logic

Volume 108

Semitopology

Decentralised collaborative action
via topology, algebra, and logic

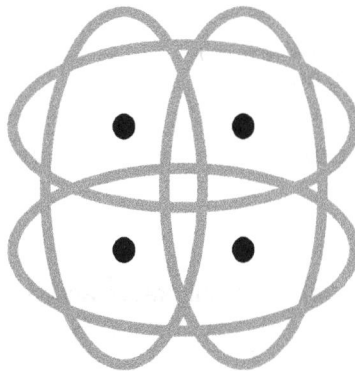

Volume 101
The Logic of Partitions. With Two Major Applications
David Ellerman

Volume 102
Bounded Reasoning Volume 1: Classical Propositional Logic
Marcello D'Agostino, Dov Gabbay, Costanza Larese, Sanjay Modgil

Volume 103
The Fertile Debate. Affective Exploration of a Controversy
Claire Polo

Volume 104
Argument, Sex and Logic
Dov Gabbay, Gadi Rozenberg and Lydia Rivlin

Volume 105
Logic as a Tool. A Guide to Formal Logical Reasoning
Valentin Goranko

Volume 106
New Directions in Term Logic
George Englebretsen, ed

Volume 107
Non-commutative Algebras. Pseudo-BCK Algebreas versus m-pseudo-BCK Algebras
Afrodita Iorgulescu

Volume 108
Semitopology: decentralised collaborative action via topology, algebra, and logic
Murdoch J. Gabbay

Studies in Logic Series Editor
Dov Gabbay dov.gabbay@kcl.ac.uk

Semitopology

Decentralised collaborative action via topology, algebra, and logic

Murdoch J. Gabbay

ISBN 978-1-84890-465-1

College Publications
Scientific Director: Dov Gabbay
Managing Director: Jane Spurr

http://www.collegepublications.co.uk

Contents

Contents v

1 Introduction 1
1.1 What is 'decentralised collaborative action'? 1
1.2 What is a semitopology? 5
1.3 Who should read this document? 7
1.4 Why did I write it? 7
1.5 Map of the work 8

I Point-set semitopologies 13

2 Semitopology 15
2.1 Definitions, examples, and some discussion 15
2.1.1 Definitions 15
2.1.2 Examples 16
2.1.3 Why the name 'semitopologies', and other discussion 19
2.2 Continuity, and its interpretation 21
2.3 Neighbourhoods of a point 23

3 Transitive sets & topens 25
3.1 Some background on sets intersection 25
3.2 Transitive open sets and value assignments 26
3.3 Examples and discussion of transitive sets and topens 29
3.4 Closure properties of transitive sets 30
3.5 Closure properties of topens 32
3.6 Intertwined points 33
3.6.1 The basic definition, and some lemmas 33
3.6.2 Pointwise characterisation of transitive sets 35
3.7 Strong topens: topens that are also subspaces 36
3.7.1 Definition and main result 36

3.7.2 Connection to lattice theory . . . 38
3.7.3 Topens in topologies . . . 40

4 Interiors, communities & regular points **43**
4.1 Community of a (regular) point . . . 43
4.2 Further exploration of (quasi-/weak) regularity and topen sets . . . 46
4.3 Intersection and partition properties of regular spaces . . . 50
4.4 Examples of communities and (ir)regular points . . . 51

5 Closed sets **55**
5.1 Closed sets . . . 55
5.2 Duality between closure and interior . . . 58
5.3 Transitivity and closure . . . 59
5.4 Closed neighbourhoods and intertwined points . . . 60
5.4.1 Definition and basic properties . . . 60
5.4.2 Application to characterise (quasi/weak) regularity . . . 62
5.5 Intersections of communities with open sets . . . 64
5.6 Regularity, maximal topens, & minimal closed neighbourhoods . . . 65
5.7 More on minimal closed neighbourhoods . . . 69
5.7.1 Regular open/closed sets . . . 69
5.7.2 Intersections of regular open sets . . . 71
5.7.3 Minimal nonempty regular closed sets are precisely the minimal closed neighbourhoods . . . 72
5.8 How are $p_{\between}$ and $|p|$ related? . . . 73

6 (Un)conflicted points: transitivity of $\between$ **75**
6.1 The basic definition . . . 75
6.2 Regular = weakly regular + unconflicted . . . 77
6.3 The boundary of $p_{\between}$ . . . 78
6.4 The intertwined preorder . . . 80
6.4.1 Definition and properties . . . 80
6.4.2 Application to quasiregular conflicted spaces . . . 82
6.4.3 (Un)conflicted points and boundaries of closed sets . . . 84
6.5 Regular = quasiregular + hypertransitive . . . 86
6.5.1 Hypertransitivity . . . 86
6.5.2 The equivalence . . . 87

7 The product semitopology **91**
7.1 Basic definitions and results (shared with topologies) . . . 91
7.2 Componentwise composition of semitopological properties . . . 92
7.3 Minimal closed neighbourhoods, and a counterexample . . . 95

8 The witnesses semitopology **99**
8.1 Discussion . . . 99
8.2 The witness function and semitopology . . . 100
8.3 Examples . . . 103

8.4 Computing open and closed sets in witness semitopologies 104
8.4.1 Computing open sets: X is open when $X \prec X$ 104
8.4.2 Computing closed sets using limit points: $|P| = lim(P)$ 106
8.5 Declarative content of witness semitopologies 108
8.5.1 Witnessed sets and Horn clause theories 108
8.5.2 Witnessed sets and topologies 110

9 (Strongly) chain-complete semitopologies 113
9.1 Definition and discussion 113
9.2 Elementary properties of the definition 115
9.3 Consequences of being strongly chain-complete 115
9.3.1 Strongly chain-complete implies $\between$-complete 116
9.3.2 Indirectly regular points: inherent properties 116
9.3.3 Indirectly regular points in the context of other regularity properties . 117
9.4 Witness semitopologies are chain-complete 119
9.5 Minimal sets: open covers and atoms 122
9.5.1 Open covers (minimal open neighbourhoods) 122
9.5.2 Atoms (minimal nonempty open sets) 124
9.5.3 Discussion 125

10 Kernels: the atoms in a community 127
10.1 Definition and examples 127
10.2 Characterisations of the kernel 130
10.3 Further properties of kernels 133
10.3.1 Intersections between the kernel of p and its open neighbourhoods . . 133
10.3.2 Idempotence properties of the kernel and community 135

11 Dense subsets & continuous extensions 139
11.1 Definition and basic properties 139
11.2 Dense subsets of topen sets 141
11.3 Explaining kernels 142
11.4 Unifying is-transitive and is-strongly-dense-in 143
11.5 Towards a continuous extension result 145
11.6 Kernels determine values of continuous extensions 149

II Semiframes: algebra and duality 153

12 Semiframes: compatible complete semilattices 155
12.1 Complete join-semilattices, and morphisms between them 155
12.2 The compatibility relation 157
12.3 The definition of a semiframe 159

13 Semifilters & abstract points 163
13.1 The basic definition, and discussion 163
13.2 Properties of semifilters 165

13.2.1 Things that are familiar from filters 165
13.2.2 Things that are different from filters 167
13.3 Sets of abstract points 169
13.4 The semitopology of abstract points 170

14 Spatial semiframes & sober semitopologies 173
14.1 Definition of spatial semiframes 173
14.2 The neighbourhood semifilter $nbhd(p)$ 175
14.2.1 The definition and basic lemma 175
14.2.2 Application to semiframes of open sets 176
14.2.3 Application to characterise T_0 spaces 176
14.3 Sober semitopologies 177
14.3.1 The definition and a key result 177
14.3.2 Sober topologies contrasted with sober semitopologies 179

15 Four categories & functors between them 183
15.1 The categories Sober/SemiTop of (sober) semitopologies 183
15.2 The categories Spatial/SemiFrame of (spatial) semiframes 185
15.3 Functoriality of the maps 185
15.4 Sober semitopologies are dual to spatial semiframes 188

16 Well-behavedness conditions, dually 191
16.1 (Maximal) semifilters and transitive elements 191
16.2 The compatibility system x^* 193
16.3 The compatibility system F^* 194
16.3.1 Basic definitions and results 194
16.3.2 Strong compatibility: when F^* is a semifilter 195
16.4 Semiframe characterisation of community 197
16.5 Semiframe characterisation of regularity 199
16.6 Semiframe characterisation of (quasi/weak)regularity 200
16.7 Characterisation of being intertwined 201
16.8 Strong compatibility in semitopologies 203

17 Graph representation of semitopologies 207
17.1 From a semitopology to its intersection graph 207
17.1.1 The basic definition 207
17.1.2 The node preorder $\leq$ 208
17.1.3 Transitive elements 209
17.2 From a semiframe to its subintersection graph 211
17.2.1 The subintersection relation $\ltimes$ 212
17.2.2 Recovering $\leq$ and $*$ from $\ltimes$ 213

III Logic and computation 217

18 Three-valued logic 219

18.1 Three-valued logic, valuations, and continuity . . . 219
18.1.1 The semitopology **3** . . . 220
18.1.2 Indicator functions and characteristic sets . . . 221
18.2 Three-valued truth-tables . . . 222
18.2.1 Truth-tables of connectives . . . 224
18.2.2 Conjunction and disjunction . . . 225
18.2.3 Implication(s) . . . 225
18.3 Predicates . . . 226
18.4 Validity . . . 228
18.4.1 The definition . . . 228
18.4.2 $f^{\vDash}$ and $f^{\vDash\neg}$: the designated sets of a valuation . . . 229
18.4.3 Validity of conjunction and quantification . . . 229
18.4.4 Logical implications . . . 230
18.5 Logical equivalence . . . 232
18.6 A sequent system . . . 232

19 Axiomatisations **237**
19.1 Theory arising from a witness function . . . 237
19.2 Axiomatisation of continuity . . . 239
19.3 Quantifying over continuous valuations, within the logic . . . 240
19.3.1 Basic properties . . . 240
19.3.2 Valuation quantification treated as a modality . . . 241

20 Intertwined-ness & regularity via three-valued logic **243**
20.1 Logical characterisation of being intertwined . . . 243
20.2 Logical characterisation of topological indistinguishability . . . 246
20.3 Logical characterisation of $interior(f^{\vDash}) \neq \varnothing$ . . . 247
20.4 Logical characterisation of being quasiregular . . . 248
20.5 Logical characterisation of being weakly regular . . . 249
20.6 Logical characterisation of being unconflicted . . . 250
20.7 Logical characterisation of being regular . . . 250

21 Computational complexity & logic programming **253**
21.1 Translation of SAT to intertwinedness problem . . . 253
21.2 Horn satisfiability over Bool and **3** . . . 256
21.2.1 The (standard) HORNSAT algorithm for Boolean logic . . . 256
21.2.2 A proposal for HORNSAT over **3** . . . 257

22 Extremal valuations **259**
22.1 Definition of an extremal valuation . . . 259
22.2 Topological characterisation of extremal valuations . . . 260
22.3 Maximal disjoint pairs of open sets . . . 262
22.4 Logical characterisation of extremal valuations . . . 265
22.5 Extremal valuations and regular points . . . 267
22.6 Characterisations of intertwinedness properties . . . 268
22.6.1 Studying $p \between p'$ and $p{\equiv}p'$ . . . 268

22.6.2 Topological indistinguishability $\mathrel{\mathring{=}}$, and the empty relation $\varnothing$ 270
22.6.3 Consensus equivalence $\between^{=}$, hypertwined $\between^{\mathsf{T}}$, and hyperdefinite 271

IV Conclusions **277**

23 Conclusions **279**
23.1 Topology vs. semitopology 279
23.2 Related work 281
23.3 Future work 284
23.4 Open problems 289
23.5 Final comments 289

List of Figures **291**

Bibliography **293**

Index **301**

To my wife and son. Maths is delightful, and so are you.

Introduction 1

1.1 What is 'decentralised collaborative action'?

This book reports on the outcome of a mathematical investigation into what I will call *decentralised collaborative action*, motivated by recent developments in *decentralised permissionless heterogeneous computing systems*. Let's unpack the jargon:

- A system is *decentralised* when it is distributed over several machines, and the system as a whole is not centrally controlled.
  Most blockchain systems and peer-to-peer networks are decentralised in the sense we intend, or at least they are supposed to be. They are distributed over multiple participants, and no single entity controls the system. The internet is also (mostly) decentralised, at least in principle.[1]
- A system is *permissionless* (or *unpermissioned*) when participants can leave and join the system at any time.
  Nature is naturally permissionless (living things do not need permission to be born or die). National voting systems *are* permissioned (because citizens require certification from the government to be allowed to vote).
- A system is *heterogeneous* when participants may be following different rules.[2]
  Ethereum and Tezos are decentralised and permissionless, but they are not heterogeneous in the sense we intend. If you are running a Tezos or Ethereum node, then you are not forced to follow the rules, but if you do not then by definition you are not following the rules.

[1] The internet was designed to be an information network that would be resilient to nuclear attack. It did this by being 'centrifugal'; emphasising node-to-node actions instead of centre-to-centre actions. See [Rya10], summarised by Ars Technica [Tec11].

[2] By 'different rules' we include the situation where an algorithm (such as a consensus algorithm) is agreed between participants but a critical parameter may vary substantively across them, e.g. imagine a blockchain in which some participants require a $>2/3$ majority to act, and others require just a $>1/2$ majority.

In contrast, consider the combination of Tezos and Ethereum as a single system connected by a *blockchain bridge*.[3] This is heterogeneous, because Tezos nodes and Ethereum nodes have different rules and different consensus mechanisms. A Tezos node is not a bad node just because it is not following the rules of Ethereum, and vice-versa, but because of the blockchain bridge, they can be considered to be operating within a single (heterogeneous) combined system.

So a decentralised heterogeneous permissionless system consists of *some* participants communicating to do *something*, with no *a priori* restrictions on who, what, or how.

If you are a blockchain engineer then this scenario — with its weak well-behavedness assumptions that do not even assume all participants share a common ruleset — might seem a terrible idea which we should not allow, because it admits crazy networks with bad behaviour. But here the generality is a feature, not a bug:

1. As mathematicians, we *want* to admit general models, including bad ones, so that we can formalise their good and bad behaviour[4] and express conditions to include or exclude it.
2. Surprisingly, it will turn out that there is still a lot that we can say even about the general case, and we shall see that much useful structure will emerge *even from very weak assumptions* (a detailed summary is listed in Subsection 1.5).

So granted that the generality of decentralised collaborative action is a feature, not a bug; but how should we approach this mathematical generality? The key is to look at how groups of participants can *progress their local state*. To see this we need a little more discussion.

In a decentralised system, a participant must store state locally — if there were a global source of truth for state then whoever controls that truth would *de facto* control the system — and communicate with other participants to decide on how their local states evolve. There must be *some* rules about how this state should be updated, even if these rules may differ across participants in the system, and even though the rules may not always be followed. It turns out that one common feature of decentralised heterogeneous permissionless systems is a notion of what I will call an *actionable coalition*, by which I intend

> *a set of participants who are legally entitled (but not obliged) to collaborate to update their local state (possibly but not necessarily in identical ways).*

We will consider three examples:

1. Ethereum.
   Ethereum's consensus protocol is proof-of-stake, so an actionable coalition on Ethereum is any group of participants who hold a majority stake of tokens (this is a bit of a simplification, but it will do).
2. Ethereum and Tezos with a blockchain bridge between them.
   Tezos's consensus protocol is also proof-of-stake. An actionable coalition in this system is either an actionable coalition of Ethereum, or one of Tezos, *or* the sets union of an

[3] See https://ethereum.org/en/bridges/ (permalink: https://web.archive.org/web/20240324090911/https://ethereum.org/en/bridges/).

[4] ...which will vary by application; e.g. sometimes all participants should play by the same rules, but in the case of a blockchain bridge we specifically want to *admit* different rules.

actionable coalition from each, along with the bridging node (again, a simplification, but it will do).[5]

3. A Tango dance evening where men will only dance with women and vice-versa.[6]
   An actionable coalition is any set containing equal numbers of male leads and female followers.

If an actionable coalition can communicate to agree on a set of local state updates, e.g. if the Tango lead leads a move and the Tango follower chooses to follow it, then the participants in this coalition are entitled to update their local states accordingly. Note that local state updates need not be literally identical across participants; they just need to be mutually agreed upon and then actioned.

Some important notes:

1. The actionable coalition can progress *without* consulting the rest of the system.
2. Being in an actionable coalition does not imply control. This set describes a potential legal collaboration, but participants can choose what actionable coalition to work with, if any, and they can also choose not to follow the rules.[7]
3. If O is an actionable coalition for $p \in O$, and $p' \in O$ is another participant in O, then O is also an actionable coalition for p'. Note that this makes actionable coalitions look a bit like open sets in a topology.

So we can now introduce our first mathematical abstraction: we identify participants as *points*, and we let let *open sets* be *actionable coalitions*. An actionable coalition is a *coalition of participants with the capacity to act*. They are not obliged to act, and if they do their action need not be identical across all participants, but the potential exists for this set to collaborate to progress their states.

1. With reference to our couples dance example: an example of an actionable coalition that is not minimal is a set containing two male leads and two female followers. There are two ways for the participants to pair off to collaborate (i.e. dance).
2. With reference to our bridged blockchain example: an example of a set that contains an actionable coalition but is not one itself is an actionable coalition from Ethereum, along with the bridging node. The Ethereum coalition on its own is actionable, but the bridging node cannot take any action without also collaborating with an actionable coalition from Tezos.

To get a flavour of our mathematical results, consider a fundamental problem in any decentralised system: ensuring that its participants remain in agreement, for some suitable sense of 'agree'. To take a simple example from blockchain: if we reach a situation where half of the nodes say that we have paid for a service, and the other half say that we have not — then

[5]Typically, participants can update their state if they held a majority of the stake at some time in the past (e.g. two weeks ago) — the idea being that all participants have reached agreement on, and learned, the state of the network two weeks in the past, so this can be treated as immutable common knowledge without undermining the decentralised nature of the system in the present [Goo14, Subsection 3.2.1, final paragraph].

[6]Many dancers can lead as well as follow, but for the sake of the mathematics we will simplify.

[7]If you put your elbow into your dance partner's eye, or simply deliver a poor lead or a poor follow, then the other dancer might stop dancing with you or turn you down if you ask for another dance. But neither of you are *compelled* to dance with one another, and if you do, you are not *compelled* to dance well.

everyone has a problem, because the system has become incoherent and it is not clear how the system as a whole can restore coherence and progress.[8] This phenomenon is called *forking*, and blockchain designers really want to avoid it!

We will call our mathematical abstraction of agreement, *antiseparation*. In a little more detail, antiseparation properties are coherence properties that are guaranteed to hold of a decentralised system *just* by analysing the structure of its actionable coalitions. It turns out that we can get surprisingly detailed information about agreement/antiseparation properties, even working from quite weak and abstract mathematical assumptions on the actionable coalitions.

We emphasise this point: sometimes we can predict important macro properties of a system's behaviour without knowing anything about its specifics, so long as we have certain good properties on its actionable coalitions.

Let us start by considering a simple situation where participants are trying to agree on a binary consensus problem: whether to announce a single value 'true' or 'false'. Continuing the theme of simplicity, assume some finite nonempty set of participants $\mathbb{E}$ and let their actionable coalitions be just any set of participants that forms a majority (so it contains strictly more than half of the set of all participants). Now suppose that the participants in some actionable coalition $O \subseteq \mathbb{E}$ have communicated and have agreed on 'true'. Because they form an actionable coalition, they are entitled to act and to announce 'true', and so they do. They have now all committed to this state update and they cannot change their minds.

So: can this system fork? Consider some participant $p \notin O$. If p wants to make progress, is must also agree on 'true', because all of its actionable coalitions intersect with O and so contain at least one participant that has committed to 'true' and cannot change its mind. This does not mean that p has to agree on 'true'; it could choose not to progress, or it could break the rules. But, by definition if p does want to progress legally, then the decision has been made and it must eventually go along with the majority. Thus, we have proved that any progress that is made by one participant within the rules (... must be shared with some actionable coalition of that participant, and since all such coalitions intersect it ...) must eventually be followed any other participant that also progresses. Thus forking is impossible.

The reader may already be familiar with this example, but note that this antiseparation property comes simply *from the structure of the actionable coalitions*. There is no need to consider the protocol, or even how values are interpreted. It turns out that antiseparation-style behaviour is common, and arises even if we do not require actionable coalitions that are simple majorities. For example, let participants be $\mathbb{Z} = \{0, 1, \ 1, 2, \ 2, \dots\}$ and let actionable coalitions be generated by sets of three consecutive numbers starting at an even number $\{2i, 2i{+}1, 2i{+}2\}$, and suppose again that we are trying to agree on 'true' or 'false'. Note that in contrast to the previous example, actionable coalitions need not intersect. Yet, the moment one triplet of participants commits to 'true', the rest of the system is obliged to eventually agree, if all participants play by the rules. Now this example system is not necessarily particularly safe or desirable in practice, because we can imagine that $\{0, 1, 2\}$ agree on 'true', and $\{4, 5, 6\}$ acting independently but in good faith agree on 'false', and then 3 cannot legally progress, because within $\{2, 3, 4\}$, 2 has announced 'true' and 4 has announced 'false' and 3 cannot agree with both. But, we know that *if* all participants do legally progress, then they announce the same

[8] coherent (adj.) 1550s, "harmonious;" 1570s, "sticking together," also "connected, consistent" (of speech, thought, etc.), from French cohérent (16c.), from Latin cohaerentem (nominative cohaerens), present participle of cohaerere "cohere," from assimilated form of com "together" (see co-) + haerere "to adhere, stick" (etymologyonline: https://www.etymonline.com/word/coherent).

value. So this example illustrates how antiseparation can arise even when actionable coalitions are rather small.[9]

The two examples above are quite different. In one, all actionable coalitions intersect, and in the other they mostly do not. This suggests that a 'general mathematics of (anti)separation' is possible, based on the study of actionable coalitions. In a nutshell, that mathematical story is what we will develop.

1.2 What is a semitopology?

So at a high level, what do we have?

1. There is a notion of what we can call an *actionable coalition* (or just: *open set*). This is a set $O \subseteq \mathsf{P}$ of participants with the capability, though not the obligation, to act collaboratively to advance (= update / transition) the local state of the elements in O, possibly but not necessarily in the same way for every $p \in O$.
2. $\varnothing$ is trivially an actionable coalition. Also we assume that P is actionable, effectively assuming that every point is a member of at least one actionable coalition.
3. A sets union of actionable coalitions, is an actionable coalition.

This leads us to the definition of a semitopology.

NOTATION 1.2.1. Suppose P is a set. Write $pow(\mathsf{P})$ for the powerset of P (the set of subsets of P); there will be more on this in Notation 8.2.1.

DEFINITION 1.2.2. A **semitopological space**, or **semitopology** for short, consists of a pair $(\mathsf{P}, \mathsf{Open}(\mathsf{P}))$ of

- a (possibly empty) set P of **points**, and
- a set $\mathsf{Open}(\mathsf{P}) \subseteq pow(\mathsf{P})$ of **open sets**,

such that:

1. $\varnothing \in \mathsf{Open}(\mathsf{P})$ and $\mathsf{P} \in \mathsf{Open}(\mathsf{P})$.
2. If $X \subseteq \mathsf{Open}(\mathsf{P})$ then $\bigcup X \in \mathsf{Open}(\mathsf{P})$.[10]

We may write $\mathsf{Open}(\mathsf{P})$ just as Open, if P is irrelevant or understood, and we may write $\mathsf{Open}_{\neq\varnothing}$ for the set of nonempty open sets.

The reader will recognise a semitopology as being like a *topology* on P [Eng89, Wil70], but without the condition that the intersection of two open sets necessarily be an open set. This reflects the fact that the intersection of two actionable coalitions need not itself be an actionable coalition.

Armed with this simple definition and bearing in mind its modern relevance as noted above, we introduce and survey semitopologies and their properties. There is an emphasis (though not an exclusive one) on studying decentralised collaborative actions, which (broadly speaking) amounts to studying antiseparation properties of points, and how this interacts with topological

[9] See also Remark 3.2.6.

[10] There is a little overlap between this clause and the first one: if $X = \varnothing$ then by convention $\bigcup X = \varnothing$. Thus, $\varnothing \in \mathsf{Open}(\mathsf{P})$ follows from both clause 1 and clause 2. If desired, the reader can just remove the condition $\varnothing \in \mathsf{Open}(\mathsf{P})$ from clause 1, and no harm would come of it.

continuity of functions out of semitopologies. The details of what this means are unpacked below.

We will proceed in three parts, by which we hope to give a comprehensive overview of our approach to decentralised collaborative action:

> point-set semitopologies, algebra, and logic.

The point-set semitopologies are most pertinent to concrete models (i.e. real networks), the algebra is most pertinent to clarifying an abstract (algebraic) view of what are the essential structures in play, and the logic is pertinent to specifying properties and — because logic is a portal to computation — computing/checking these properties.

REMARK 1.2.3. Traditional notions of consensus and voting can be understood in a semitopological framework. For instance, a committee may make a decision by two-thirds majority vote; the set of all 2/3 majorities of some **P** is a semitopology (note that it is not a topology). Also, concrete algorithms to attain consensus often use a notion of *quorum* [Lam98, LSP82] as a set of participants whose unanimous adoption of a value guarantees that other (typically all other) participants will eventually also adopt this value. Social choice theorists have a similar notion called a *winning coalition* [Rik62, Item 5, page 40]. If the reader has a background in logic then they may be reminded of a whole field of *generalised quantifiers* (a good survey is in [Wes11]).

The reader should just note that these examples have a synchronous, centralised flavour.

For instance: a vote in the typical democratic sense is a synchronous, global operation (unless the result is disputed): votes are cast, collected, and then everyone gets together — e.g. in a vote counting hall — to count the votes and agree on who won and so certify the outcome.[11] This is certainly a collaborative action, but it is centralised.

Our semitopological framework adds to the above by allowing us to study *decentralised* collaborative action, which can progress by local state updates on actionable coalitions (which certainly do not need to be simple majorities), and they can so act without *necessarily* having to synchronise step-by-synchronous-step on global state updates.

REMARK 1.2.4. Let us take a moment to give a high-level motivation for semitopologies, in the style of (very abstract) justifications that have been given for topologies. A classic text on topology [Vic89] justifies topology as follows:

1. Logically, open sets model *affirmations*:[12] an open set O corresponds to an affirmation (of O) [Vic89, page 10].[13]
2. Computationally, open sets model *semidecidable properties*. See the first page of the preface in [Vic89].

The notion of an actionable coalition justifies semitopologies in similar terms.

---

[11] I have seen this happen; votes being tallied up while under supervision by representatives of all parties on the ballot. It is a moving sight. But it is not decentralised.

[12] *Affirmation:* Something declared to be true; a positive statement or judgment. https://www.wordnik.com/words/affirmation (permalink: https://web.archive.org/web/20230608073651/https://www.wordnik.com/words/affirmation).

[13] *Side note:* I sometimes get asked here why in topology, open sets are closed under *arbitrary* unions but only *finite* intersections. Surely an infinite collection of positive affirmations is still a positive affirmation? No, not quite: consider the affirmation 'I am a nonempty open neighbourhood of π in $\mathbb{Q}$'. This is closed under finite intersections, but not infinite intersections.

1. Logically, open sets model *collaborative* affirmations. An open set corresponds to an affirmation (by collaboration between the elements of O).
2. Computationally, open sets model *actionable properties*, this being a (semidecidable) property that furthermore *enough participants agree on that it can be acted on.*

Alternatively, we could just read this document in the spirit of pure mathematics — I am sympathetic to this view — in which case a test for whether semitopologies are an interesting definition is just whether we get interestingly more structure out of it than we put in to the definition. We shall see that this is indeed the case.

1.3 Who should read this document?

1. *Practitioners,* especially those using quorum systems and fail-prone systems, looking for a mathematical framework that subsumes what they are already doing, puts it in a broader context, creates a common language to speak with one another and with mathematicians, and suggests new engineering options.

2. *Theoreticians,* looking to use semitopologies to construct the next generation of advanced decentralised computer systems — especially where the maths makes it easier to reuse existing tools whose applicability would not otherwise be clear (e.g. using SAT solvers to compute semitopological antiseparation properties from Part I using the logic in Part III).

3. *Pure mathematicians,* who might be delighted to discover a new topology-adjacent field and might see it as a fresh research opportunity.[14]

   Pure mathematicians can also learn a lot from this material regarding what things are important and interesting to look at, and what distinctions make a difference in practice; I know that I have.

4. *Mathematicians looking to get into practical systems.*

   Real systems can be messy because they have to accommodate a messy reality. Semitopologies provide a useful abstraction that can help understand what is going on at a high level.

1.4 Why did I write it?

1. Numerous authors have recently studied designing systems where participants have different opinions on who is part of the system or on who is trustworthy or not [ACTZ24, SWRM21, CLZ23, LCL23, BKK22, GPG18, LLM+19, LGM19, FHNS22, LL23]. These systems go by names such as *(permissionless) fail-prone systems* and *(heterogeneous) quorum systems* (more discussion, with more references, is in Subsection 23.2).
   Most of these systems are (or to be more precise: they directly give rise to) semitopologies, and we would argue that the literature above is, in fact, *rediscovering topology through semitopology*, but they did not know it. This document makes the connection to classical mathematics explicit, and builds on it in interesting ways.

[14] E.g. *algebraic semitopology* does not exist yet, nor does *semitopological epistemological logic*; but perhaps somebody might read this work and be inspired to create them.

2. Like topology, semitopology is not a single theorem; it is a method and a research topic. As such, the topic needs a text to describe its current scope, and show how the parts of the theory fit together.[15]
   For instance, the algebraic theory is motivated by the point-set semitopologies; and conversely, to fully understand what point-set semitopologies really are, we need the algebraic theory. Similarly for the logic. Here, we have the space to tell a proper story arc.
3. By design, this document is interdisciplinary, speaking to two communities:
   - *pure mathematicians*, who probably know lots about point-set topology but maybe know less about modern consensus or blockchain systems, and
   - *practitioners*, who conversely are very familiar with the engineering, and may have already reinvented (semi)topologies but not realise it, but may be unfamiliar with the terminology and methods of pure mathematics.[16]

   I intend this maths to be accessible and useful to both types of reader, and hope thereby to make a constructive difference to the development of both the theory and practice of decentralised systems.

1.5 Map of the work

As already mentioned, this work is in three parts:

1. We consider point-set semitopologies in Sections 2 to 11.
2. We then take an algebraic, point-free, categorical approach by studying semiframes in Sections 12 to 17.
3. Finally, we consider logic over semitopologies in Sections 18 to 22.

In a nutshell, we will study the topology, algebra, and logic of semiframes, as follows:

1. Section 1 is the Introduction. You Are Here.

SEMITOPOLOGIES

2. In Section 2 we show how **continuity corresponds to local agreement** (Definition 1.2.2 and Lemma 2.2.4).

3. In Section 3 we discuss **transitive sets**, **topens**, and **intertwined points**. These are all different views on the anti-separation well-behavedness properties that will interest us. Most of Section 3 is concerned with showing how these different views relate and in what senses they are equivalent (e.g. Proposition 3.6.9). Transitive sets are guaranteed to be in agreement (in a sense made precise in Theorem 3.2.2 and Corollary 3.2.3), and we take a first step to understanding the fine structure of semitopologies by proving that every semitopology partitions into topen sets (Theorem 3.5.4), plus other kinds of points which we classify in the next Section.

[15] ...if only so that others can build on this and do even more, and even better!
[16] This is not a hypothetical.

4. In Section 4 we start to classify points in more detail, introducing notions of **regularity** for points in Definition 4.1.4. This is the start of a classification of good properties for points in semitopologies, including: regular, weakly regular, indirectly regular, quasiregular, unconflicted, hypertransitive, and more.

5. In Section 5 we study **closed sets**, and in particular the interaction between intertwined points, topens, and closures. Typical results are Proposition 5.4.3 and Theorem 5.6.2 which characterise sets of intertwined points as minimal closures. The significance to consensus is discussed in Remarks 5.5.1 and 5.5.5.

6. In Section 6 we study **unconflicted** and **hypertransitive** points, leading to two useful characterisations of regularity in Theorems 6.2.2 and 6.5.8.

7. In Section 7 we consider **product semitopologies**. These are defined just as for topologies (Definition 7.1.2) but we study how the semitopological properties we have considered above — like being intertwined, topen, regular, conflicted, and so forth — interact with taking products. This is also useful for building large complex counterexamples out of smaller simpler ones (examples in Corollary 7.2.6 or Theorem 7.3.4).

8. In Section 8 we construct a novel theory of computationally tractable semitopologies, based on **witness functions** (Definition 8.2.2(1)). We call semitopologies generated by witness functions *witness semitopologies*. These display excellent algorithmic behaviour (Remarks 8.4.6 and 8.4.14) and we note deep reasons why this is so by showing that witness functions correspond to Horn clause theories, and that open and closed sets in the witness semitopology are related to answer sets to those theories; see Subsection 8.5.

9. In Section 9 we introduce **(strongly) chain-complete** semitopologies. We argue in Remark 9.4.6 that these have properties making them a suitable abstraction of finite semitopologies — finite semitopologies are of particular interest because these are the ones that we can build. We study their properties and prove a key result that witness semitopologies are chain-complete (Theorem 9.4.1), even if they are infinite.[17]

10. A key property in a strongly chain-complete semitopology is that the poset of open sets is *atomic*, i.e. minimal nonempty open sets always exist. In Section 10 we study **kernels** — unions of atomic transitive open sets — especially in strongly chain-complete semitopologies where atoms are guaranteed to exist. We will see that the kernel dictates behaviour in a sense we make formal (see discussion in Remark 10.1.1).

11. In Section 11 we study notions of **dense subset of** from topology and see that this splits into two notions: *weakly dense in* and *strongly dense in* (Definition 11.1.2). Transitivity turns out to be closely related to denseness (Proposition 11.4.6). We prove a continuous extension result and show that this leads naturally back to the notion of regular point and topen set which we developed to begin with (Remark 11.5.1).

[17]We discuss why infinite semitopologies matter, even in a world of finite implementations, in Remark 9.4.7. Note also that in a real system there may be hostile participants who report an unbounded space of 'phantom' points, either for denial-of-service or to create 'extra voters'. So even a system that is physically finite may present itself as infinite.

SEMIFRAMES

12. In Section 12 we introduce **semiframes**. These are the algebraic version of semitopologies, and they are to semitopologies as frames are to topologies. We discover that semiframes are not just join-semilattices; **semiframes are *compatible* semilattices**, which include a *compatibility relation* $*$ to abstract the property of sets intersection $\between$ (see Remark 12.1.2).

13. In Section 13 we introduce **semifilters**. These play a similar role as filters do in topologies, except that semifilters have a *compatibility condition* instead of closure under finite meets. We develop the notion of abstract points (completely prime semifilters), and show how to build a semitopology out of the abstract points of a semiframe.

14. In Section 14 we introduce **sober semitopologies** and **spatial semiframes**. The reader familiar with categorical duality will know these conditions. Some of the details are significantly different (see for instance the discussion in Subsection 14.3.2) but at a high level these conditions work in the proofs just as they do for the topological duality.

15. In Section 15 we consider the **duality** between suitable categories of (sober) semitopologies and (spatial) semiframes.

16. In Section 16 we **dualise the well-behavedness conditions** from Section 3 to algebraic versions. The correspondence is good (Proposition 16.6.2) but also imperfect in some interesting ways (Remark 16.8.8).

17. In Section 17 we briefly consider alternative **graph-based representations** of semitopologies.

LOGIC

18. In Section 18 we introduce a **three-valued modal logic** to describe properties of semitopologies. Logic is closely related to computation, so what we can describe we can — by passing it to a solver or a prover — also compute.

19. In Section 19 we use our logic to start **axiomatising**. In particular, in Definition 19.1.1 we carry out the most basic task and write down axioms in our logic that correspond to continuity.

20. In Section 20 we **axiomatise the regularity properties** that we have already investigated.

21. In Section 21 we look at computing in more detail. We show how to **convert a witness function into a logical theory** suitable for passing to a SAT solver, we investigate notions of Horn clause programming (both for two-valued and three-valued logic), and we show in Theorem 21.1.6 that **determining whether two points are intertwined is NP-complete** in general.

22. In Section 22 we open up a fresh line of inquiry and consider **extremal valuations**, which roughly speaking correspond to final states of a system in which every participant who could return a definite value, has returned a definite value. This turns up to open a new design space which we put in the context of the maths thus far.

CONCLUSIONS

23. In Section 23 we conclude and discuss related and future work.

REMARK 1.5.1. Algebraic topology has been applied to the solvability of distributed-computing tasks in various computational models (e.g. the impossibility of wait-free k-set consensus using read-write registers and the Asynchronous Computability Theorem [HS93, BG93a, SZ93]; see [HKR13] for a survey). Semitopology is not topology, and this work is not about algebraic topology applied to the solvability of distributed-computing tasks!

We are interested in the mathematics of actionable coalitions, as made precise by point-set semitopologies; their antiseparation properties; and the implications to partially continuous functions on of them. If we discuss distributed systems, it is by way of providing motivating examples or noting applicability.

Part I

Point-set semitopologies

Semitopology 2

2.1 Definitions, examples, and some discussion

2.1.1 Definitions

Recall from Definition 1.2.2 the definition of a semitopology.

REMARK 2.1.1.

1. As a sets structure, a semitopology on P is like a *topology* on P, but without the condition that the intersection of two open sets be an open set.
2. As a lattice structure, a semitopology on P is a bounded complete join-subsemilattice of $pow(\mathsf{P})$.[1]
3. Every semitopology $(\mathsf{P}, \mathsf{Open})$ induces two natural topological completions: the least topology that contains Open, and the greatest topology contained in Open. But there is more to semitopologies than just their topological completions, because:
    a) We are explicitly interested in situations where intersections of open sets need *not* be open.
    b) Completing to a topology loses information. For example: the 'many', 'all-but-one', and 'more-than-one' semitopologies in Example 2.1.4 express three distinct notions of quorum, yet if P is infinite then for all three, the least topology containing them is the discrete semitopology (Definition 2.1.3(1)), and the greatest topology that they contain is the trivial topology $\{\varnothing, \mathsf{P}\}$ (Example 2.1.4(4)). See also the overview in Subsection 23.1.

Semitopologies are not topologies. We take a moment to spell out one concrete difference:

[1] *Bounded* means closed under empty intersections and unions, i.e. containing the empty and the full set of points. *Complete* means closed under arbitrary (possibly empty, possibly infinite) sets unions. The reader may know that a complete lattice is also co-complete: if we have all joins, then we also have all meets. However, note that there is no reason for the meets in Open to coincide with the meets in $pow(\mathsf{P})$, i.e. for them to be sets intersections.

Also, note that this does not mean that semitopologies are 'just' bounded complete join-subsemilattices. They are in fact *compatible* bounded complete join-semilattices. See Section 12.

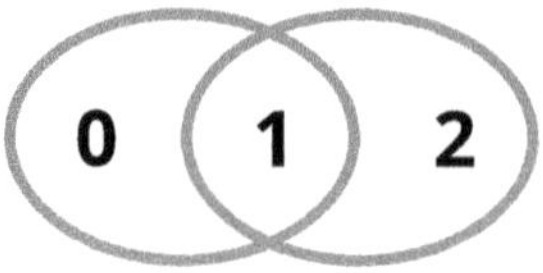

Figure 2.1: An example of a point with two minimal open neighbourhoods (Lemma 2.1.2)

LEMMA 2.1.2. *In topologies, if a point p has a minimal open neighbourhood then it is least (= unique minimal). In semitopologies, a point may have multiple distinct minimal open neighbourhoods.*[2]

Proof. To see that in a topology every minimal open neighbourhood is least, just note that if $p \in A$ and $p \in B$ then $p \in A \cap B$. So if A and B are two minimal open neighbourhoods then $A \cap B$ is contained in both and by minimality is equal to both.

To see that in a semitopology a minimal open neighbourhood need not be least, it suffices to provide an example. Consider $(\mathsf{P}, \mathsf{Open})$ defined as follows, as illustrated in Figure 2.1:

- $\mathsf{P} = \{0, 1, 2\}$
- $\mathsf{Open} = \{\varnothing, \{0,1\}, \{1,2\}, \{0,1,2\}\}$

Note that 1 has two minimal open neighbourhoods: $\{0, 1\}$ and $\{1, 2\}$. □

2.1.2 Examples

As standard, we can make any set Val into a semitopology (indeed, it is also a topology) just by letting open sets be the powerset:

DEFINITION 2.1.3.

1. Call $(\mathsf{P}, pow(\mathsf{P}))$ the **discrete semitopology on P**.
   We may call a set with the discrete semitopology a **semitopology of values**, and when we do we will usually call it Val. We may identify Val-the-set and Val-the-discrete-semitopology; meaning will always be clear.
2. When $(\mathsf{P}, \mathsf{Open})$ is a semitopology and Val is a semitopology of values, we may call a function $f : \mathsf{P} \to \mathsf{Val}$ a **value assignment**.
   Note that a value just assigns values to points, and in particular we do not assume *a priori* that it is continuous, where continuity is defined just as for topologies (see Definition 2.2.1).

EXAMPLE 2.1.4. We consider further examples of semitopologies:

1. Every topology is also a semitopology; intersections of open sets are allowed to be open in a semitopology, they are just not constrained to be open. In particular, the discrete topology is also a discrete semitopology (Definition 2.1.3(1)).

2. The **initial semitopology** $(\varnothing, \{\varnothing\})$ and the **final semitopology** $(\{*\}, \{\varnothing, \{*\}\})$ are semitopologies.

[2] We study minimal open neighbourhoods in detail, starting from Definition 9.5.2.

3. An important discrete semitopological space is

$$\mathbb{B} = \{\bot, \top\} \quad \text{with the discrete semitopology} \quad \mathsf{Open}(\mathbb{B}) = \{\varnothing, \{\bot\}, \{\top\}, \{\bot, \top\}\}.$$

We may silently treat $\mathbb{B}$ as a (discrete) semitopological space henceforth.

4. Take P to be any nonempty set. Let the **trivial semitopology** (this is also a topology) on P have

$$\mathsf{Open} = \{\varnothing, \mathsf{P}\}.$$

So (as usual) there are only two open sets: the one containing nothing, and the one containing every point.[3]

The only nonempty open is P itself, reflecting a notion of actionable coalition that requires unanimous agreement.

5. Suppose P is a set and $\mathcal{F} \subseteq pow(\mathsf{P})$ is nonempty and up-closed (so if $P \in \mathcal{F}$ and $P \subseteq P' \subseteq \mathsf{P}$ then $P' \in \mathcal{F}$, then $(\mathsf{P}, \mathcal{F})$ is a semitopology. This is not necessarily a topology, because we do not insist that $\mathcal{F}$ is a filter (i.e. is closed under intersections).

We give four sub-examples for different choices of $\mathcal{P} \subseteq pow(\mathsf{P})$. Partly this is to illustrate how varying $\mathcal{F}$ can encode different systems, but also, these variations can have substantially different behavior, as Lemmas 8.3.3 and 8.3.5 will illustrate later.

a) Take P to be any finite nonempty set. Let the **supermajority semitopology** have

$$\mathsf{Open} = \{\varnothing\} \cup \{O \subseteq \mathsf{P} \mid cardinality(O) > 2/3 * cardinality(\mathsf{P})\}.$$

So O is open when it contains more than two-thirds of the points.
Two-thirds is a common threshold used for making progress in consensus and voting algorithms.[4]

b) Take P to be any nonempty set. Let the **many semitopology** have

$$\mathsf{Open} = \{\varnothing\} \cup \{O \subseteq \mathsf{P} \mid cardinality(O) = cardinality(\mathsf{P})\}.$$

For example, if $\mathsf{P} = \mathbb{N}$ then open sets include $evens = \{2 * n \mid n \in \mathbb{N}\}$ and $odds = \{2 * n{+}1 \mid n \in \mathbb{N}\}$.
Its notion of open set captures an idea that an actionable coalition is a set that may not be all of P, but does at least biject with it.

c) Take P to be any nonempty set. Let the **all-but-one semitopology** have

$$\mathsf{Open} = \{\varnothing,\ \mathsf{P}\} \cup \{\mathsf{P} \setminus \{p\} \mid p \in \mathsf{P}\}.$$

This semitopology is not a topology. See also Lemma 8.3.3.
The notion of actionable coalition here is that there may be at most one objector (but not two).

---

[3]According to Wikipedia, this space is also called *indiscrete*, *anti-discrete*, *concrete*, and *codiscrete* (https://en.wikipedia.org/wiki/Trivial_topology).

[4]In the context of consensus algorithms, we usually take a strict inequality $cardinality(O) > 2/3 * cardinality(\mathsf{P})$ — to guarantee that the intersection of any three nonempty open sets is nonempty. However, in the context of voting (as in e.g. the rules for the US Senate to amend the Constitution) the majority may be taken to have the form $cardinality(O) \geq 2/3 * cardinality(\mathsf{P})$ — to guarantee that votes *for* outnumber votes *against*, by at least two-to-one.

d) Take P to be any set with cardinality at least 2. Let the **more-than-one semitopology** have

$$\mathsf{Open} = \{\varnothing\} \cup \{O \subseteq \mathsf{P} \mid \mathit{cardinality}(O) \geq 2\}.$$

This semitopology is not a topology. See also Lemma 8.3.5.

This notion of actionable coalition reflects a security principle in banking and accounting (and elsewhere) of *separation of duties*, that functional responsibilities be separated such that at least two people are required to complete an action — so that errors (or worse) cannot be made without being discovered by another person.

6. Take $\mathsf{P} = \mathbb{R}$ (the set of real numbers) and let open sets be generated by intervals of the form $[0, r)$ or $(\text{-}r, 0]$ for any strictly positive real number $r > 0$.

   This semitopology is not a topology, since (for example) $(1, 0]$ and $[0, 1)$ are open, but their intersection $\{0\}$ is not open.

7. In [NW94] a notion of *quorum system* is discussed, defined as any collection of pairwise intersecting sets. Quorum systems are a field of study in their own right, especially in the theory of concrete consensus algorithms.

   Every quorum system gives rise naturally to a semitopology, just by closing under arbitrary unions. We obtain what we will call an *intertwined space* (Notation 3.6.5; a semitopology all of whose nonempty open sets intersect)[5] and this is also clearly related to the *compatibility condition* in the notions of *semifilter* and *abstract point* in Section 13.

   Going in the other direction is interesting for a different reason, that it is slightly less canonical: of course every intertwined space is already a quorum system; but (for the finite case) we can also map to the set of all open covers of all points (Definition 9.5.2(2); in the notation of that Definition, we would write this as $\bigcup_{p\in\mathsf{P}}\{O\in\mathsf{Open} \mid O \gtrdot p\}$).

   To give one specific example of a quorum system from [NW94], consider $n \times n$ grid of cells with quorums being sets consisting of any full row and a full column; note that any two quorums must intersect in at least two points. We obtain a semitopology just by closing under arbitrary unions.

Remark 2.1.5 (Logical models of semitopologies). One class of examples of semitopologies deserves its own discussion. Consider an arbitrary logical system with predicates Pred and entailment relation $\vdash$.[6] Call $\Phi \subseteq \mathsf{Pred}$ **deductively closed** when $\Phi \vdash \phi$ implies $\phi \in \Phi$. Then take

- $\mathsf{P} = \mathsf{Pred}$, and
- let $O \in \mathsf{Open}$ be Pred or the complement to a deductively closed set Φ, so $O = \mathsf{Pred} \setminus \Phi$.

Note that an arbitrary union of open sets is open (because an arbitrary intersection of deductively closed sets is deductively closed), but an intersection of open sets need not be open (because the union of deductively closed sets need not be deductively closed). This is a semitopology. This example will be important to us and we will return to it in Subsection 8.5.

[5] A topologist would call this a *hyperconnected space*, but be careful! There are multiple such notions in semitopologies, so intuitions need not transfer over. See the discussion in Subsection 3.7.3.

[6] A validity relation $\models$ would also work.

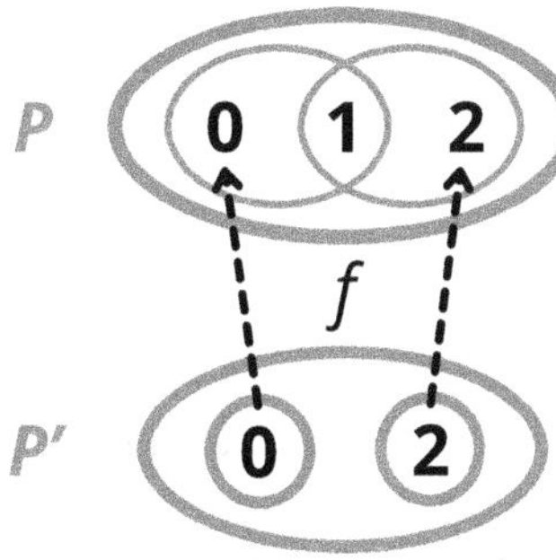

Figure 2.2: Two nonidentical semitopologies (Remark 2.1.7)

2.1.3 Why the name 'semitopologies', and other discussion

REMARK 2.1.6 (Why the name 'semitopologies'). When we give a name 'semitopologies' to things that are like topologies but without intersections, this is a riff on

- 'semilattices', for things that are like lattices with joins but without meets (or vice-versa), and
- 'semigroups', for things that are like groups but without inverses.

But, this terminology also reflects a real mathematical connection, because semitopologies *are* semilattices *are* semigroups, in standard ways which we take a moment to spell out:

- A semitopology $(\mathsf{P}, \mathsf{Open})$ is a bounded join subsemilattice of the powerset $pow(\mathsf{P})$, by taking the join V to be sets union $\cup$ and the bounds $\bot$ and $\top$ to be $\varnothing$ and P respectively.
- A semilattice is an idempotent commutative monoid, which is an idempotent commutative semigroup with an identity, by taking the multiplication $\circ$ to be V and the identity element to be $\bot$ ($\top$ becomes what is called a *zero* or *absorbing* element, such that $\top \circ x = \top$ always).

REMARK 2.1.7 (Semitopologies are not *just* semilattices). We noted in Remark 2.1.6 that every semitopology is a semilattice. This is true, but the reader should not read this statement as reductive: semitopologies are not *just* semilattices.

To see why, consider the following two simple semitopologies, as illustrated in Figure 2.2:

1. $(\mathsf{P}, \mathsf{Open})$ where $\mathsf{P} = \{0, 1, 2\}$ and $\mathsf{Open} = \{\varnothing, \{0, 1\}, \{1, 2\}, \{0, 1, 2\}\}$.
2. $(\mathsf{P}', \mathsf{Open}')$ where $\mathsf{P} = \{0, 2\}$ and $\mathsf{Open}' = \{\varnothing, \{0\}, \{2\}, \{0, 2\}\}$.

Note that the semilattices of open sets Open and Open' are isomorphic — so, when viewed as semilattices these two semitopologies are the same (up to isomorphism).

However, $(\mathsf{P}, \mathsf{Open})$ is not the same semitopology as $(\mathsf{P}', \mathsf{Open}')$. There is more than one way to see this, but perhaps the simplest indication is that for every continuous $f : (\mathsf{P}, \mathsf{Open}) \to (\mathsf{P}', \mathsf{Open}')$, there is no continuous map $g : (\mathsf{P}', \mathsf{Open}') \to (\mathsf{P}, \mathsf{Open})$ such that $g \circ f$ is the identity (we will define continuity in a moment in Definition 2.2.1(2) but it is just as for topologies, so we take the liberty of using it here). There are a limited number of possibilities for f and g, and we can just enumerate them and check:

- If $f(0) = 0$ and $f(2) = 2$ and $g(1) = 0$, then $g^{-1}(\{2\}) = \{2\} \notin \mathsf{Open}$, and if $g(1) = 1$ then $g^{-1}(\{0\}) = \{0\} \notin \mathsf{Open}$.
- If $f(0) = 0$ and $f(2) = 1$ and $g(1) = 0$, then $g^{-1}(\{2\}) = \{1\} \notin \mathsf{Open}$, and if $g(1) = 2$ then $g^{-1}(\{0\}) = \{0\} \notin \mathsf{Open}$.
- Other possibilities are no harder.

A similar observation holds for *topologies*: for example, if we write $(\mathbb{Q}, \mathsf{Open}_{\mathbb{Q}})$ for the rational numbers with their usual open set topology, and $(\mathbb{R}, \mathsf{Open}_{\mathbb{R}})$ for the real numbers with their usual open set topology, then their topologies are isomorphic as lattices, with one direction of the isomorphism given just by $O \in \mathsf{Open}_{\mathbb{R}}$ maps to $O \cap \mathbb{Q} \in \mathsf{Open}_{\mathbb{Q}}$. This counterexample works for semitopologies too since every topology is also a semitopology.

However, we would still argue that the counterexample in Figure 2.2 is inherently stronger; not just because it is smaller (two and three points instead of countably and uncountably many) but also because — while we can recover $\mathbb{R}$ from $\mathbb{Q}$ in a natural and canonical way by forming a completion — the upper semitopology in Figure 2.2 is not *a priori* canonically derived from the lower one. The two semitopologies in Figure 2.2 seem to be distinct in some structural way, yet they still corresponding to the same semilattice, so we see that there is other structure here, which is not reflected by the pure semilattice derived from their open sets.

REMARK 2.1.8 ('Stronger' does not necessarily equal 'better'). We conclude with some easy predictions about the theory of semitopologies, made just from general mathematical principles. Fewer axioms means:

1. *more* models,
2. *finer discrimination* between definitions, and
3. (because there are more models) *more counterexamples*.

So we can expect a theory with the look-and-feel of topology, but with new models, new distinctions between definitions that in topology may be equivalent, and some new definitions, theorems, and counterexamples.

Note that fewer axioms does not necessarily mean fewer interesting properties. On the contrary: if we can make finer distinctions, there may also be more interesting things to prove; and assumptions can become *more* impactful in a weaker system, because they may exclude more models than would have been the case with more powerful axioms.

For example consider semigroup theory and group theory: every group is a semigroup, but both groups and semigroups have their own distinct character, literature, and applications. To take this to an extreme, consider the *terminal* theory, which has just one first-order axiom: $\exists x.\forall y.x = y$. This 'subsumes' groups, lattices, graphs, and much besides, in the sense that every model of the terminal theory *is* a group, a lattice, and a graph, in a natural way. But models of this theory are so restricted (just the singleton model with one element) that there is not much left to say about it. Additional assumptions add nothing of value, because there was only one element to begin with!

REMARK 2.1.9 (Counting semitopologies). The number of semitopologies with n points follows OEIS sequence A102894 (https://oeis.org/A102894): the *"number of families of subsets of* $\{1, \dots, n\}$ *that are closed under intersection and contain both the universe and the empty set."* (The sequence counts closed sets rather than open sets, but these are in bijection.)

2.2 Continuity, and its interpretation

The topological notion of continuity works fine in semitopologies, and the fact that there are no surprises is a feature. In Remark 2.2.5 we explain how these notions matter to us:

DEFINITION 2.2.1. We import standard topological notions of inverse image and continuity:

1. Suppose P and P' are any sets and $f : \mathsf{P} \to \mathsf{P}'$ is a function. Suppose $O' \subseteq \mathsf{P}'$. Then write $f^{\text{-}1}(O')$ for the **inverse image** or **preimage** of O', defined by

$$f^{\text{-}1}(O') = \{p \in \mathsf{P} \mid f(p) \in O'\}.$$

2. Suppose $(\mathsf{P}, \mathsf{Open})$ and $(\mathsf{P}', \mathsf{Open}')$ are semitopological spaces (Definition 1.2.2). Call a function $f : \mathsf{P} \to \mathsf{P}'$ **continuous** when the inverse image of an open set is open. In symbols:

$$\forall O' \in \mathsf{Open}'. f^{\text{-}1}(O') \in \mathsf{Open}.$$

3. Call a function $f : \mathsf{P} \to \mathsf{P}'$ **continuous at** $p \in \mathsf{P}$ when

$$\forall O' \in \mathsf{Open}'. f(p) \in O' \Longrightarrow \exists O_{p,O'} \in \mathsf{Open}. p \in O_{p,O'} \wedge O_{p,O'} \subseteq f^{\text{-}1}(O').$$

   In words: f is continuous at p when the inverse image of every open neighbourhood of $f(p)$ contains an open neighbourhood of p.

4. Call a function $f : \mathsf{P} \to \mathsf{P}'$ **continuous on** $P \subseteq \mathsf{P}$ when f is continuous at every $p \in P$.

LEMMA 2.2.2. *Suppose* $(\mathsf{P}, \mathsf{Open})$ *and* $(\mathsf{P}', \mathsf{Open}')$ *are semitopological spaces (Definition 1.2.2) and suppose* $f : \mathsf{P} \to \mathsf{P}'$ *is a function. Then the following are equivalent:*

1. *f is continuous (Definition 2.2.1(2)).*
2. *f is continuous at every $p \in \mathsf{P}$ (Definition 2.2.1(3)).*

Proof. The top-down implication is immediate, taking $O = f^{\text{-}1}(O')$.

For the bottom-up implication, given p and an open neighbourhood $O' \ni f(p)$, we write

$$O = \bigcup\{O_{p,O'} \in \mathsf{Open} \mid p \in \mathsf{P},\ f(p) \in O'\}.$$

Above, $O_{p,O'}$ is the open neighbourhood of p in the preimage of O', which we know exists by Definition 2.2.1(3).

It is routine to check that $O = f^{\text{-}1}(O')$, and since this is a union of open sets, it is open. □

DEFINITION 2.2.3. Suppose that:

- $(\mathsf{P}, \mathsf{Open})$ is a semitopology and
- Val is a semitopology of values (Definition 2.1.3(1)) and
- $f : \mathsf{P} \to \mathsf{Val}$ is a value assignment (Definition 2.1.3(2); an assignment of a value to each element in P).

Then:

1. Call f **locally constant at** $p \in \mathsf{P}$ when there exists $p \in O_p \in \mathsf{Open}$ such that

$$\forall p' \in O_p. f(p) = f(p').$$

   So f is locally constant at p when it is constant on some open neighbourhood O_p of p.
2. Call f **locally constant** when it is locally constant at every $p \in \mathsf{P}$.

LEMMA 2.2.4. *Suppose* $(\mathsf{P}, \mathsf{Open})$ *is a semitopology and* Val *is a semitopology of values and* $f : \mathsf{P} \to \mathsf{Val}$ *is a value assignment. Then the following are equivalent:*

- *f is locally constant / locally constant at $p \in \mathsf{P}$ (Definition 2.2.3).*
- *f is continuous / continuous at $p \in \mathsf{P}$ (Definition 2.2.1).*

Proof. This is just by pushing around definitions, but we spell it out:

- Suppose f is continuous, consider $p \in \mathsf{P}$, and write $v = f(p)$. By our assumptions we know that $f^{-1}(v)$ is open, and $p \in f^{-1}(v)$. This is an open neighbourhood O_p on which f is constant, so we are done.

- Suppose f is locally constant, consider $p \in \mathsf{P}$, and write $v = f(p)$. By assumption we can find $p \in O_p \in \mathsf{Open}$ on which f is constant, so that $O_p \subseteq f^{-1}(v)$. □

REMARK 2.2.5 (Continuity = agreement). Lemma 2.2.4 tells us that we can view the problem of attaining agreement across an actionable coalition (as discussed in Subsection 1.1) as being the same thing as computing a value assignment that is continuous on that coalition (and possibly elsewhere).

To see why, consider a semitopology $(\mathsf{P}, \mathsf{Open})$ and following the intuitions discussed in Subsection 1.1 view points $p \in \mathsf{P}$ as *participants*; and view open neighbourhoods $p \in O \in \mathsf{Open}$ as **actionable coalitions** that include p. Then to say "f is a value assignment that is continuous at p" is to say that:

- f assigns a value or belief to $p \in \mathsf{P}$, and
- p is part of a (by Lemma 2.2.4 continuity) set of peers that agrees with p and (being open) can progress to act on this agreement.

Conceptually and mathematically this reduces the general question

How can we model collaborative action?

(which, to be fair, has more than one possible answer!) to a more specific research question

Understand continuous value assignments on semitopologies.

We then devote ourselves to elaborating (some of) a body of mathematics that we can pull out of this idea.

2.3 Neighbourhoods of a point

Definition 2.3.1 is a standard notion from topology, and Lemma 2.3.2 is a (standard) characterisation of openness, which will be useful later:

DEFINITION 2.3.1. Suppose $(\mathsf{P}, \mathsf{Open})$ is a semitopology and $p \in \mathsf{P}$ and $O \in \mathsf{Open}$. Then call O an **open neighbourhood** of p when $p \in O$.

In other words: an open set is (by definition) an *open neighbourhood* precisely for the points that it contains.

LEMMA 2.3.2. *Suppose* $(\mathsf{P}, \mathsf{Open})$ *is a semitopology and suppose* $P \subseteq \mathsf{P}$ *is any set of points. Then the following are equivalent:*

- $P \in \mathsf{Open}$.
- *Every point* p *in* P *has an open neighbourhood in* P.

In symbols we can write:

$$\forall p \in P. \exists O \in \mathsf{Open}. (p \in O \land O \subseteq P) \quad \textit{if and only if} \quad P \in \mathsf{Open}$$

Proof. If P is open then P itself is an open neighbourhood for every point that it contains.

Conversely, if every $p \in P$ contains some open neighbourhood $p \in O_p \subseteq P$ then $P = \bigcup \{O_p \mid p \in P\}$ and this is open by condition 2 of Definition 1.2.2. □

REMARK 2.3.3. An initial inspiration for modelling collaborative action using semitopologies, came from noting that the standard topological property described above in Lemma 2.3.2, corresponds to the *quorum sharing* property in [LGM19, Property 1]; the connection to topological ideas had not been noticed in [LGM19].

Transitive sets & topens 3

3.1 Some background on sets intersection

Some notation will be convenient:

NOTATION 3.1.1. Suppose X, Y, and Z are sets.

1. Write
$$X \between Y \quad \text{when} \quad X \cap Y \neq \varnothing.$$
When $X \between Y$ holds then we say (as standard) that X and Y **intersect**.
2. We may chain the $\between$ notation, writing for example
$$X \between Y \between Z \quad \text{for} \quad X \between Y \wedge Y \between Z$$
3. We may write $X \not\between Y$ for $\neg(X \between Y)$, thus $X \not\between Y$ when $X \cap Y = \varnothing$.

REMARK 3.1.2. *Note on design in Notation 3.1.1:* It is uncontroversial that if $X \neq \varnothing$ and $Y \neq \varnothing$ then $X \between Y$ should hold precisely when $X \cap Y \neq \varnothing$ — but there is an edge case! What truth-value should $X \between Y$ return when X or Y is empty?

1. It might be nice if $X \subseteq Y$ would imply $X \between Y$. This argues for setting
$$(X = \varnothing \vee Y = \varnothing) \Longrightarrow X \between Y.$$
2. It might be nice if $X \between Y$ were monotone on both arguments (i.e. if $X \between Y$ and $X \subseteq X'$ then $X' \between Y$). This argues for setting
$$(X = \varnothing \vee Y = \varnothing) \Longrightarrow X \not\between Y.$$
3. It might be nice if $X \between X$ always — after all, should a set *not* intersect itself? — and this argues for setting
$$\varnothing \between \varnothing,$$
even if we also set $\varnothing \not\between Y$ for nonempty Y.

All three choices are defensible, and they are consistent with the following nice property:

$$X \between Y \Longrightarrow (X \between X \vee Y \between Y).$$

We choose the second — if X or Y is empty then $X \not\between Y$ — because it gives the simplest definition that $X \between Y$ precisely when $X \cap Y \neq \varnothing$.

We list some elementary properties of $\between$ from Notation 3.1.1(1):

LEMMA 3.1.3.

1. *$X \between X$ if and only if $X \neq \varnothing$.*
2. *$X \between Y$ if and only if $Y \between X$.*
3. *$X \between (Y \cup Z)$ if and only if $(X \between Y) \vee (X \between Z)$.*
4. *If $X \subseteq X'$ and $X \neq \varnothing$ then $X \between X'$.*
5. *Suppose $X \between Y$. Then $X \subseteq X'$ implies $X' \between Y$, and $Y \subseteq Y'$ implies $X \between Y'$.*
6. *If $X \between Y$ then $X \neq \varnothing$ and $Y \neq \varnothing$.*

Proof. By facts of sets intersection. □

3.2 Transitive open sets and value assignments

DEFINITION 3.2.1. Suppose $(\mathsf{P}, \mathsf{Open})$ is a semitopology. Suppose $T \subseteq \mathsf{P}$ is any set of points.

1. Call T **transitive** when

$$\forall O, O' \in \mathsf{Open}. O \between T \between O' \Longrightarrow O \between O'.$$

2. Call T **topen** when T is nonempty transitive and open.[1]
   We may write

$$\mathsf{Topen} = \{T \in \mathsf{Open}_{\neq\varnothing} \mid T \text{ is transitive}\}.$$

3. Call S a **maximal topen** when S is a topen that is not a subset of any strictly larger topen.[2]

Theorem 3.2.2 clarifies why transitivity is interesting: continuous value assignments are constant — if we think of points as participants, 'constant function' here means 'in agreement' — across transitive sets.

THEOREM 3.2.2. *Suppose that:*

- $(\mathsf{P}, \mathsf{Open})$ *is a semitopology.*
- Val *is a semitopology of values (a nonempty set with the discrete semitopology; see Definition 2.1.3(1)).*
- $f : \mathsf{P} \to \mathsf{Val}$ *is a value assignment (Definition 2.1.3(2)).*

[1]The empty set is trivially transitive and open, so it would make sense to admit it as a (degenerate) topen. However, it turns out that we mostly need the notion of 'topen' to refer to certain kinds of neighbourhoods of points (we will call them *communities*; see Definition 4.1.4). It is therefore convenient to exclude the empty set from being topen, because while it is the neighbourhood of every point that it contains, it is not a neighbourhood of any point.

[2]'Transitive open' → 'topen', like 'closed and open' → 'clopen'.
For convenient reference, note that related notions of *strong* transitivity and topen are in Definition 3.7.5.

- $T \subseteq \mathsf{P}$ *is a transitive set (Definition 3.2.1) — in particular this will hold if* T *is topen — and* $p, p' \in T$.

Then:

1. *If* f *is continuous at* p *and* p' *then* $f(p) = f(p')$.
2. *As a corollary, if* f *is continuous on* T*, then* f *is constant on* T.

In words we can say:

Continuous value assignments are constant across transitive sets.

Proof. Part 2 follows from part 1 since if $f(p) = f(p')$ for *any* $p, p' \in T$, then by definition f is constant on T. So we now just need to prove part 1 of this result.

Consider $p, p' \in T$. By continuity on T, there exist open neighbourhoods $p \in O \subseteq f^{-1}(f(p))$ and $p' \in O' \subseteq f^{-1}(f(p'))$. By construction $O \between T \between O'$ (because $p \in O \cap T$ and $p' \in T \cap O'$). By transitivity of T it follows that $O \between O'$. Thus, there exists $p'' \in O \cap O'$, and by construction $f(p) = f(p'') = f(p')$. □

Corollary 3.2.3 is an easy and useful consequence of Theorem 3.2.2:

COROLLARY 3.2.3. *Suppose that:*

- $(\mathsf{P}, \mathsf{Open})$ *is a semitopology.*
- $f : \mathsf{P} \to \mathsf{Val}$ *is a value assignment to some set of values* Val *(Definition 2.1.3).*
- f *is continuous on topen sets* $T, T' \in \mathsf{Topen}$.

Then

$$T \between T' \quad \textit{implies} \quad \forall p \in T, p' \in T'. f(p) = f(p').$$

Proof. By Theorem 3.2.2 f is constant on T and T'. We assumed that T and T' intersect, and the result follows. □

A converse to Theorem 3.2.2 also holds:

PROPOSITION 3.2.4. *Suppose that:*

- $(\mathsf{P}, \mathsf{Open})$ *is a semitopology.*
- Val *is a semitopology of values with at least two elements (to exclude a degenerate case that no functions exist, or they exist but there is only one because there is only one value to map to).*
- $T \subseteq \mathsf{P}$ *is any set.*

Then

- if *for every* $p, p' \in T$ *and every value assignment* $f : \mathsf{P} \to \mathsf{Val}$, f *continuous at* p *and* p' *implies* $f(p) = f(p')$,
- then T *is transitive.*

Proof. We prove the contrapositive. Suppose T is not transitive, so there exist $O, O' \in \mathsf{Open}$ such that $O \between T \between O'$ and yet $O \cap O' = \varnothing$. We choose two distinct values $v \neq v' \in \mathsf{Val}$ and define f to map any point in O to v and any point in $\mathsf{P} \setminus O$ to v'.

Choose some $p \in O$ and $p' \in O'$. It does not matter which, and some such p and p' exist, because O and O' are nonempty by Lemma 3.1.3(6), since $O \between T$ and $O' \between T$).

We note that $f(p) = v$ and $f(p') = v'$ and f is continuous at $p \in O$ and $p' \in O' \subseteq \mathsf{P} \setminus O$, yet $f(p) \neq f(p')$. □

We can sum up what Theorem 3.2.2 and Proposition 3.2.4 mean, as follows:

Remark 3.2.5. Suppose $(\mathsf{P}, \mathsf{Open})$ is a semitopology and Val is a semitopology of values with at least two elements. Say that a value assignment $f : \mathsf{P} \to \mathsf{Val}$ **splits** a set $T \subseteq \mathsf{P}$ when there exist $p, p' \in T$ such that f is continuous at p and p' and $f(p) \neq f(p')$. Then Theorem 3.2.2 and Proposition 3.2.4 together say in words that:

> $T \subseteq \mathsf{P}$ is transitive if and only if it cannot be split by a value assignment that is continuous on T.

Intuitively, transitive sets characterise areas of guaranteed agreement.

This reminds us of a basic result in topology about *connected spaces* [Wil70, Chapter 8, section 26]. Call a topological space $(\mathsf{T}, \mathsf{Open})$ **disconnected** when there exist open sets $O, O' \in \mathsf{Open}$ such that $O \cap O' = \varnothing$ (in our notation: $O \not\between O'$) and $O \cup O' = \mathsf{T}$; otherwise call $(\mathsf{T}, \mathsf{Open})$ **connected**. Then $(\mathsf{T}, \mathsf{Open})$ is disconnected if and only if (in our terminology above) it can be split by a value assignment. Theorem 3.2.2 and Proposition 3.2.4 are not identical to that result, but they are in the same spirit.

Remark 3.2.6. The notion of transitive set gives us enough to comment on the two examples in Subsection 1.1. Recall that we considered:

1. A nonempty finite set $\mathbb{E}$ with open sets $\mathsf{Open}(\mathbb{E})$ ('actionable coalitions') being majority subsets $O \subseteq \mathbb{E}$.
2. Integers $\mathbb{Z}$ with open sets $\mathsf{Open}(\mathbb{Z})$ generated by triplets $\{2i, 2i{+}1, 2i{+}2\}$.

The reader can check that in $(\mathbb{E}, \mathsf{Open}(\mathbb{E}))$ *every* set is transitive, because every pair of nonempty open sets intersect; thus, no $T \subseteq \mathbb{E}$ can be split by a value assignment that is continuous on T. In contrast, the reader can check that in $(\mathbb{Z}, \mathsf{Open}(\mathbb{Z}))$, most sets are not transitive, including (for example) $\{0, 4\}$. This lack of transitivity reflects an intuitive observation we made in Subsection 1.1 that our second example was 'not necessarily particularly safe or desirable in practice'; in our more technical language, we can now note that there exists a value assignment that splits $\{0, 4\}$, yet is continuous at 0 and 4. What $(\mathbb{Z}, \mathsf{Open}(\mathbb{Z}))$ does satisfy is the weaker (but still useful!) safety property that any continuous value assignment that is continuous everywhere, is constant (corresponding to our informal observation that "*if* all participants do legally progress, then they announce the same value").[3] This reflects a useful intuition, that the topological notion of 'continuity at a point', corresponds to an intuition of p as a participant 'following the rules'.

[3]We can be more precise if we like: e.g. T cannot be split by a value assignment that is continuous on a contiguous segment of $\mathbb{Z}$ that includes T. Continuity on all of $\mathbb{Z}$ is one sufficient condition for this, which corresponds (in the language of consensus) to assuming that all participants are correct. But we digress.

3.3 Examples and discussion of transitive sets and topens

We may routinely order sets by subset inclusion; including open sets, topens, closed sets, and so on, and we may talk about maximal, minimal, greatest, and least elements. We include the (standard) definition for reference:

NOTATION 3.3.1. Suppose $(\mathsf{P}, \leq)$ is a poset. Then:

1. Call $p \in \mathsf{P}$ **maximal** when $\forall p'.p \leq p' \Longrightarrow p' = p$ and **minimal** when $\forall p'.p' \leq p \Longrightarrow p' = p$.
2. Call $p \in \mathsf{P}$ **greatest** when $\forall p.p' \leq p$ and **least** when $\forall p'.p \leq p'$.

EXAMPLE 3.3.2 (Examples of transitive sets).

1. $\{p\}$ is transitive, for any single point $p \in \mathsf{P}$.
2. The empty set $\varnothing$ is (trivially) transitive. It is not topen because we insist in Definition 3.2.1(2) that topens are nonempty.
3. Call a set $P \subseteq \mathsf{P}$ *topologically indistinguishable* when (using Notation 3.1.1) for every open set O,
$$P \between O \Longleftrightarrow P \subseteq O.$$
It is easy to check that if P is topologically indistinguishable, then it is transitive.

EXAMPLE 3.3.3 (Examples of topens).

1. Take $\mathsf{P} = \{0, 1, 2\}$, with open sets $\varnothing$, P, $\{0\}$, and $\{2\}$. This has two maximal topens $\{0\}$ and $\{2\}$ as illustrated in Figure 3.1 (top-left diagram).
2. Take $\mathsf{P} = \{0, 1, 2\}$, with open sets $\varnothing$, P, $\{0\}$, $\{0, 1\}$, $\{2\}$, $\{1, 2\}$, and $\{0, 2\}$. This has two maximal topens $\{0\}$ and $\{2\}$, as illustrated in Figure 3.1 (top-right diagram).
3. Take $\mathsf{P} = \{0, 1, 2, 3, 4\}$, with open sets generated by $\{0, 1\}$, $\{1\}$, $\{3\}$, and $\{3, 4\}$. This has two maximal topens $\{0, 1\}$ and $\{3, 4\}$, as illustrated in Figure 3.1 (lower-left diagram).
4. Take $\mathsf{P} = \{0, 1, 2, *\}$, with open sets generated by $\{0\}$, $\{1\}$, $\{2\}$, $\{0, 1, *\}$, and $\{1, 2, *\}$. This has three maximal topens $\{0\}$, $\{1\}$, and $\{2\}$, as illustrated in Figure 3.1 (lower-right diagram).
5. Take the all-but-one semitopology from Example 2.1.4(5c) on $\mathbb{N}$: so $\mathsf{P} = \mathbb{N}$ with opens $\varnothing$, $\mathbb{N}$, and $\mathbb{N} \setminus \{x\}$ for every $x \in \mathbb{N}$. This has a single maximal topen $\mathbb{N}$.
6. The semitopology in Figure 5.2 has no topen sets at all ($\varnothing$ is transitive and open, but by definition in Definition 3.2.1(2) topens have to be nonempty).

REMARK 3.3.4 (Discussion). We take a moment for a high-level discussion of where we are going.

The semiopologies in Example 3.3.3 invite us to ask what makes these examples different (especially parts 1 and 2). Clearly they are not equal, but that is a superficial answer in the sense that it is valid just in the world of sets, and it ignores semitopological structure.

For comparison: if we ask what makes 0 and 1 different in $\mathbb{N}$, we could just to say that $0 \neq 1$, but this ignores what makes them different *as numbers*. For more insight, we could note that 0 is the additive unit whereas 1 is the multiplicative unit of $\mathbb{N}$ as a semiring; or that 0 is a least element and 1 is the unique atom of $\mathbb{N}$ as a well-founded poset; or that 1 is the successor of 0 of $\mathbb{N}$ as a well-founded inductive structure. Each of these answers gives us more understanding, not only into 0 and 1 but also into the structures that can be given to $\mathbb{N}$ itself.

So we can ask:

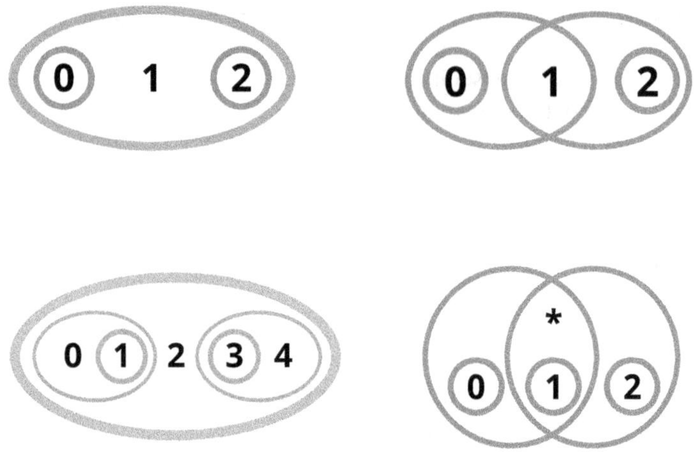

Here and elsewhere, we might omit open sets that are unions of open sets that are illustrated. For example, we explicitly draw the universal open set in the left-hand diagrams above, but not in the right-hand diagrams above. Meaning is clear and we get cleaner diagrams.

Figure 3.1: Examples of topens (Example 3.3.3)

> *What semitopological property or properties on points can identify the essential nature of the differences between the semitopologies in Example 3.3.3?*

There would be some truth to saying that the rest of our investigation is devoted to developing and understanding answers to this question! In particular, we will shortly define the set of *intertwined points* $p_{\between}$ in Definition 3.6.1. Example 3.6.2 will note that $1_{\between} = \{0, 1, 2\}$ in Example 3.3.3(1), whereas $1_{\between} = \{1\}$ in Example 3.3.3(2), and $x_{\between} = \mathbb{N}$ for every x in Example 3.3.3(3).

3.4 Closure properties of transitive sets

REMARK 3.4.1. Transitive sets have some nice closure properties which we treat in this Subsection — here we mean 'closure' in the sense of "the set of transitive sets is closed under various operations", and not in the topological sense of 'closed sets'.

Topens — nonempty transitive *open* sets — will have even better closure properties, which emanate from the requirement in Lemma 3.4.3 that at least one of the transitive sets T or T' is open. See Subsection 3.5.

LEMMA 3.4.2. *Suppose* $(\mathsf{P}, \mathsf{Open})$ *is a semitopology and* $T \subseteq \mathsf{P}$. *Then:*

1. *If* T *is transitive and* $T' \subseteq T$, *then* T' *is transitive.*
2. *If* T *is topen and* $\varnothing \neq T' \subseteq T$ *is nonempty and open, then* T' *is topen.*

Proof.

1. By Definition 3.2.1 it suffices to consider open sets O and O' such that $O \between T' \between O'$, and prove that $O \between O'$. But this is simple: by Lemma 3.1.3(5) $O \between T \between O'$, so $O \between O'$ follows by transitivity of T.

2. Direct from part 1 of this result and Definition 3.2.1(2). □

LEMMA 3.4.3. *Suppose that:*

- $(\mathsf{P}, \mathsf{Open})$ *is a semitopology.*
- $T, T' \subseteq \mathsf{P}$ *are transitive.*
- *At least one of T and T' is open.*

Then:

1. $\forall O, O' \in \mathsf{Open}. O \between T \between T' \between O' \Longrightarrow O \between O'$.
2. *If $T \between T'$ then $T \cup T'$ is transitive.*

Proof.

1. We simplify using Definition 3.2.1 and our assumption that one of T and T' is open. We consider the case that T' is open:

$$\begin{aligned} O \between T \between T' \between O' &\Longrightarrow O \between T' \between O' && T \text{ transitive, } T' \text{ open} \\ &\Longrightarrow O \between O' && T' \text{ transitive.} \end{aligned}$$

The argument for when T is open, is precisely similar.

2. Suppose $O \between T \cup T' \between O'$. By Lemma 3.1.3(3) (at least) one of the following four possibilities must hold:

$$O \between T \wedge T \between O', \quad O \between T' \wedge T \between O', \quad O \between T \wedge T' \between O', \quad \text{or} \quad O \between T' \wedge T' \between O'.$$

If $O \between T \wedge T' \between O'$ then by part 1 of this result we have $O \between O'$ as required. The other possibilities are no harder. □

DEFINITION 3.4.4 (Ascending/descending chain). A **chain** of sets $\mathcal{X}$ is a collection of sets that is totally ordered by subset inclusion $\subseteq$.[4]

We may call a chain **ascending** or **descending** if we want to emphasise that we are thinking of the sets as 'going up' or 'going down'.

LEMMA 3.4.5. *Suppose $(\mathsf{P}, \mathsf{Open})$ is a semitopology and suppose $\mathcal{T}$ is a chain of transitive sets (Definition 3.4.4). Then $\bigcup \mathcal{T}$ is a transitive set.*

Proof. Suppose $O \between \bigcup \mathcal{T} \between O'$. Then there exist $T, T' \in \mathcal{T}$ such that $O \between T$ and $T' \between O'$. But $\mathcal{T}$ is totally ordered, so either $T \subseteq T'$ or $T \supseteq T'$. In the former case it follows that $O \between T' \between O'$ so that $O \between O'$ by transitivity of T'; the latter case is precisely similar. □

[4]A total order is reflexive, transitive, antisymmetric, and total.

3.5 Closure properties of topens

Definition 3.5.1 will be useful in Proposition 3.5.2(2):

DEFINITION 3.5.1. Suppose $(\mathsf{P}, \mathsf{Open})$ is a semitopology. Call a set of nonempty open sets $\mathcal{O} \subseteq \mathsf{Open}_{\neq\varnothing}$ a **clique** when its elements pairwise intersect.[5] In symbols:

$$\mathcal{O} \subseteq \mathsf{Open} \text{ is a clique} \quad \text{when} \quad \forall O, O' \in \mathcal{O}. O \between O'.$$

Note that if $\mathcal{O}$ is a clique then every $O \in \mathcal{O}$ is nonempty, since if $O = \varnothing$ then $O \not\between O$ by Lemma 3.1.3(1).

PROPOSITION 3.5.2. *Suppose* $(\mathsf{P}, \mathsf{Open})$ *is a semitopology. Then:*

1. *If T and T' are an intersecting pair of topens (i.e. $T \between T'$), then $T \cup T'$ is topen.*
2. *If $\mathcal{T}$ is a clique of topens (Definition 3.5.1), then $\bigcup \mathcal{T}$ is topen.*
3. *If $\mathcal{T}$ is a nonempty ascending chain of topens then $\bigcup \mathcal{T}$ is topen.*

Proof.

1. $T \cup T'$ is open because by Definition 1.2.2(2) open sets are closed under arbitrary unions, and by Lemma 3.4.3(2) $T \cup T'$ is transitive.
2. $\bigcup \mathcal{T}$ is open by Definition 1.2.2(2). Also, if $O \between \bigcup \mathcal{T} \between O'$ then there exist $T, T' \in \mathcal{T}$ such that $O \between T$ and $T' \between O'$. We assumed $T \between T'$, so by Lemma 3.4.3(1) (since T and T' are open) we have $O \between O'$ as required.
3. Any chain is pairwise intersecting. We use part 2 of this result.[6] □

COROLLARY 3.5.3. *Suppose* $(\mathsf{P}, \mathsf{Open})$ *is a semitopology. Then every topen T is contained in a unique maximal topen.*

Proof. Consider $\mathcal{T}$ defined by

$$\mathcal{T} = \{T \cup T' \mid T' \text{ topen} \wedge T \between T'\}.$$

By Proposition 3.5.2(1) this is a set of topens. By construction they all contain T, and by our assumption that $T \neq \varnothing$ they pairwise intersect (since they all contain T, at least), so by Proposition 3.5.2(2) $\bigcup \mathcal{T}$ is topen. It is easy to check that this is the unique maximal transitive open set that contains T. □

THEOREM 3.5.4. *Suppose* $(\mathsf{P}, \mathsf{Open})$ *is a semitopology. Then any $P \subseteq \mathsf{P}$, and in particular P itself, can be partitioned into:*

- *Some disjoint collection of maximal topens.*

---

[5] We call this a *clique*, because if we form the *intersection graph* with nodes elements of $\mathcal{O}$ and with an (undirected) edge between O and O' when $O \between O'$, then $\mathcal{O}$ is a clique precisely when its intersection graph is indeed a clique. See also Definition 3.7.12. We will return to intersection graphs in Subsection 17.1.1.

[6] We could also use Lemma 3.4.5. The chain needs to be nonempty because $\bigcup \varnothing = \varnothing$ and this is open but not topen (= nonempty, transitive, and open). The reader might ask why Lemma 3.4.5 was not derived directly from Lemma 3.4.3(2); this is because (interestingly) Lemma 3.4.5 does not require openness.

- *A set of other points, which are not contained in any topen.*

Proof. Routine from Corollary 3.5.3. □

REMARK 3.5.5. It may be useful to put Theorem 3.5.4 in the context of the terminology, results, and examples that will follow below. We will have Definition 4.1.4(3&6) and Theorem 4.2.6. These will allow us to call a point p contained in some maximal topen T *regular*, to call the maximal topen T of a regular point its *community*, and a point that is not contained in any topen *irregular*. Then Theorem 3.5.4 says that a semitopology P can be partitioned into:

- Disjoint maximal communities of regular points which, in a sense made formal in Theorem 3.2.2, are a coalition acting together — and
- a set of irregular points, which are in no community and so are not members of any coalition.

We give examples in Example 3.3.3 and Figure 3.1, and we will see more elaborate examples below (see in particular the collection in Example 5.6.10).

In the special case that the entire space consists of a single topen community, there are no irregular points and all participants are guaranteed to agree, where algorithms succeed. For the application of a single blockchain trying to arrive at consensus, this discussion tells us that we want it to consist of a single topen.

3.6 Intertwined points

3.6.1 The basic definition, and some lemmas

DEFINITION 3.6.1. Suppose $(\mathsf{P}, \mathsf{Open})$ is a semitopology and $p, p' \in \mathsf{P}$.

1. Call p and p' **intertwined** when $\{p, p'\}$ is transitive. Unpacking Definition 3.2.1 this means:
$$\forall O, O' \in \mathsf{Open}.(p \in O \wedge p' \in O') \Longrightarrow O \between O'.$$
By a mild abuse of notation, write
$$p \between p' \quad \text{when} \quad p \text{ and } p' \text{ are intertwined.}$$
2. Define $p_{\between}$ (read 'intertwined of p') to be the set of points intertwined with p. In symbols:
$$p_{\between} = \{p' \in \mathsf{P} \mid p \between p'\}.$$

EXAMPLE 3.6.2. We return to the examples in Example 3.3.3. There we note that:

1. $1_{\between} = \{0, 1, 2\}$ and $0_{\between} = \{0, 1\}$ and $2_{\between} = \{1, 2\}$.
2. $1_{\between} = \{1\}$ and $0_{\between} = \{0\}$ and $2_{\between} = \{2\}$.
3. $0_{\between} = 1_{\between} = \{0, 1, 2\}$ and $3_{\between} = 4_{\between} = \{2, 3, 4\}$ and $2_{\between} = \mathsf{P}$.
4. $0_{\between} = \{0\}$ and $1_{\between} = *_{\between} = \{1, *\}$ and $2_{\between} = \{2\}$.
5. $x_{\between} = \mathsf{P}$ for every x.
6. $x_{\between} = \{x\}$ for every x.

Here is one reason to care about intertwined points; a value assignment is constant on a pair of intertwined points, where it is continuous:

LEMMA 3.6.3. *Suppose* Val *is a semitopology of values and* $f : \mathsf{P} \to \mathsf{Val}$ *is a value assignment (Definition 2.1.3) and* $p, p' \in \mathsf{P}$ *and* $p \between p'$. *Then if* f *is continuous at* p *and* p', *then* $f(p) = f(p')$.

Proof. $\{p, p'\}$ is transitive by Definition 3.6.1(1). we use Theorem 3.2.2. □

We might suppose that being intertwined is transitive. Lemma 3.6.4 shows that this is not necessarily the case (the case when $\between$ *is* transitive at p is an important well-behavedness property, which we will call being *unconflicted*; see Subsection 6.1 and Definition 6.1.1):

LEMMA 3.6.4. *Suppose* $(\mathsf{P}, \mathsf{Open})$ *is a semitopology. Then:*

1. *The 'is intertwined' relation* $\between$ *is reflexive and symmetric.*
2. $\between$ *is not necessarily transitive. That is:* $p' \between p \between p''$ *does not necessarily imply* $p' \between p''$.

Proof. Reflexivity and symmetry are clear from Definition 3.6.1(1) and Lemma 3.1.3(3).

To show that transitivity need not hold, it suffices to provide a counterexample. The semitopology from Example 3.3.3(1) (illustrated in Figure 3.1, top-left diagram) will do. Take

$$\mathsf{P} = \{0, 1, 2\} \quad \text{and} \quad \mathsf{Open} = \{\varnothing, \mathsf{P}, \{0\}, \{2\}\}.$$

Then

$$0 \between 1 \text{ and } 1 \between 2, \quad \text{but} \quad \neg(0 \between 2).$$

□

We conclude with an easy observation:

NOTATION 3.6.5. Suppose $(\mathsf{P}, \mathsf{Open})$ is a semitopology. Call P **intertwined** when

$$\forall p, p' \in \mathsf{P}. p \between p'.$$

In words: P is intertwined when all of its points are pairwise intertwined.

Lemma 3.6.6 will be useful later, notably for Lemma 4.1.9:

LEMMA 3.6.6. *Suppose* $(\mathsf{P}, \mathsf{Open})$ *is a semitopology. Then the following conditions are equivalent:*

1. P *is an intertwined space.*
2. P *is a transitive set in the sense of Definition 3.2.1(1).*
3. *All nonempty open sets intersect.*
4. *Every nonempty open set is topen.*

Proof. Routine by unpacking the definitions. □

REMARK 3.6.7. A topologist would call an intertwined space *hyperconnected* (see Definition 3.7.12 and the following discussion). This is also — modulo closing under arbitrary unions — what an expert in the classical theory of consensus might call a *quorum system* [NW94].

3.6.2 Pointwise characterisation of transitive sets

LEMMA 3.6.8. *Suppose* $(\mathsf{P}, \mathsf{Open})$ *is a semitopology and* $T \subseteq \mathsf{P}$. *Then the following are equivalent:*

1. T *is transitive.*
2. $p \between p'$ *(meaning by Definition 3.6.1 that* $\{p, p'\}$ *is transitive) for every* $p, p' \in T$.

Proof. Suppose T is transitive. Then by Lemma 3.4.2(1), $\{p, p'\}$ is transitive for every $p, p' \in T$.

Suppose $\{p, p'\}$ is transitive for every $p, p' \in T$. Consider open sets O and O' such that $O \between T \between O'$. Choose $p \in O \cap T$ and $p' \in O \cap T'$. By construction $\{p, p'\} \subseteq T$ so this is transitive. It follows that $O \between O'$ as required. □

The special case of Lemma 3.6.8 where T is an open set will be particularly useful:

PROPOSITION 3.6.9. *Suppose* $(\mathsf{P}, \mathsf{Open})$ *is a semitopology and* $T \subseteq \mathsf{P}$. *Then the following are equivalent:*

1. T *is topen.*
2. $T \in \mathsf{Open}_{\neq \varnothing}$ *and* $\forall p, p' \in T. p \between p'$.

In words we can say:

> *A topen is a nonempty open set of intertwined points.*

Proof. By Definition 3.2.1(2), T is topen when it is nonempty, open, and transitive. By Lemma 3.6.8 this last condition is equivalent to $p \between p'$ for every $p, p' \in T$. □

REMARK 3.6.10 (Intertwined as 'non-Hausdorff').
Recall that we call a topological space $(\mathsf{P}, \mathsf{Open})$ **Hausdorff** (or T_2) when any two points can be separated by pairwise disjoint open sets. Using the $\between$ symbol from Notation 3.1.1, we rephrase the Hausdorff condition as

$$\forall p, p'. p \neq p' \Longrightarrow \exists O, O'. (p \in O \wedge p' \in O' \wedge \neg(O \between O')),$$

we can simplify to

$$\forall p, p'. p \neq p' \Longrightarrow p \not\between p',$$

and thus we simplify the Hausdorff condition just to

$$\forall p. p_{\between} = \{p\}. \tag{3.1}$$

Note how distinct p and p' being intertwined is the *opposite* of being Hausdorff: $p \between p'$ when $p' \in p_{\between}$, and they *cannot* be separated by pairwise disjoint open sets. Thus the assertion $p \between p'$ in Proposition 3.6.9 is a negation to the Hausdorff property:

$$\exists p. p_{\between} \neq \{p\}.$$

This is useful because for semitopologies as applied to consensus,

- being Hausdorff means that the space is separated (which is probably a bad thing, if we are looking for a system with lots of points in consensus), whereas

- being full of intertwined points means by Theorem 3.2.2 that the system will (where algorithms succeed) be full of points whose value assignment agrees (which is a good thing).

In the blockchain literature, we say that a blockchain *forks* when it partitions into two sets of participants with incompatible beliefs about the state of the system. In this light, we can view Theorem 3.2.2 as a result making precise sufficient conditions to ensure that this does not happen.

3.7 Strong topens: topens that are also subspaces

3.7.1 Definition and main result

Let us take stock and recall that:

- T is *topen* when it is a nonempty open transitive set (Definition 3.2.1).
- T is *transitive* when $O \between T \between O'$ implies $O \between O'$ for all $O, O' \in \mathsf{Opens}$ (Definition 3.2.1).
- $O \between O'$ means that $O \cap O' \neq \varnothing$ (Notation 3.1.1).

But, note above that if T is topen and $O \between T \between O'$ then $O \cap O'$ need not intersect *inside* T. It could be that O and O' intersect outside of T (an example is in the proof Lemma 3.7.2 below).

Definition 3.7.1 spells out a standard topological construction in the language of semitopologies:

Definition 3.7.1 (Subspaces). Suppose $(\mathsf{P}, \mathsf{Open})$ is a semitopology and suppose $T \subseteq \mathsf{P}$ is a set of points. Write $(T, \mathsf{Open} \cap T)$ for the semitopology such that:

- The points are T.
- The open sets have the form $O \cap T$ for $O \in \mathsf{Open}$.

We say that $(T, \mathsf{Open} \cap T)$ is T with the **semitopology induced by** $(\mathsf{P}, \mathsf{Open})$.

We may call $(T, \mathsf{Open} \cap T)$ a **subspace** of $(\mathsf{P}, \mathsf{Open})$, and if the open sets are understood then we may omit mention of them and just write:

A subset $T \subseteq \mathsf{P}$ is naturally a **(semitopological) subspace** of P.

Lemma 3.7.2. *The property of being a (maximal) topen is not necessarily closed under taking subspaces.*

Proof. It suffices to exhibit a semitopology $(\mathsf{P}, \mathsf{Open})$ and a subset $T \subseteq \mathsf{P}$ such that T is topen in $(\mathsf{P}, \mathsf{Open})$ but T is not topen in $(T, \mathsf{Open} \cap T)$. We set:

$$\mathsf{P} = \{0,1,2\} \qquad \mathsf{Open} = \{\varnothing,\ \{0,2\},\ \{1,2\},\ \{0,1\},\ \mathsf{P}\} \qquad T = \{0,1\}$$

as illustrated in Figure 3.2 (left-hand diagram). Now:

- T is topen in $(\mathsf{P}, \mathsf{Open})$, because every open neighbourhood of 0 — that is $\{0,2\}$, $\{0,1\}$, and P — intersects with every open neighbourhood of 1 — that is $\{1,2\}$, $\{0,1\}$, and P.
- T is not topen in $(T, \mathsf{Open} \cap T)$, because $\{0\}$ is an open neighbourhood of 0 and $\{1\}$ is an open neighbourhood of 1 and these do not intersect. □

(a) A topen that is not strong (Lemma 3.7.2)

(b) A transitive set that is not strongly transitive (Lemma 3.7.6(2))

Figure 3.2: Two counterexamples for (strong) transitivity

Lemma 3.7.2 motivates the following definitions:

DEFINITION 3.7.3. Suppose X, Y, and Z are sets. Write $X \between_Y Z$, and say that X and Z **meet** or **intersect in** Y, when $(X \cap Y) \between (Z \cap Y)$.

LEMMA 3.7.4. *Suppose X, Y, and Z are sets. Then:*

1. *The following are equivalent:*

$$X \cap Y \cap Z \neq \varnothing \quad \Longleftrightarrow \quad X \between_Y Z \quad \Longleftrightarrow \quad Y \between_X Z \quad \Longleftrightarrow \quad X \between_Z Y.$$

2. *$X \between_Y Y$ if and only if $X \between Y$.*
3. *If $X \between_Y Z$ then $X \between Z$.*

Proof. From Definition 3.7.3, by elementary sets calculations. □

DEFINITION 3.7.5. Suppose $(\mathsf{P}, \mathsf{Open})$ is a semitopology and recall from Definition 3.2.1 the notions of *transitive set* and *topen*.

1. Call $T \subseteq \mathsf{P}$ **strongly transitive** when

$$\forall O, O' \in \mathsf{Open}. O \between T \between O' \Longrightarrow O \between_T O'.$$

2. Call T a **strong topen** when T is nonempty open and strongly transitive,

LEMMA 3.7.6. *Suppose $(\mathsf{P}, \mathsf{Open})$ is a semitopology and $T \subseteq \mathsf{P}$. Then:*

1. *If T is strongly transitive then it is transitive.*
2. *The reverse implication need not hold (even if $(\mathsf{P}, \mathsf{Open})$ is a topology): it is possible for T to be transitive but not strongly transitive.*

Proof. We consider each part in turn:

1. Suppose T is strongly transitive and suppose $O \between T \between O'$. By Lemma 3.7.4(2) $O \between_T T \between_T O'$. By strong transitivity $O \between_T O'$. By Lemma 3.7.4(3) $O \between O'$. Thus T is transitive.

2. It suffices to provide a counterexample. This is illustrated in Figure 3.2 (right-hand diagram). We set:

   - $\mathsf{P} = \{0, 1, 2\}$, and
   - $\mathsf{Open} = \{\varnothing, \{1\}, \{0,1\}, \{1,2\}, \{0,1,2\}\}$.
   - We set $T = \{0, 2\}$.

   We note that $(\mathsf{P}, \mathsf{Open})$ is a topology, and it is easy to check that T is transitive — we just note that $\{0,1\} \between T \between \{1,2\}$ and $\{0,1\} \between \{1,2\}$. However, T is not strongly transitive, because $\{0,1\} \cap \{1,2\} = \{1\} \not\between T$. □

PROPOSITION 3.7.7. *Suppose* $(\mathsf{P}, \mathsf{Open})$ *is a semitopology and suppose* $T \in \mathsf{Open}$. *Then the following are equivalent:*

1. *T is a strong topen.*
2. *T is a topen in $(T, \mathsf{Open} \cap T)$ (Definition 3.7.1).*

Proof. Suppose T is a strong topen; thus T is nonempty, open, and strongly transitive in $(\mathsf{P}, \mathsf{Open})$. Then by construction T is open in $(T, \mathsf{Open} \cap T)$, and the strong transitivity property of Definition 3.7.5 asserts precisely that T is transitive as a subset of $(T, \mathsf{Open} \cap T)$.

Now suppose T is a topen in $(T, \mathsf{Open} \cap T)$; thus T is nonempty, open, and transitive in $(T, \mathsf{Open} \cap T)$. Then T is nonempty and by assumption above $T \in \mathsf{Open}$.[7] Now suppose $O, O' \in \mathsf{Open}$ and $O \between T \between O'$. Then by Lemma 3.7.4(2) $O \between_T T \between_T O'$, so by transitivity of T in $(T, \mathsf{Open} \cap T)$ also $O \between_T O'$, and thus by Lemma 3.7.4(3) also $O \between O'$. □

3.7.2 Connection to lattice theory

There is a notion from order-theory of a *join-irreducible* element (see for example in [DP02, Definition 2.42]), and a dual notion of *meet-irreducible* element:

DEFINITION 3.7.8. Call an element s in a lattice $\mathcal{L}$

- **join-irreducible** when s is not a bottom element, and s is not a join of two strictly smaller elements: if $x \vee y = s$ then $x = s$, or $y = s$, and
- **meet-irreducible** when s is not a top element, and s is not a meet of two strictly greater elements: if $x \wedge y = s$ then $x = s$ or $y = s$.

This definition is typically given for lattices, but it makes just as much sense for semilattices as well.

EXAMPLE 3.7.9.

1. Consider the lattice of finite (possibly empty) subsets of $\mathbb{N}$, with $\mathbb{N}$ adjoined as a top element. Then $\mathbb{N}$ is join-irreducible; $\mathbb{N} \subseteq \mathbb{N}$ is not a bottom element, and if $x \cup y = \mathbb{N}$ then either $x = \mathbb{N}$ or $y = \mathbb{N}$.

[7] It does not follow from T being open in $(T, \mathsf{Open} \cap T)$ that T is open in $(\mathsf{P}, \mathsf{Open})$, which is why we included an assumption that this holds in the statement of the result.

2. Consider $\mathbb{N}$ with the **final segment semitopology** such that opens are either $\varnothing$ or sets $n_{\geq} = \{n' \in \mathbb{N} \mid n' \geq n\}$.
   Then $\varnothing$ is meet-irreducible; $\varnothing$ is not a top element, and if $x \cap y = \varnothing$ then either $x = \varnothing$ or $y = \varnothing$.
3. Consider the integers with the lattice structure in which meet is minimum and join is maximum. Then every element is join- and meet-irreducible; if $x \vee y = z$ then $x = z$ or $y = z$, and similarly for $x \wedge y$.

We spell out how this is related to our notions of transitivity from Definitions 3.2.1 and 3.7.5:

LEMMA 3.7.10. *Suppose* $(\mathsf{P}, \mathsf{Open})$ *is a semitopology and* $T \subseteq \mathsf{P}$. *Then:*

1. *T is strongly transitive if and only if $\varnothing$ is meet-irreducible in $(T, \mathsf{Open} \cap T)$ (Definition 3.7.1).*
2. *T is transitive if $\varnothing$ is meet-irreducible in $(T, \mathsf{Open} \cap T)$.*
3. *If T is transitive it does not necessarily follow that $\varnothing$ is meet-irreducible in $(T, \mathsf{Open} \cap T)$.*

Proof. We reason as follows:

1. $\varnothing$ is meet-irreducible in $(T, \mathsf{Open} \cap T)$ means that $(O \cap T) \cap (O' \cap T) = \varnothing$ implies $O \cap T = \varnothing$ or $O \cap T' = \varnothing$.

   T is strongly transitive when (taking the contrapositive in Definition 3.7.5(1)) $(O \cap T) \cap (T \cap O') = \varnothing$ implies $O \cap T = \varnothing$ or $T \cap O' = \varnothing$.

   That these conditions are equivalent follows by straightforward sets manipulations.

2. We can use part 1 of this result and Lemma 3.7.6(1), or give a direct argument by sets calculations: if $O \cap O' = \varnothing$ then $(O \cap T) \cap (T \cap O') = \varnothing$ and by meet-irreducibility $O \cap T = \varnothing$ or $T \cap O' = \varnothing$ as required.

3. Figure 3.2 (left-hand diagram) provides a counterexample, taking $T = \{0, 1\}$ and $O = \{0, 2\}$ and $O' = \{1, 2\}$. Then $(O \cap T) \cap (T \cap O') = \varnothing$ but it is not the case that $O \cap T = \varnothing$ or $O' \cap T = \varnothing$. □

REMARK 3.7.11. The proof of Lemma 3.7.10 not hard, but the result is interesting for what it says, and also for what it does not say:

1. The notion of being a strong topen maps naturally to something in order theory; namely that $\varnothing$ is meet-irreducible in the induced poset $\{O \cap T \mid O \in \mathsf{Open}\}$ which is the set of open sets of the subspace $(T, \mathsf{Open} \cap T)$ of $(\mathsf{P}, \mathsf{Open})$.

2. However, this mapping is imperfect: the poset is not a lattice, and it is also not a sub-poset of Open — even if T is topen. If Open were a topology and closed under intersections then we would have a lattice — but it is precisely the point of difference between semitopologies vs. topologies that open sets need not be closed under intersections.

3. Being transitive does not correspond to meet-irreducibility; there is an implication in one direction, but certainly not in the other.

So, Lemma 3.7.10 says that (strong) transitivity has a flavour of meet-irreducibility, but in a way that also illustrates — as did Proposition 5.6.6(2) — how semitopologies are different, because they are not closed under intersections, and have their own behaviour.

See also the characterisation of strong transitivity in Lemma 11.4.2 and the surrounding discussion.

3.7.3 Topens in topologies

We conclude by briefly looking at what 'being topen' means if our semitopology is actually a topology. We recall a standard definition from topology:

DEFINITION 3.7.12. Suppose $(\mathsf{P}, \mathsf{Open})$ is a semitopology. Call $T \subseteq \mathsf{P}$ **hyperconnected** when all nonempty open subsets of T intersect.[8] In symbols:

$$\forall O, O' \in \mathsf{Open}_{\neq\varnothing}. O, O' \subseteq T \Longrightarrow O \between O'.$$

LEMMA 3.7.13. *Suppose* $(\mathsf{P}, \mathsf{Open})$ *is a semitopology. Then if* $T \subseteq \mathsf{P}$ *is transitive then it is hyperconnected.*

Proof. Suppose $\varnothing \neq O, O' \subseteq T$. Then $O \between T \between O'$ and by transitivity $O \between O'$ as required. □

What is arguably particularly interesting about Lemma 3.7.13 is that its reverse implication does *not* hold, and in quite a strong sense:

LEMMA 3.7.14. *Suppose* $(\mathsf{P}, \mathsf{Open})$ *is a semitopology and* $T \subseteq \mathsf{P}$. *Then:*

1. *T can be hyperconnected but not transitive, even if* $(\mathsf{P}, \mathsf{Open})$ *is a topology (not just a semitopology).*
2. *T can be hyperconnected but not transitive, even if T is an open set.*

Proof. It suffices to provide counterexamples:

1. Consider the semitopology illustrated in the lower-left diagram in Figure 3.1 (which is a topology), and set $T = \{0, 4\}$. This has no nonempty open subsets so it is trivially hyperconnected. However, T is not transitive because $\{0, 1\} \between T \between \{3, 4\}$ yet $\{0, 1\} \not\between \{3, 4\}$.

2. Consider the semitopology illustrated in the top-right diagram in Figure 3.1, and set $T = \{0, 1\}$. This has two nonempty open subsets, $\{0\}$ and $\{0, 1\}$, so it is hyperconnected. However, T is not transitive, because $\{0\} \between T \between \{1, 2\}$ yet $\{0\} \not\between \{1, 2\}$. □

We know from Lemma 3.7.6(2) that 'transitive' does not imply 'strongly transitive' for an arbitrary subset $T \subseteq \mathsf{P}$, even in a topology. When read together with Lemmas 3.7.13 and 3.7.14, this invites the question of what happens when

- $(\mathsf{P}, \mathsf{Open})$ is a topology, and *also*
- T is an open set.

[8] Calling this *hyperconnected* is a slight but natural generalisation of the usual definition: in topology, 'hyperconnected' is typically used to refer to an entire space rather than a subset of it. In the case that $T = \mathsf{P}$, our definition specialises to the usual one.

In this natural special case, strong transitivity, transitivity, and being hyperconnected, all become equivalent:

LEMMA 3.7.15. *Suppose* $(\mathsf{P}, \mathsf{Open})$ *is a topology and suppose* $T \in \mathsf{Open}$ *is an open set. Then the following are equivalent:*

- *T is a strong topen (Definition 3.7.5(2)).*
- *T is a topen.*
- *T is hyperconnected.*

Proof. We assumed T is open, so the equivalence above can also be thought of as

$$\text{strongly transitive} \Longleftrightarrow \text{transitive} \Longleftrightarrow \text{all nonempty open subsets intersect.}$$

We prove a chain of implications:

- If T is a strong topen then it is a topen by Lemma 3.7.6(1).
- If T is a topen then we use Lemma 3.7.13.
- Suppose T is hyperconnected, so every pair of nonempty open subsets of T intersect; and suppose $O, O' \in \mathsf{Open}_{\neq\varnothing}$ and $O \between T \between O'$. Then also $(O \cap T) \between T \between (O' \cap T)$. Now $O \cap T$ and $O' \cap T$ are open: because T is open; and P is a topology (not just a semitopology), so intersections of open sets are open. By transitivity of T we have $O \cap T \between O' \cap T$. Since O and O' were arbitrary, T is strongly transitive. □

Interiors, communities & regular points 4

4.1 Community of a (regular) point

Definition 4.1.1 is standard:

DEFINITION 4.1.1 (Open interior). Suppose $(\mathsf{P}, \mathsf{Open})$ is a semitopology and $P \subseteq \mathsf{P}$. Define $interior(P)$ the **(open) interior of** P by

$$interior(P) = \bigcup\{O \in \mathsf{Open} \mid O \subseteq P\}.$$

LEMMA 4.1.2. *Suppose* $(\mathsf{P}, \mathsf{Open})$ *is a semitopology and* $P \subseteq \mathsf{P}$. *Then* $interior(P)$ *from Definition 4.1.1 is the greatest open subset of* P.

Proof. Routine by the construction in Definition 4.1.1 and closure of open sets under unions (Definition 1.2.2(2)). □

COROLLARY 4.1.3. *Suppose* $(\mathsf{P}, \mathsf{Open})$ *is a semitopology and* $P, P' \subseteq \mathsf{P}$. *Then if* $P \subseteq P'$ *then* $interior(P) \subseteq interior(P')$.

Proof. Routine using Lemma 4.1.2. □

DEFINITION 4.1.4 (Community of a point, and regularity). Suppose $(\mathsf{P}, \mathsf{Open})$ is a semitopology and $p \in \mathsf{P}$. Then:

1. Define $K(p)$ the **community of** p by

$$K(p) = interior(p_{\between}).$$

2. Extend K to subsets $P \subseteq \mathsf{P}$ by taking a sets union:

$$K(P) = \bigcup\{K(p) \mid p \in P\}.$$

3. Call p a **regular point** when its community is a topen neighbourhood of p. In symbols:

$$p \text{ is regular} \quad \text{when} \quad p \in K(p) \in \mathsf{Topen}.$$

4. Call p a **weakly regular point** when its community is an open (but not necessarily topen) neighbourhood of p. In symbols:

$$p \text{ is weakly regular} \quad \text{when} \quad p \in K(p) \in \mathsf{Open}.$$

5. Call p a **quasiregular point** when its community is nonempty. In symbols:

$$p \text{ is quasiregular} \quad \text{when} \quad \varnothing \neq K(p) \in \mathsf{Open}.$$

6. If p is not regular then we may call it an **irregular point**, or just say that it is not regular.
7. If $P \subseteq \mathsf{P}$ and every $p \in P$ is regular/weakly regular/quasiregular/irregular then we may call P a **regular/weakly regular/quasiregular/irregular set** respectively (see also Definition 6.1.1(2)).

Remark 4.1.5. Lemmas 4.1.6 and 4.1.8 give an overview of the relationships between the properties in Definition 4.1.4. For reference, two more regularity-flavoured conditions will appear later: *indirect regularity* in Definition 9.3.2, and an *MCN* property is mentioned in Remark 22.6.18. See also Remark 9.3.6.

Lemma 4.1.6. *Suppose* $(\mathsf{P}, \mathsf{Open})$ *is a semitopology and* $p \in \mathsf{P}$. *Then:*

1. *If* p *is regular, then* p *is weakly regular.*
2. *If* p *is weakly regular, then* p *is quasiregular.*

Proof. We consider each part in turn:

1. If p is regular then by Definition 4.1.4(3) $p \in K(p) \in \mathsf{Topen}$, so certainly $p \in K(p)$ and by Definition 4.1.4(4) p is weakly regular.

2. If p is weakly regular then by Definition 4.1.4(4) $p \in K(p) \in \mathsf{Open}$, so certainly $K(p) \neq \varnothing$ and by Definition 4.1.4(5) p is quasiregular. □

Example 4.1.7.

1. In Figure 3.2 (left-hand diagram), 0, 1, and 2 are three intertwined points and the entire space $\{0, 1, 2\}$ consists of a single topen set. It follows that 0, 1, and 2 are all regular and their community is $\{0, 1, 2\}$.
2. In Figure 3.1 (top-left diagram), 0 and 2 are regular and 1 is weakly regular but not regular ($1 \in K(1) = \{0, 1, 2\}$ but $\{0, 1, 2\}$ is not topen).
3. In Figure 3.1 (lower-right diagram), 0, 1, and 2 are regular and $*$ is quasiregular ($K(*) = \{1\}$).
4. In Figure 3.1 (top-right diagram), 0 and 2 are regular and 1 is neither regular, weakly regular, nor quasiregular ($K(1) = \varnothing$).
5. In a semitopology of values $(\mathsf{Val}, pow(\mathsf{Val}))$ (Definition 2.1.3) every value $v \in \mathsf{Val}$ is regular, weakly regular, and unconflicted.

6. In $\mathbb{R}$ with its usual topology (which is also a semitopology), every point is unconflicted because the topology is Hausdorff and by Equation 3.1 in Remark 3.6.10 this means precisely that $p_{\between} = \{p\}$ so p is intertwined just with itself. Furthermore p is not (quasi/weakly)regular, because $K(p) = interior(p_{\between}) = \varnothing$.

LEMMA 4.1.8. *Suppose* $(\mathsf{P}, \mathsf{Open})$ *is a semitopology and* $p \in \mathsf{P}$. *Then:*

1. *p might not be quasiregular (i.e.* $K(p) = \varnothing$*); thus by Lemma 4.1.6 it is also not weakly regular and not regular.*
2. *p might be quasiregular but not weakly regular (i.e.* $K(p) \neq \varnothing$ *but* $p \notin K(p)$*); and*
3. *p might be weakly regular but not regular (i.e.* $p \in K(p) \notin \mathsf{Topen}$*).*

Proof. We consider each part in turn:

1. Point $0 \in \mathbb{R}$ in Example 4.1.7(6) is not quasiregular.

2. Point 1 in Example 4.1.7(2) (illustrated in Figure 3.1, top-left diagram) is weakly regular ($K(1) = \{0, 1, 2\}$) but not regular ($K(1)$ is open but not topen).

3. Point $*$ in Example 4.1.7(3) (illustrated in Figure 3.1, lower-right diagram) is quasiregular ($K(*) = \{1\}$ is nonempty but does not contain $*$). □

LEMMA 4.1.9. *Suppose* $(\mathsf{P}, \mathsf{Open})$ *is a semitopology. Then:*

1. *If all nonempty open sets intersect then* $(\mathsf{P}, \mathsf{Open})$ *is regular (meaning that every* $p \in \mathsf{P}$ *is regular).*
2. *The reverse implication need not hold: it is possible for* $(\mathsf{P}, \mathsf{Open})$ *to be regular but not all open sets intersect (cf. Corollary 4.3.3).*

Proof. We consider each part in turn:

1. By Lemma 3.6.6(2) $\mathsf{P} \in \mathsf{Topen}$ (since it is transitive and open). By Lemma 3.6.6(1) $p_{\between} = \mathsf{P}$ for every $p \in \mathsf{P}$, thus $K(p) = interior(p_{\between}) = \mathsf{P}$. Thus $p \in K(p) \in \mathsf{Topen}$ for every $p \in \mathsf{P}$, so P is regular.

2. It suffices to provide a counterexample. We take any discrete semitopology with at least two elements; e.g. $(\{0, 1\}, pow(\{0, 1\}))$. Then $\{0\} \not\between \{1\}$, but by Corollary 4.2.7 0 and 1 are both regular. □

EXAMPLE 4.1.10. When we started looking at semitopologies we gave some examples in Example 2.1.4. These may seem quite elementary now, but we run through them commenting on which spaces are regular, weakly regular, or quasiregular:

- Any discrete semitopology is regular; topen neighbourhoods are just the singleton sets.
- The initial semitopology is regular: it has no topen neighbourhoods, but also no points. The final semitopology is regular: it has one topen neighbourhood, containing one point. The trivial topology is regular; it has a single topen neighbourhood that is P itself.
- The supermajority semitopology is regular. It has one topen neighbourhood containing all of P.

- The many semitopology is regular if P is finite (because it is equal to the trivial semitopology), and not even quasiregular if P is infinite, because (for infinite P) $p_{\between} = \varnothing$ for every point. For example, if $\mathsf{P} = \mathbb{N}$ and p is even and p' is odd, then $evens = \{2 * n \mid n \in \mathbb{N}\}$ and $odds = \{2 * n+1 \mid n \in \mathbb{N}\}$ are disjoint open neighbourhoods of p and p' respectively.
- The all-but-one semitopology is regular for P having cardinality of 3 or more, since all points are intertwined so there is a single topen neighbourhood which is the whole space. If P has cardinality 2 or 1 then we have a discrete semitopology (on two points or one point) and these too are regular, with two or one topen neighbourhoods.
- The more-than-one semitopology is not even quasiregular for P having cardinality of 4 or more. If P has cardinality 3 then we get the left-hand topology in Figure 3.2, which is regular. If P has cardinality 2 then we get the trivial semitopology, which is regular.
- Take $\mathsf{P} = \mathbb{R}$ (the set of real numbers) and let open sets be generated by intervals of the form $[0, r)$ or $(\text{-}r, 0]$ for any strictly positive real number $r > 0$. The reader can check that this semitopology is regular.
- Any quorum system induces an intertwined semitopology, as outlined in Example 2.1.4(7). By Lemmas 4.1.9(1) and 3.6.6 this is a regular semitopology, and every nonempty open set is a topen neighbourhood.

REMARK 4.1.11. We pause to recap:

1. $K(p)$ always exists and always is open. It may or may not be empty, may or may not be topen, and may or may not contain p.

2. When $p \in K(p) \in \mathsf{Topen}$ we call p 'regular', which suggests that non-regular behaviour — $p \notin K(p)$ and/or $K(p) \notin \mathsf{Topen}$, or even $K(p) = \varnothing$ — is 'bad behaviour', and being regular 'good behaviour'.

   But what is this good behaviour that regularity implies? Theorem 3.2.2 (continuous value assignments are constant on topens) tells us that a regular p is surrounded by a topen neighbourhood of points $K(p) = interior(p_{\between})$ that must agree with it under continuous value assignments. Using our terminology *community* and *regular*, we can say that *the community of a regular p shares its values.*

3. We can sum up the above intuitively as follows:

   a) We care about transitivity because it implies agreement.
   b) We care about being open, because it implies actionability.
   c) Thus, a regular point is interesting because it is a participant in a maximal topen neighbourhood and therefore can *i)* come to agreement and *ii)* take action on that agreement.

4. The question then arises how the community of p can be (semi)topologically characterised. We will explore, notably in Theorem 4.2.6, Proposition 5.4.10, and Theorem 5.6.2; see also Remark 5.6.1.

4.2 Further exploration of (quasi-/weak) regularity and topen sets

REMARK 4.2.1. Recall three common separation axioms from topology:

1. T_0: if $p_1 \neq p_2$ then there exists some $O \in \mathsf{Open}$ such that $(p_1 \in O)$ xor $(p_2 \in O)$, where xor denotes *exclusive or*.
2. T_1: if $p_1 \neq p_2$ then there exist $O_1, O_2 \in \mathsf{Open}$ such that $p_i \in O_j \Longleftrightarrow i = j$ for $i, j \in \{1, 2\}$.
3. T_2, or the *Hausdorff condition*: if $p_1 \neq p_2$ then there exist $O_1, O_2 \in \mathsf{Open}$ such that $p_i \in O_j \Longleftrightarrow i = j$ for $i, j \in \{1, 2\}$, and $O_1 \cap O_2 = \varnothing$. Cf. the discussion in Remark 3.6.10.

Even the weakest of the well-behavedness property for semitopologies that we consider in Definition 4.1.4 — quasiregularity — is in some sense strongly opposed to the space being Hausdorff/T_2 (though not to being T_1), as Lemma 4.2.2 makes formal.

LEMMA 4.2.2.

1. *Every quasiregular Hausdorff semitopology is discrete.*
   In more detail: if $(\mathsf{P}, \mathsf{Open})$ *is a semitopology that is quasiregular (Definition 4.1.4(5)) and Hausdorff (equation 3.1 in Remark 3.6.10), then* $\mathsf{Open} = pow(\mathsf{P})$.
2. *There exists a (quasi)regular* T_1 *semitopology that is not discrete.*

Proof. We consider each part in turn:

1. By the Hausdorff property, $p_{\between} = \{p\}$. By the quasiregularity property, $K(p) \neq \varnothing$. It follows that $K(p) = \{p\}$. But by construction in Definition 4.1.4(1), $K(p)$ is an open interior. Thus $\{p\} \in \mathsf{Open}$. The result follows.

2. It suffices to provide an example. We use the left-hand semitopology in Figure 3.2. Thus $\mathsf{P} = \{0, 1, 2\}$ and Open is generated by $\{0, 1\}$, $\{1, 2\}$, and $\{2, 0\}$. All nonempty open sets intersect, so by Lemma 4.1.9(1) P is regular. It is also T_1 (Remark 4.2.1). □

Lemma 4.2.3 confirms in a different way that regularity (Definition 4.1.4(3)) is non-trivially distinct from weak regularity and quasiregularity:

LEMMA 4.2.3. *Suppose* $(\mathsf{P}, \mathsf{Open})$ *is a semitopology and* $p \in \mathsf{P}$. *Then:*

1. $K(p) \in \mathsf{Open}$.
2. $K(p)$ *is not necessarily topen; equivalently* $K(p)$ *is not necessarily transitive. (More on this later in Subsection 4.4.)*

Proof. $K(p)$ is open by construction in Definition 4.1.4(1), since it is an open interior.

For part 2, it suffices to provide a counterexample. We consider the semitopology from Example 3.3.3(1) (illustrated in Figure 3.1, top-left diagram). We calculate that $K(1) = \{0, 1, 2\}$ so that $K(1)$ is an open neighbourhood of 1 — but it is not transitive, and thus not topen, since $\{0\} \cap \{2\} = \varnothing$.

Further checking reveals that $\{0\}$ and $\{2\}$ are two maximal topens within $K(1)$. □

So what is $K(p)$? We start by characterising $K(p)$ as the *greatest* topen neighbourhood of p, if this exists:

LEMMA 4.2.4. *Suppose* $(\mathsf{P}, \mathsf{Open})$ *is a semitopology and recall from Definition 4.1.4(3) that* p *is regular when* $K(p)$ *is a topen neighbourhood of* p.

1. *If* $K(p)$ *is a topen neighbourhood of* p *(i.e. if* p *is regular) then* $K(p)$ *is a maximal topen.*
2. *If* $p \in T \in \mathsf{Topen}$ *is a maximal topen neighbourhood of* p *then* $T = K(p)$.

Proof.

1. Since p is regular, by definition, $K(p)$ is topen and is a neighbourhood of p. It remains to show that $K(p)$ is a maximal topen.

   Suppose T is a topen neighbourhood of p; we wish to prove $T \subseteq K(p) = interior(p_{\between})$. Since T is open it would suffice to show that $T \subseteq p_{\between}$. By Proposition 3.6.9 $p \between p'$ for every $p' \in T$, and it follows immediately that $T \subseteq p_{\between}$.

2. Suppose T is a maximal topen neighbourhood of p.

   First, note that T is open, and by Proposition 3.6.9 $T \subseteq p_{\between}$, so $T \subseteq K(p)$.

   By assumption $p \in T \cap K(p)$ and both are topen so by Proposition 3.5.2(1) $T \cup K(p)$ is topen, and by maximality $K(p) \subseteq T$. □

Remark 4.2.5. We can use Lemma 4.2.4 to characterise regularity in five equivalent ways: see Theorem 4.2.6 and Corollary 4.2.8. Other characterisations will follow but will require additional machinery to state (the notion of *closed neighbourhood*; see Definition 5.4.1). See Corollary 5.4.13 and Theorem 5.6.2.

Theorem 4.2.6. *Suppose* $(\mathsf{P}, \mathsf{Open})$ *is a semitopology and* $p \in \mathsf{P}$. *Then the following are equivalent:*

1. *p is regular, or in full:* $p \in K(p) \in \mathsf{Topen}$.
2. *$K(p)$ is the greatest topen neighbourhood of p.*
3. *$K(p)$ is a maximal topen neighbourhood of p.*
4. *p has a maximal topen neighbourhood.*
5. *p has some topen neighbourhood.*

Proof. We prove a cycle of implications:

1. If $K(p)$ is a topen neighbourhood of p then it is maximal by Lemma 4.2.4(1). Furthermore this maximal topen neighbourhood of p is necessarily greatest, since if we have two maximal topen neighbourhoods of p then their union is a larger topen neighbourhood of p by Proposition 3.5.2(1) (union of intersecting topens is topen).

2. If $p_{\between}$ is the greatest topen neighbourhood of p, then certainly it is a maximal topen neighbourhood of p.

3. If $p_{\between}$ is a maximal topen neighbourhood of p, then certainly p has a maximal topen neighbourhood.

4. If p has a maximal topen neighbourhood then certainly p has a topen neighbourhood.

5. Suppose p has a topen neighbourhood T. By Corollary 3.5.3 we may assume without loss of generality that T is a maximal topen. We use Lemma 4.2.4(2). □

Theorem 4.2.6 has numerous corollaries:

Corollary 4.2.7. *Suppose* $(\mathsf{P}, \mathsf{Open})$ *is a semitopology and* $p \in \mathsf{P}$ *and* $\{p\} \in \mathsf{Open}$. *Then p is regular.*

Proof. We noted in Example 3.3.2(1) that a singleton $\{p\}$ is always transitive, so if $\{p\}$ is also open, then it is topen, so that p has a topen neighbourhood and by Theorem 4.2.6(5) p is topen.[1] □

COROLLARY 4.2.8. *Suppose* $(\mathsf{P}, \mathsf{Open})$ *is a semitopology and* $p \in \mathsf{P}$. *Then the following are equivalent:*

1. p *is regular.*
2. p *is weakly regular and* $K(p) = K(p')$ *for every* $p' \in K(p)$.

Proof. We prove two implications, using Theorem 4.2.6:

- Suppose p is regular. By Lemma 4.1.6(1) p is weakly regular. Now consider $p' \in K(p)$. By Theorem 4.2.6 $K(p)$ is topen, so it is a topen neighbourhood of p'. By Theorem 4.2.6 $K(p')$ is a greatest topen neighbourhood of p'. But by Theorem 4.2.6 $K(p)$ is also a greatest topen neighbourhood of p, and $K(p) \between K(p')$ since they both contain p'. By Proposition 3.5.2(1) and maximality, they are equal.

- Suppose p is weakly regular and suppose $K(p) = K(p')$ for every $p' \in K(p)$, and consider $p', p'' \in K(p)$. Then $p' \between p''$ holds, since $p'' \in K(p') = K(p)$. By Proposition 3.6.9 $K(p)$ is topen, and by weak regularity $p \in K(p)$, so by Theorem 4.2.6 p is regular as required. □

REMARK 4.2.9. With regards to Corollary 4.2.8, it might be useful to look at Example 3.3.3(2) and Figure 3.1 (top-right diagram). In that example the point 1 is *not* regular, and its community $\{0, 1, 2\}$ is not a community for 0 or 2.

COROLLARY 4.2.10. *Suppose* $(\mathsf{P}, \mathsf{Open})$ *is a semitopology and* $p, p' \in \mathsf{P}$. *Then if* p *is regular and* $p' \in K(p)$ *then* p' *is regular and has the same community.*

Proof. Suppose p is regular — so by Definition 4.1.4(3) $p \in K(p) \in \mathsf{Topen}$ — and suppose $p' \in K(p)$. Then by Corollary 4.2.8 $K(p) = K(p')$, so $p' \in K(p') \in \mathsf{Topen}$ and by Theorem 4.2.6 p' is regular. □

COROLLARY 4.2.11. *Suppose* $(\mathsf{P}, \mathsf{Open})$ *is a semitopology. Then the following are equivalent for* $T \subseteq \mathsf{P}$:

- T *is a maximal topen.*
- $T \neq \varnothing$ *and* $T = K(p)$ *for every* $p \in T$.

Proof. If T is a maximal topen and $p \in T$ then T is a maximal topen neighbourhood of p. By Theorem 4.2.6(2&5) $T = K(p)$.

If $T \neq \varnothing$ and $T = K(p)$ for every $p \in T$, then $K(p) = K(p')$ for every $p' \in K(p)$ and by Corollary 4.2.8 p is regular, so that by Definition 4.1.4(3) $T = K(p) \in \mathsf{Topen}$ as required. □

[1] It does not follow from $p \in \{p\} \in \mathsf{Topen}$ that $K(p) = \{p\}$: consider $\mathsf{P} = \{0, 1\}$ and $\mathsf{Open} = \{\varnothing, \{0\}, \{0, 1\}\}$ and $p = 0$; then $\{p\} \in \mathsf{Topen}$ yet $K(p) = \{0, 1\}$.

4.3 Intersection and partition properties of regular spaces

Proposition 4.3.1 is useful for consensus in practice. Suppose we are a regular point q and we have reached consensus with some topen neighbourhood $O \ni q$. Suppose further that our topen neighbourhood O intersects with the maximal topen neighbourhood $K(p)$ of some other regular point p. Then Proposition 4.3.1 tells us that we were inside $K(p)$ all along. See also Remark 5.3.3.

PROPOSITION 4.3.1. *Suppose* $(\mathsf{P}, \mathsf{Open})$ *is a semitopology and* $p \in \mathsf{P}$ *is regular and* $O \in \mathsf{Topen}$ *is topen. Then*

$$O \between K(p) \quad \text{if and only if} \quad O \subseteq K(p).$$

Proof. The right-to-left implication is immediate from Notation 3.1.1(1), given that topens are nonempty by Definition 3.2.1(2).

For the left-to-right implication, suppose $O \between K(p)$. By Theorem 4.2.6 $K(p)$ is a maximal topen, and by Proposition 3.5.2(1) $O \cup K(p)$ is topen. Then $O \subseteq K(p)$ follows by maximality. □

PROPOSITION 4.3.2. *Suppose* $(\mathsf{P}, \mathsf{Open})$ *is a semitopology and suppose* $p, p' \in \mathsf{P}$ *are regular. Then*

$$K(p) \between K(p') \quad \Longleftrightarrow \quad K(p) = K(p')$$

(See also Corollary 6.4.6, which considers similar properties for p and p' that are not necessarily regular.)

Proof. We prove two implications.

- Suppose there exists $p'' \in K(p) \cap K(p')$. By Corollary 4.2.10 (p'' is regular and) $K(p) = K(p'') = K(p')$.
- Suppose $K(p) = K(p')$. By assumption $p \in K(p)$, so $p \in K(p')$. Thus $p \in K(p) \cap K(p')$. □

Corollary 4.3.3 is a simple characterisation of regular semitopological spaces (it is also a kind of continuation to Lemma 4.1.9(2)):

COROLLARY 4.3.3. *Suppose* $(\mathsf{P}, \mathsf{Open})$ *is a semitopology. Then the following are equivalent:*

1. $(\mathsf{P}, \mathsf{Open})$ *is regular.*
2. P *partitions into topen sets: there exists some set of topen sets* $\mathcal{T}$ *such that* $T \not\between T'$ *for every* $T, T' \in \mathcal{T}$ *and* $\mathsf{P} = \bigcup \mathcal{T}$.
3. *Every* $X \subseteq \mathsf{P}$ *has a cover of topen sets: there exists some set of topen sets* $\mathcal{T}$ *such that* $X \subseteq \bigcup \mathcal{T}$.

Proof. The proof is routine from the machinery that we already have. We prove equivalence of parts 1 and 2:

1. Suppose $(\mathsf{P}, \mathsf{Open})$ is regular, meaning by Definition 4.1.4(7&3) that $p \in K(p) \in \mathsf{Topen}$ for every $p \in \mathsf{P}$. We set $\mathcal{T} = \{K(p) \mid p \in \mathsf{P}\}$. By assumption this covers P in topens, and by Proposition 4.3.2 the cover is a partition.

2. Suppose $\mathcal{T}$ is a topen partition of P. By definition for every point p there exists $T \in \mathcal{T}$ such that $p \in T$ and so p has a topen neighbourhood. By Theorem 4.2.6(5&1) p is regular.

We prove equivalence of parts 2 and 3:

1. Suppose $\mathcal{T}$ is a topen partition of P, and suppose $X \subseteq \mathcal{P}$. Then trivially $X \subseteq \bigcup \mathcal{T}$.

2. Suppose every $X \subseteq \mathsf{P}$ has a cover of topen sets. Then P has a cover of topen sets; write it $\mathcal{T}$. By Corollary 3.5.3 we may assume without loss of generality that $\mathcal{T}$ is a partition, and we are done. □

REMARK 4.3.4. The moral we take from the results and examples above (and those to follow) is that the world we are entering has rather different well-behavedness criteria than those familiar from the study of typical Hausdorff topologies like $\mathbb{R}$. Put crudely:

1. 'Bad' spaces are spaces that are not regular.
   $\mathbb{R}$ with its usual topology (which is also a semitopology) is an example of a 'bad' semitopology; it is not even quasiregular.
2. 'Good' spaces are spaces that are regular.
   The supermajority and all-but-one semitopologies from Example 2.1.4(5a&5c) are typical examples of 'good' semitopologies; both are intertwined spaces (Notation 3.6.5).
3. Corollary 4.3.3 shows that the 'good' spaces are just the (disjoint, possibly infinite) unions of intertwined spaces.

4.4 Examples of communities and (ir)regular points

By Definition 4.1.4 a point p is regular when its community is a topen neighbourhood. Then a point is *not* regular when its community is *not* a topen neighbourhood of p. We saw one example of this in Lemma 4.2.3. In this subsection we take a moment to investigate the possible behaviour in more detail.

EXAMPLE 4.4.1.

1. We noted in Example 4.4.1(6) and Lemma 4.1.8(1) that for $\mathbb{R}$ the real numbers with its usual topology, every $p \in \mathbb{R}$ is not regular. Then $x_{\between} = \{x\}$ and $K(x) = \varnothing$ for every $x \in \mathbb{R}$.

2. We continue the semitopology from Example 3.3.3(1) (illustrated in Figure 3.1, top-left diagram), as used in Lemma 4.2.3:

   - $\mathsf{P} = \{0, 1, 2\}$.
   - Open is generated by $\{0\}$ and $\{2\}$.

   Then:

   - $0_{\between} = \{0, 1\}$ and $K(0) = \mathit{interior}(0_{\between}) = \{0\}$.
   - $2_{\between} = \{1, 2\}$ and $K(2) = \mathit{interior}(2_{\between}) = \{2\}$.
   - $1_{\between} = \{0, 1, 2\}$ and $K(1) = \{0, 1, 2\}$.

3. We take, as illustrated in Figure 4.1 (left-hand diagram):

Figure 4.1: Illustration of Example 4.4.1(3&4)

- $\mathsf{P} = \{0,1,2,3,4\}$.
- Open is generated by $\{1,2\}$, $\{0,1,3\}$, $\{0,2,4\}$, $\{3\}$, and $\{4\}$.

Then:

- $x_{\between} = \{0,1,2\}$ and $K(x) = \mathit{interior}(x_{\between}) = \{1,2\}$ for $x \in \{0,1,2\}$.
- $x_{\between} = \{x\} = K(x)$ for $x \in \{3,4\}$.

(We return to this example in Example 10.1.3(3), and we will also use it in the proof of Lemma 10.2.4.)

4. We take, as illustrated in Figure 4.1 (right-hand diagram):

   - $\mathsf{P} = \{0,1,2,3,4\}$.
   - Open is generated by $\{1\}$, $\{2\}$, $\{3\}$, $\{4\}$, $\{0,1,2,3\}$, and $\{0,1,2,4\}$.

   Then:

   - $0_{\between} = \{0,1,2\}$ and $K(0) = \{1,2\}$.
   - $K(0)$ is not transitive and consists of two distinct topens $\{1\}$ and $\{2\}$.
   - $0 \notin K(0)$.

   See Remark 5.6.3 for further discussion of this example.

5. The reader can also look ahead to Example 5.6.10. In Example 5.6.10(1), every point p is regular and $K(p) = \mathbb{Q}^2$. In Example 5.6.10(2), no point p is regular and $K(p) = \varnothing \subseteq \mathbb{Q}^2$.

LEMMA 4.4.2. *Suppose* $(\mathsf{P}, \mathsf{Open})$ *is a semitopology and* $p \in \mathsf{P}$. *Then precisely one of the following possibilities must hold, and each one is possible:*

1. *p is regular: $p \in K(p)$ and $K(p)$ is topen (nonempty, open, and transitive).*
2. *$K(p)$ is topen, but $p \notin K(p)$.*
3. *$K(p) = \varnothing$.*
4. *$K(p)$ is open but not transitive. (Both $p \in K(p)$ and $p \notin K(p)$ are possible.)*

Proof.

1. To see that p can be regular, consider $\mathsf{P} = \{0\}$ with the discrete topology. Then $p \in K(p) = \{0\}$.

2. To see that it is possible for $K(p)$ to be topen but p is not in it, consider Example 4.4.1(3). There, $\mathsf{P} = \{0,1,2,3,4\}$ and $0_{\between} = \{0,1,2\}$ and $K(0) = \{1,2\}$. Then $K(0)$ is topen, but $0 \notin K(0)$.

   (Another, slightly more compact but more distant, example is $p = *$ in the lower-right semitopology in Figure 3.1.)

3. To see that $K(p) = \varnothing$ is possible, consider Example 4.4.1(1) (the real numbers $\mathbb{R}$ with its usual topology). Then by Remark 3.6.10 $r_{\between} = \{r\}$ and so $K(x) = \mathit{interior}(\{r\}) = \varnothing$. (See also Example 5.6.10(2) for a more elaborate example.)

4. To see that it is possible for $K(p)$ to be an open neighbourhood of p but not transitive, see Example 4.4.1(2). There, $\mathsf{P} = \{0,1,2\}$ and $1 \in 1_{\between} = \{0,1,2\} = K(1)$, but $\{0,1,2\}$ is not transitive (it contains two disjoint topens: $\{0\}$ and $\{2\}$).

   To see that it is possible for $K(p)$ to be open and nonempty yet not contain p and not be transitive, see Example 4.4.1(4) for $p = 0$, and see also Remark 5.6.3 for a discussion of the connection with minimal closed neighbourhoods.

The possibilities above are clearly mutually exclusive and exhaustive. □

Closed sets 5

5.1 Closed sets

REMARK 5.1.1. In Subsection 5.1 we check that some familiar properties of closures carry over from topologies to semitopologies. There are no technical surprises, but this is in itself a mathematical result that needs to be checked. From Subsection 5.3 and the following Subsections we will study the close relation between closures and sets of intertwined points.

First, we spare a few words on why closures are particularly interesting in semitopologies:

1. A participant may wish to compute a quorum that it can be confident of remaining in agreement with, where algorithms succeed. The notion of maximal topen from Definition 3.2.1(3) provides this, as discussed in Remark 3.2.5 — but computing maximal topens is hard since the definition involves quantifications over all open sets and there may be many of them.
2. However, computing *closures* is quite tractable in the right circumstances (see Section 8 and Remark 8.4.14), so a characterisation of maximal topens using closed sets and closures is of practical interest.
3. Closures are significant for other reasons too: see the discussions in Remarks 5.5.1 and 5.5.5, and see the later material in Section 11 that considers *dense sets*.

Thus, and just as is the case in topology, closures are an interesting object of study.

DEFINITION 5.1.2. Suppose $(\mathsf{P}, \mathsf{Open})$ is a semitopology and suppose $p \in \mathsf{P}$ and $P \subseteq \mathsf{P}$. Then:

1. Define $|P| \subseteq \mathsf{P}$ the **closure of** P to be the set of points p such that every open neighbourhood of p intersects P. In symbols using Notation 3.1.1:

$$|P| = \{p' \in \mathsf{P} \mid \forall O \in \mathsf{Open}. p' \in O \Longrightarrow P \between O\}.$$

2. As is standard, we may write $|p|$ for $|\{p\}|$. Unpacking definitions for reference:

$$|p| = \{p' \in \mathsf{P} \mid \forall O \in \mathsf{Open}. p' \in O \Longrightarrow p \in O\}.$$

LEMMA 5.1.3. *Suppose* $(\mathsf{P}, \mathsf{Open})$ *is a semitopology and suppose* $P, P' \subseteq \mathsf{P}$. *Then taking the closure of a set is:*

1. Monotone: *If* $P \subseteq P'$ *then* $|P| \subseteq |P'|$.
2. Increasing: $P \subseteq |P|$.
3. Idempotent: $|P| = ||P||$.

Proof. By routine calculations from Definition 5.1.2. □

LEMMA 5.1.4. *Suppose* $(\mathsf{P}, \mathsf{Open})$ *is a semitopology and* $P \subseteq \mathsf{P}$ *and* $O \in \mathsf{Open}$. *Then*

$$P \between O \quad \textit{if and only if} \quad |P| \between O.$$

Proof. Suppose $P \between O$. Then $|P| \between O$ using Lemma 5.1.3(2).

Suppose $|P| \between O$. Pick $p \in |P| \cap O$. By construction of $|P|$ in Definition 5.1.2 $p \in O \Longrightarrow P \between O$. It follows that $P \between O$ as required. □

DEFINITION 5.1.5. Suppose $(\mathsf{P}, \mathsf{Open})$ is a semitopology and suppose $C \subseteq \mathsf{P}$.

1. Call C a **closed set** when $C = |C|$.
2. Call C a **clopen set** when C is closed and open.
3. Write Closed for the set of **closed sets** (as we wrote Open for the open sets; the ambient semitopology will always be clear or understood).

LEMMA 5.1.6. *Suppose* $(\mathsf{P}, \mathsf{Open})$ *is a semitopology and suppose* $P \subseteq \mathsf{P}$. *Then* $|P|$ *is closed and contains* P. *In symbols:*

$$P \subseteq |P| \in \mathsf{Closed}.$$

Proof. From Definition 5.1.5(1) and Lemma 5.1.3(2 & 3). □

EXAMPLE 5.1.7.

1. Take $\mathsf{P} = \{0,1\}$ and $\mathsf{Open} = \{\varnothing, \{0\}, \{0,1\}\}$. Then the reader can verify that:
   - $\{0\}$ is open.
   - The closure of $\{1\}$ is $\{1\}$ and $\{1\}$ is closed.
   - The closure of $\{0\}$ is $\{0,1\}$.
   - $\varnothing$ and $\{0,1\}$ are the only clopen sets.
2. Now take $\mathsf{P} = \{0,1\}$ and $\mathsf{Open} = \{\varnothing, \{0\}, \{1\}, \{0,1\}\}$.[1] Then the reader can verify that:
   - Every set is clopen.
   - The closure of every set is itself.

REMARK 5.1.8. There are two standard definitions for when a set is closed: when it is equal to its closure (as per Definition 5.1.5(1)), and when it is the complement of an open set. In topology these are equivalent. We do need to check that the same holds in semitopology, but as it turns out the proof is routine:

[1]Following Definition 2.1.3 and Example 2.1.4(3), this is just $\{0,1\}$ with the *discrete semitopology*.

LEMMA 5.1.9. *Suppose* $(\mathsf{P}, \mathsf{Open})$ *is a semitopology. Then:*

1. *Suppose* $C \in \mathsf{Closed}$ *is closed (by Definition 5.1.5:* $C = |C|$*). Then* $\mathsf{P} \setminus C$ *is open.*
2. *Suppose* $O \in \mathsf{Open}$ *is open. Then* $\mathsf{P} \setminus O$ *is closed (by Definition 5.1.5:* $|\mathsf{P} \setminus O| = \mathsf{P} \setminus O$*).*

Proof.

1. Suppose $p \in \mathsf{P} \setminus C$. Since $C = |C|$, we have $p \in \mathsf{P} \setminus |C|$. Unpacking Definition 5.1.2, this means precisely that there exists $O_p \in \mathsf{Open}$ with $p \in O_p \not\between C$. We use Lemma 2.3.2.
2. Suppose $O \in \mathsf{Open}$. Combining Lemma 2.3.2 with Definition 5.1.2 it follows that $O \not\between |\mathsf{P} \setminus O|$ so that $|\mathsf{P} \setminus O| \subseteq \mathsf{P} \setminus O$. Furthermore, by Lemma 5.1.3(2) $\mathsf{P} \setminus O \subseteq |\mathsf{P} \setminus O|$. □

COROLLARY 5.1.10. *If* $C \in \mathsf{Closed}$ *then* $\mathsf{P} \setminus C = \bigcup_{O \in \mathsf{Open}} O \not\between C$.

Proof. By Lemma 5.1.9(1) $\mathsf{P} \setminus C \subseteq \bigcup_{O \in \mathsf{Open}} O \not\between C$. Conversely, if $O \not\between C$ then $O \subseteq \mathsf{P} \setminus C$ by Definition 5.1.2(1). □

COROLLARY 5.1.11. *Suppose* $(\mathsf{P}, \mathsf{Open})$ *is a semitopology and* $P \subseteq \mathsf{P}$ *and* $\mathcal{C} \subseteq pow(\mathsf{P})$*. Then:*

1. $\varnothing$ *and* P *are closed.*
2. *If every* $C \in \mathcal{C}$ *is closed, then* $\bigcap \mathcal{C}$ *is closed. Or succinctly in symbols:*
$$\mathcal{C} \subseteq \mathsf{Closed} \Longrightarrow \bigcap \mathcal{C} \in \mathsf{Closed}.$$
3. $|P|$ *is equal to the intersection of all the closed sets that contain it. In symbols:*
$$|P| = \bigcap \{C \in \mathsf{Closed} \mid P \subseteq C\}.$$

Proof.

1. Immediate from Lemma 5.1.9(2).
2. From Lemma 5.1.9 and Definition 1.2.2(1&2).
3. By Lemma 5.1.6 $\bigcap \{C \in \mathsf{Closed} \mid P \subseteq C\} \subseteq |P|$. By construction $P \subseteq \bigcap \{C \in \mathsf{Closed} \mid P \subseteq C\}$, and using Lemma 5.1.3(1) and part 3 of this result we have
$$|P| \overset{L5.1.3(1)}{\subseteq} |\bigcap \{C \in \mathsf{Closed} \mid P \subseteq C\}| \overset{pt.2}{=} \bigcap \{C \in \mathsf{Closed} \mid P \subseteq C\}.$$
□

The usual characterisation of continuity in terms of inverse images of closed sets being closed, remains valid:

COROLLARY 5.1.12. *Suppose* $(\mathsf{P}, \mathsf{Open})$ *and* $(\mathsf{P}', \mathsf{Open}')$ *are semitopological spaces (Definition 1.2.2) and suppose* $f : \mathsf{P} \to \mathsf{P}'$ *is a function. Then the following are equivalent:*

1. f *is continuous, meaning by Definition 2.2.1(2) that* $f^{-1}(O') \in \mathsf{Open}$ *for every* $O' \in \mathsf{Open}'$.
2. $f^{-1}(C') \in \mathsf{Closed}$ *for every* $C' \in \mathsf{Closed}'$.

Proof. By routine calculations as for topologies, using Lemma 5.1.9 and the fact that the inverse image of a complement is the complement of the inverse image; see [Wil70, Theorem 7.2, page 44] or [Eng89, Proposition 1.4.1(iv), page 28]. □

5.2 Duality between closure and interior

The usual dualities between closures and interiors remain valid in semitopologies. There are no surprises but this still needs to be checked, so we spell out the details:

LEMMA 5.2.1. *Suppose* $(\mathsf{P}, \mathsf{Open})$ *is a semitopology and* $O \in \mathsf{Open}$ *and* $C \in \mathsf{Closed}$. *Then:*

1. $O \subseteq interior(|O|)$. *The inclusion may be strict.*
2. $|interior(C)| \subseteq C$. *The inclusion may be strict.*
3. $interior(\mathsf{P} \setminus O) = \mathsf{P} \setminus |O|$.
4. $|\mathsf{P} \setminus C| = \mathsf{P} \setminus interior(C)$.

Proof. The reasoning is just as for topologies, but we spell out the details:

1. By Lemma 5.1.3(2) $O \subseteq |O|$. By Corollary 4.1.3 $interior(O) \subseteq interior(|O|)$. By Lemma 4.1.2 $O = interior(O)$, so we are done.

   For an example of the strict inclusion, consider $\mathbb{R}$ with the usual topology (which is also a semitopology) and take $O = (0,1) \cup (1,2)$. Then $O \subsetneq interior(|O|) = (0,2)$.

2. By Lemma 4.1.2 $interior(C) \subseteq C$. By Lemma 5.1.3(1) $|interior(C)| \subseteq |C|$. By Definition 5.1.5(1) (since we assumed $C \in \mathsf{Closed}$) $|C| = C$, so we are done.

   For an example of the strict inclusion, consider $\mathbb{R}$ with the usual topology and take $C = \{0\}$. Then $|interior(C)| = \varnothing \subsetneq C$.

3. Consider some $p' \in \mathsf{P}$. By Definition 4.1.1 $p' \in interior(\mathsf{P} \setminus O)$ when there exists some $O' \in \mathsf{Open}$ such that $p' \in O' \not\between O$. By definition in Definition 5.1.2(1) this happens precisely when $p' \notin |O|$.

4. By Definition 5.1.2(1), $p' \notin |\mathsf{P} \setminus C|$ precisely when there exists some $O' \in \mathsf{Open}$ such that $p' \in O' \not\between \mathsf{P} \setminus C$. By facts of sets this means precisely that $p' \in O' \subseteq C$. By Definition 4.1.1 this means precisely that $p' \in interior(C)$. □

COROLLARY 5.2.2. *Suppose* $(\mathsf{P}, \mathsf{Open})$ *is a semitopology and* $O \in \mathsf{Open}$ *and* $C \in \mathsf{Closed}$. *Then:*

1. $|O| = |interior(|O|)|$.
2. $interior(C) = interior(|interior(C)|)$.

Proof. We use Lemma 5.2.1(1&3) along with Lemma 5.1.3(1) and Corollary 4.1.3:

$$|O| \overset{L5.2.1(1)\&L5.1.3(1)}{\subseteq} |interior(|O|)| \overset{L5.2.1(2)}{\subseteq} interior(|O|)$$
$$interior(C) \overset{L5.2.1(1)}{\subseteq} interior(|interior(C)|) \overset{L5.2.1(2)\&C4.1.3}{\subseteq} interior(C)$$

□

5.3 Transitivity and closure

We explore how the topological closure operation interacts with taking transitive sets.

LEMMA 5.3.1. *Suppose* $(\mathsf{P}, \mathsf{Open})$ *is a semitopology and* $T \subseteq \mathsf{P}$ *is transitive and* $O \in \mathsf{Open}$. *Then*

$$T \between O \quad \textit{implies} \quad |T| \subseteq |O|.$$

Proof. Unpacking Definition 5.1.2 we have:

$$p' \in |T| \iff \forall O' \in \mathsf{Open}.p' \in O' \Longrightarrow O' \between T \qquad \text{and}$$
$$p' \in |O| \iff \forall O' \in \mathsf{Open}.p' \in O' \Longrightarrow O' \between O.$$

It would suffice to prove $O' \between T \Longrightarrow O' \between O$ for any $O' \in \mathsf{Open}$.

So suppose $O' \between T$. By assumption $T \between O$ and by transitivity of T (Definition 3.2.1) $O' \between O$. □

PROPOSITION 5.3.2. *Suppose* $(\mathsf{P}, \mathsf{Open})$ *is a semitopology and* $T \in \mathsf{Topen}$ *and* $O \in \mathsf{Open}$. *Then the following are equivalent:*

$$T \between O \quad \textit{if and only if} \quad T \subseteq |T| \subseteq |O|.$$

Proof. We prove two implications:

- Suppose $T \between O$. By Lemma 5.3.1 $|T| \subseteq |O|$. By Lemma 5.1.3(2) (as standard) $T \subseteq |T|$.
- Suppose $T \subseteq |T| \subseteq |O|$. Then $T \between |O|$ and by Lemma 5.1.4 (since T is nonempty (and transitive) and open) also $T \between O$. □

REMARK 5.3.3. In retrospect we can see the imprint of topens (Definition 3.2.1) in previous work, if we look at things in a certain way. Many consensus algorithms have the property that once consensus is established in a quorum O, it propagates to $|O|$.

This is apparent (for example) in the Grade-Cast algorithm [FM88], in which participants assign a confidence grade of 0, 1 or 2 to their output and must ensure that if any participant outputs v with grade 2 then all must output v with grade at least 1. In this algorithm, if a participant finds that all its quorums intersect some set S that unanimously supports value v, then the participant assigns grade at least 1 to v. From our point of view here, this is just taking a closure in the style we discussed in Remark 5.1.1. If T unanimously supports v and participants communicate enough, then eventually every member of $|T|$ assigns grade at least 1 to v. Thus, Proposition 5.3.2 suggests that, to convince a topen to agree on a value, we can first convince an open neighbourhood that intersects the topen, and then use Grade-Cast to convince the closure of that open set and thus in particular the topen which we know must be contained in that closure.

REMARK 5.3.4. Later on we will revisit these ideas and fit them into a nice general framework having to do with dense subsets. See Lemma 11.4.5 and Proposition 11.4.6.

We conclude with an easy observation which will be useful later. Recall from Notation 3.6.5 the notion of an intertwined space being one such that all nonempty open sets intersect. Then we have:

LEMMA 5.3.5. *Suppose* $(\mathsf{P}, \mathsf{Open})$ *is a semitopology and suppose* $T \in \mathsf{Topen}$. *Then the following are equivalent:*

1. P *is intertwined.*
2. $|T| = \mathsf{P}$.

Proof. Suppose $|T| = \mathsf{P}$ and consider any $O, O' \in \mathsf{Open}$. Unpacking Definition 5.1.2(1) it follows that $O \between T \between O'$. By transitivity of T (Definition 3.2.1(1)) $O \between O'$ as required.

Suppose $(\mathsf{P}, \mathsf{Open})$ is intertwined. By Lemma 3.6.6 every nonempty open set is topen, thus P is topen, and $\mathsf{P} = |T|$ follows by Lemma 5.3.1. □

5.4 Closed neighbourhoods and intertwined points

5.4.1 Definition and basic properties

Definition 5.4.1. Suppose $(\mathsf{P}, \mathsf{Open})$ is a semitopology. We generalise Definition 2.3.1 as follows:

1. Call $P \subseteq \mathsf{P}$ a **neighbourhood** when it contains an open set (i.e. when $interior(P) \neq \varnothing$), and call P a **neighbourhood of** p when $p \in \mathsf{P}$ and P contains an open neighbourhood of p (i.e. when $p \in interior(P)$). In particular:
2. $C \subseteq \mathsf{P}$ is a **closed neighbourhood of** $p \in \mathsf{P}$ when C is closed and $p \in interior(C)$.
3. $C \subseteq \mathsf{P}$ is a **closed neighbourhood** when C is closed and $interior(C) \neq \varnothing$.

Remark 5.4.2.

1. If C is a closed neighbourhood of p in the sense of Definition 5.4.1(2) then C is a closed neighbourhood in the sense of Definition 5.4.1(3), just because if $p \in interior(C)$ then $interior(C) \neq \varnothing$.
2. $p \in C$ is not enough for C to be a closed neighbourhood of p; we require the stronger condition $p \in interior(C)$.

   For instance take $\mathsf{P} = \{0,1\}$ and $\mathsf{Open} = \{\varnothing, \{1\}, \mathsf{P}\}$ (the Sierpiński space; see Figure 5.3), and consider $p = 0$ and $C = \{0\}$. Then $p \in C$ but $p \notin interior(C) = \varnothing$, so that C is not a closed neighbourhood of p.

Recall from Definition 3.6.1 the notions of $p \between p'$ and $p_\between$. Proposition 5.4.3 packages up our material for convenient use in later results.

Proposition 5.4.3. *Suppose* $(\mathsf{P}, \mathsf{Open})$ *is a semitopology and* $p, p' \in \mathsf{P}$. *Then:*

1. *We can characterise when* p' *is intertwined with* p *as follows:*

$$p \between p' \quad \textit{if and only if} \quad \forall O \in \mathsf{Open}. p \in O \Longrightarrow p' \in |O|.$$

2. *As a corollary,*

$$p_\between = \bigcap \{|O| \mid p \in O \in \mathsf{Open}\}.$$

3. *Equivalently:*

$$\begin{aligned} p_\between &= \bigcap \{C \in \mathsf{Closed} \mid p \in interior(C)\} \\ &= \bigcap \{C \in \mathsf{Closed} \mid C \textit{ a closed neighbourhood of } p\} \qquad \textit{Definition 5.4.1.} \end{aligned}$$

Thus in particular, if C *is a closed neighbourhood of* p *then* $p_\between \subseteq C$.

4. $p_{\between}$ *is closed and* $\mathsf{P}\setminus p_{\between}$ *is open.*

Proof.

1. We just rearrange Definition 3.6.1. So

$$\forall O,O'\in\mathsf{Open}.((p\in O \wedge p'\in O') \Longrightarrow O \between O')$$

rearranges to

$$\forall O\in\mathsf{Open}.(p\in O \Longrightarrow \forall O'\in\mathsf{Open}.(p'\in O' \Longrightarrow O \between O')).$$

We now observe from Definition 5.1.2 that this is precisely

$$\forall O\in\mathsf{Open}.(p\in O \Longrightarrow p'\in|O|).$$

2. We just rephrase part 1 of this result.

3. Using part 2 of this result it would suffice to prove

$$\bigcap\{|O| \mid p\in O\in\mathsf{Open}\} = \bigcap\{C\in\mathsf{Closed} \mid p\in \mathit{interior}(C)\}.$$

We will do this by proving that for each O-component on the left there is a C on the right with $C\subseteq|O|$; and for each C-component on the right there is an O on the left with $|O|\subseteq C$:

- Consider some $O\in\mathsf{Open}$ with $p\in O$.
We set $C=|O|$, so that trivially $C\subseteq|O|$. By Lemma 5.2.1(1) $O\subseteq \mathit{interior}(|O|)$, so $p\in \mathit{interior}(C)$.
- Consider some $C\in\mathsf{Closed}$ such that $p\in\mathit{interior}(C)$.
We set $O=\mathit{interior}(C)$. Then $p\in O$, and by Lemma 5.2.1(2) $|O|\subseteq C$.

4. Part 3 of this result exhibits $p_{\between}$ as an intersection of closed sets, and by Corollary 5.1.11(2) this is closed. By Lemma 5.1.9(1) its complement $\mathsf{P}\setminus p_{\between}$ is open. □

Definition 5.4.4. Suppose $(\mathsf{P},\mathsf{Open})$ is a semitopology and $p\in\mathsf{P}$.

1. Write $\mathit{nbhd}(p)=\{O\in\mathsf{Open} \mid p\in\mathsf{Open}\}$ and call this the **open neighbourhood system** of $p\in\mathsf{P}$.
2. Write $\mathit{nbhd}^c(p)=\{C\in\mathsf{Closed} \mid p\in\mathsf{Closed}\}$ and call this the **closed neighbourhood system** of $p\in\mathsf{P}$.

Remark 5.4.5. As standard, we can use Definition 5.4.4 to rewrite the definition of f being continuous at p (Definition 2.2.1(3)) as

$$\forall O'\in \mathit{nbhd}(f(p)).\exists O\in\mathit{nbhd}(p).O\subseteq f^{-1}(O').$$

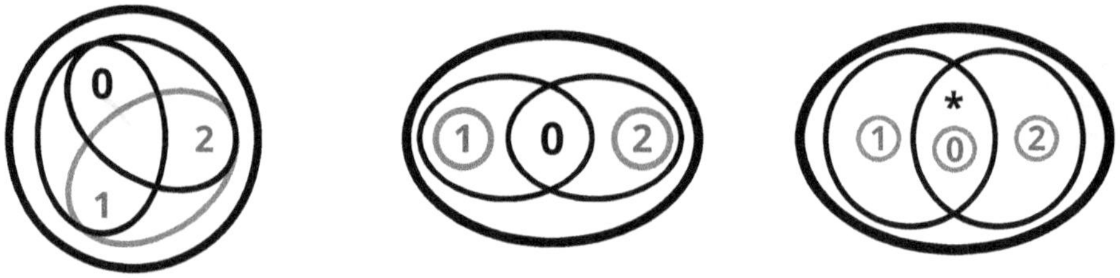

Figure 5.1: Examples of open neighbourhoods (Remarks 5.4.6 and 13.1.3)

REMARK 5.4.6. If $(\mathsf{P}, \mathsf{Open})$ is a topology, then $nbhd(p)$ is a filter (a nonempty up-closed down-directed set) and this is often called the *neighbourhood filter* of p.

We are working with semitopologies, so Open is not necessarily closed under intersections, and $nbhd(p)$ is not necessarily a filter. Figure 5.1 illustrates examples of this: e.g. in the left-hand example $\{0,1\}, \{0,2\} \in nbhd(0)$ but $\{0\} \notin nbhd(0)$, since $\{0\}$ is not an open set.

REMARK 5.4.7. We can relate Proposition 5.4.3 to a concept from topology. Following standard terminology ([Bou98, Definition 2, page 69] or [Eng89, page 52]), a **cluster point** $p \in \mathsf{P}$ of $\mathcal{O} \subseteq \mathsf{Open}$ is one such that every open neighbourhood of p intersects every $O \in \mathcal{O}$. Then Proposition 5.4.3(2) identifies $p_{\between}$ as the set of cluster points of $nbhd(p) \subseteq \mathsf{Open}$.

5.4.2 Application to characterise (quasi/weak) regularity

REMARK 5.4.8. Recall that Theorem 4.2.6 characterised regularity in multiple ways, including as the existence of a greatest topen neighbourhood. Proposition 5.4.10 below does something similar, for quasiregularity and weak regularity and the existence of closed neighbourhoods (Definition 5.4.1), and Theorem 5.6.2 is a result in the same style, for regularity.

Here, for the reader's convenience, is a summary of the relevant results:

1. Proposition 5.4.9: p is quasiregular when $p_{\between}$ is a closed neighbourhood.
2. Proposition 5.4.10: p is weakly regular when $p_{\between}$ is a closed neighbourhood of p.
3. Theorem 5.6.2: p is regular when $p_{\between}$ is a closed neighbourhood of p and is a minimal closed neighbourhood.

PROPOSITION 5.4.9. *Suppose* $(\mathsf{P}, \mathsf{Open})$ *is a semitopology and* $p \in \mathsf{P}$. *Then the following are equivalent:*

1. *p is quasiregular, or in full: $K(p) \neq \varnothing$ (Definition 4.1.4(5)).*
2. *$p_{\between}$ is a closed neighbourhood (Definition 5.4.1(3)).*

Proof. By construction in Definition 4.1.4(1), $K(p) = interior(p_{\between})$. So $K(p) \neq \varnothing$ means precisely that $p_{\between}$ is a closed neighbourhood. □

PROPOSITION 5.4.10. *Suppose* $(\mathsf{P}, \mathsf{Open})$ *is a semitopology and* $p \in \mathsf{P}$. *Then the following are equivalent:*

1. *p is weakly regular, or in full: $p \in K(p)$ (Definition 4.1.4(4)).*
2. *$p_{\between}$ is a closed neighbourhood of p (Definition 5.4.1(2)).*

3. *The poset of closed neighbourhoods of p ordered by subset inclusion, has a least element.*
4. $p_{\between}$ *is least in the poset of closed neighbourhoods of p ordered by subset inclusion.*

Proof. We prove a cycle of implications:

- Suppose $p \in interior(p_{\between})$. By Proposition 5.4.3(4) $p_{\between}$ is closed, so this makes it a closed neighbourhood of p as per Definition 5.4.1.

- Suppose $p_{\between}$ is a closed neighbourhood of p. By Proposition 5.4.3(3) $p_{\between}$ is the intersection of *all* closed neighbourhoods of p, and it follows that this poset has $p_{\between}$ as a least element.

- Assume the poset of closed neighbourhoods of p has a least element; write it C. So $C = \bigcap\{C' \in \mathsf{Closed} \mid C' \text{ is a closed neighbourhood of } p\}$ and thus by Proposition 5.4.3(3) $C = p_{\between}$.

- If $p_{\between}$ is least in the poset of closed neighbourhoods of p ordered by subset inclusion, then in particular $p_{\between}$ is a closed neighbourhood of p and it follows from Definition 5.4.1 that $p \in interior(p_{\between})$. □

Recall from Definition 4.1.4 that $K(p) = interior(p_{\between})$:

LEMMA 5.4.11. *Suppose* $(\mathsf{P}, \mathsf{Open})$ *is a semitopology and* $p \in \mathsf{P}$. *Then* $|K(p)| \subseteq p_{\between}$.

Proof. By Proposition 5.4.3(4) $p_{\between}$ is closed; we use Lemma 5.2.1(2). □

THEOREM 5.4.12. *Suppose* $(\mathsf{P}, \mathsf{Open})$ *is a semitopology and* $p \in \mathsf{P}$. *Then:*

1. *If p is weakly regular then* $|K(p)| = p_{\between}$. *In symbols:*
$$p \in K(p) \quad \textit{implies} \quad |K(p)| = p_{\between}.$$
2. *As an immediate corollary, if p is regular then* $|K(p)| = p_{\between}$.

Proof. We consider each part in turn:

1. If $p \in K(p) = interior(p_{\between})$ then $|K(p)|$ is a closed neighbourhood of p, so by Proposition 5.4.3(3) $p_{\between} \subseteq |K(p)|$. By Lemma 5.4.11 $|K(p)| \subseteq p_{\between}$.

2. By Lemma 4.1.6(1) if p is regular then it is weakly regular. We use part 1 of this result. □

We can combine Theorem 5.4.12 with Corollary 4.2.8:

COROLLARY 5.4.13. *Suppose* $(\mathsf{P}, \mathsf{Open})$ *is a semitopology and* $p \in \mathsf{P}$. *Then the following are equivalent:*

1. *p is regular.*
2. *p is weakly regular and* $p_{\between} = p'_{\between}$ *for every* $p' \in K(p)$.

Proof. Suppose p is regular and $p' \in K(p)$. Then p is weakly regular by Lemma 4.1.6(1), and $K(p) = K(p')$ by Corollary 4.2.8, and $p_{\between} = p'_{\between}$ by Theorem 5.4.12.

Suppose p is weakly regular and $p_{\between} = p'_{\between}$ for every $p' \in K(p)$. By Definition 4.1.4(1) also $K(p) = interior(p_{\between}) = interior(p'_{\between}) = K(p')$ for every $p' \in K(p)$, and by Corollary 4.2.8 p is regular. □

REMARK 5.4.14. Note a subtlety to Corollary 5.4.13: it is possible for p to be regular, yet it is not the case that $p_{\between} = p'_{\between}$ for every $p' \in p_{\between}$ (rather than for every $p' \in K(p)$). For an example consider the top-left semitopology in Figure 3.1, taking $p = 0$ and $p' = 1$; then $1 \in 0_{\between}$ but $0_{\between} = \{0, 1\}$ and $1_{\between} = \{0, 1, 2\}$.

To understand why this happens the interested reader can look ahead to Subsection 6.1: in the terminology of that Subsection, p' needs to be *unconflicted* in Corollaries 4.2.8 and 5.4.13.

5.5 Intersections of communities with open sets

REMARK 5.5.1 (An observation about consensus). Proposition 5.5.3 and Lemma 5.5.2 tell us some interesting and useful things:

- Suppose a weakly regular p wants to convince its community $K(p)$ of some belief. How might it proceed?
  By Proposition 5.5.3 it would suffice to seed one of the open neighbourhoods in its community with that belief, and then compute a *topological closure* of that open set; in Remark 5.5.5 we discuss why topological closures are particularly interesting.
- Suppose p is regular, so it is a member of a transitive open neighbourhood, and p wants to convince its community $K(p)$ of some belief.
  By Lemma 5.5.2 p need only convince *some* open set that intersects its community (this open set need not even contain p), and then compute a topological closure as in the previous point.

LEMMA 5.5.2. *Suppose* $(\mathsf{P}, \mathsf{Open})$ *is a semitopology and* $p \in \mathsf{P}$ *is regular (so* $p \in K(p) \in \mathsf{Topen}$*). Suppose* $O \in \mathsf{Open}$*. Then*

$$p \in O \between K(p) \quad \textit{implies} \quad K(p) \subseteq p_{\between} \subseteq |O|.$$

In word:

> *If an open set intersects the community of a regular point, then that community is included in the closure of the open set.*

Proof. Suppose p is regular, so $p \in K(p) \in \mathsf{Topen}$, and suppose $p \in O \between K(p)$. By Proposition 5.3.2 $K(p) \subseteq |K(p)| \subseteq |O|$. By Theorem 5.4.12 $|K(p)| = p_{\between}$, and putting this together we get

$$K(p) \subseteq p_{\between} \subseteq |O|$$

as required. □

Proposition 5.5.3 generalises Theorem 5.4.12, and is proved using it. We regain Theorem 5.4.12 as the special case where $O = K(p)$:

PROPOSITION 5.5.3. *Suppose* $(\mathsf{P}, \mathsf{Open})$ *is a semitopology and* $p \in \mathsf{P}$ *is weakly regular (so* $p \in K(p) \in \mathsf{Open}$*). Suppose* $O \in \mathsf{Open}$*. Then:*

1. $p \in O \subseteq K(p)$ *implies* $p_{\between} = |O|$.
2. *As a corollary,* $p \in O \subseteq p_{\between}$ *implies* $p_{\between} = |O|$.

Proof. If $p \in O \subseteq K(p)$ then $p \in K(p)$ and using Theorem 5.4.12 $|K(p)| \subseteq p_{\between}$. Since $O \subseteq K(p)$ also $|O| \subseteq p_{\between}$. Also, by Proposition 5.4.3(2) (since $p \in O \in \mathsf{Open}$) $p_{\between} \subseteq |O|$.

For the corollary, we note that if O is open then $O \subseteq interior(p_{\between}) = K(p)$ if and only if $O \subseteq p_{\between}$. □

REMARK 5.5.4. Note in Proposition 5.5.3 that it really matters that $p \in O$ — that is, that O is an open neighbourhood *of* p and not just an open set in $p_{\between}$.

To see why, consider the example in Lemma 4.2.3 (illustrated in Figure 3.1, top-left diagram): so $\mathsf{P} = \{0, 1, 2\}$ and $\mathsf{Open} = \{\varnothing, \mathsf{P}, \{0\}, \{2\}\}$. Note that:

- $1_{\between} = \{0, 1, 2\}$.
- If we set $O = \{0\} \subseteq \{0, 1, 2\}$ then this is open, but $|O| = \{0, 1\} \neq \{0, 1, 2\}$.
- If we set $O = \{0, 1, 2\} \subseteq \{0, 1, 2\}$ then $|O| = \{0, 1, 2\}$.

REMARK 5.5.5. Topological closures will matter because we will develop a theory of computable semitopologies which will (amongst other things) deliver a distributed algorithm to compute closures (see Remark 8.4.14).

Thus, putting together the results above with the witness semitopology machinery to come in Definition 8.2.5 onwards, we can say that from the point of view of a regular participant p, Proposition 5.5.3 and Lemma 5.5.2 reduce the problem of

> p wishes to progress with value v

to the simpler problem of

> p wishes to find an open set that intersects with the community of p, and work with this open set to agree on v (which open set does not matter; p can try several until one works).

Once this is done, the distributed algorithm will safely propagate the belief across the network.

Note that no forking is possible above (this is when a distributed system that was in agreement, partitions into subsets that are committed to incompatible values); all the action is in finding and convincing the $O \between K(p)$, and then the rest is automatic.

More discussion of this when we develop the notion of a *kernel* in Section 10.

5.6 Regularity, maximal topens, & minimal closed neighbourhoods

REMARK 5.6.1. Recall we have seen an arc of results which

- started with Theorem 4.2.6 and Corollary 4.2.8 — characterisations of regularity in terms of maximal topens — and
- passed through Proposition 5.4.10 — characterisation of weak regularity $p \in K(p) \in \mathsf{Open}$ in terms of minimal closed neighbourhoods.

We are now ready to complete this arc by stating and proving Theorem 5.6.2. This establishes a pleasing — and not-at-all-obvious — duality between 'has a maximal topen neighbourhood' and 'has a minimal closed neighbourhood'.

THEOREM 5.6.2. *Suppose* $(\mathsf{P}, \mathsf{Open})$ *is a semitopology and* $p \in \mathsf{P}$. *Then the following are equivalent:*

1. *p is regular.*
2. *$K(p)$ is a maximal/greatest topen neighbourhood of p.*
3. *p is weakly regular (meaning that $p \in K(p) = interior(p_{\between})$) and $p_{\between}$ is a minimal closed neighbourhood (Definition 5.4.1).*[2]

Proof. Equivalence of parts 1 and 2 is just Theorem 4.2.6(2).

For equivalence of parts 2 and 3 we prove two implications:

- Suppose p is regular. By Lemma 4.1.6(1) p is weakly regular. Now consider a closed neighbourhood $C' \subseteq p_{\between}$. Note that C' has a nonempty interior by Definition 5.4.1(3), so pick any p' such that

  $$p' \in interior(C') \subseteq C' \subseteq p_{\between}.$$

  It follows that $p' \in interior(p_{\between}) = K(p)$, and p is regular, so by Corollary 5.4.13 $p'_{\between} = p_{\between}$, and then by Proposition 5.4.10(2&4) (since $p' \in interior(C')$) $p'_{\between} \subseteq C'$. Putting this all together we have

  $$p_{\between} = p'_{\between} \subseteq C' \subseteq p_{\between},$$

  so that $C' = p_{\between}$ as required.

- Suppose p is weakly regular and suppose $p_{\between}$ is minimal in the poset of closed neighbourhoods ordered by subset inclusion.

  Consider some $p' \in K(p)$. By Proposition 5.4.3(3) $p'_{\between} \subseteq p_{\between}$, and by minimality it follows that $p'_{\between} = p_{\between}$. Thus also $K(p') = K(p)$.

  Now $p' \in K(p)$ was arbitrary, so by Corollary 4.2.8 p is regular as required. □

REMARK 5.6.3. Recall Example 4.4.1(4), as illustrated in Figure 4.1 (right-hand diagram). This has a point 0 whose community $K(0) = \{1,2\}$ is not a single topen (it contains two topens: $\{1\}$ and $\{2\}$).

A corollary of Theorem 5.6.2 is that $0_{\between} = \{0,1,2\}$ cannot be a minimal closed neighbourhood, because if it were then 0 would be regular and $K(0)$ would be a maximal topen neighbourhood of 0.

We check, and see that indeed, $0_{\between}$ contains *two* distinct minimal closed neighbourhoods: $\{0,1\}$ and $\{0,2\}$.

REMARK 5.6.4. Theorem 5.6.2(3) looks like Proposition 5.4.10(4), but

- Proposition 5.4.10(4) regards the *poset of closed neighbourhoods of p* (closed sets with a nonempty open interior that contains p),
- Theorem 5.6.2(3) regards the *poset of all closed neighbourhoods* (closed sets with a nonempty open interior, not necessarily including p).

[2] We really do mean "$p_{\between}$ is minimal amongst closed neighbourhoods" and *not* the weaker condition "$p_{\between}$ is minimal amongst closed neighbourhoods of p"! That weaker condition is treated in Proposition 5.4.10. See Remark 5.6.4.

So the condition used in Theorem 5.6.2(3) is strictly stronger than the condition used in Proposition 5.4.10(4). Correspondingly, the regularity condition in Theorem 5.6.2(1) can be written as $p \in K(p) \in \mathsf{Topen}$, and (as noted in Lemma 4.1.6 and Example 4.1.7(2)) this is strictly stronger than the condition $p \in K(p)$ used in Proposition 5.4.10(1).

Corollary 5.6.5 makes Remark 3.6.10 (intertwined is the opposite of Hausdorff) a little more precise:

COROLLARY 5.6.5. *Suppose* $(\mathsf{P}, \mathsf{Open})$ *is a Hausdorff semitopology (so every two points have a pair of disjoint neighbourhoods). Then if* $p \in \mathsf{P}$ *is regular, then* $\{p\}$ *is clopen.*

Proof. Suppose P is Hausdorff and consider $p \in \mathsf{P}$. By Remark 3.6.10 $p_{\between} = \{p\}$. From Theorem 5.6.2(3) $\{p\}$ is closed and has a nonempty open interior which must therefore also be equal to $\{p\}$. By Corollary 4.2.7 (or from Theorem 5.6.2(2)) this interior is transitive. □

PROPOSITION 5.6.6. *Suppose* $(\mathsf{P}, \mathsf{Open})$ *is a semitopology. Then:*

1. *Every maximal topen is equal to the interior of a minimal closed neighbourhood.*
2. *The converse implication holds if* $(\mathsf{P}, \mathsf{Open})$ *is a topology, but need not hold in the more general case that* $(\mathsf{P}, \mathsf{Open})$ *is a semitopology: there may exist a minimal closed neighbourhood whose interior is not topen.*

Proof.

1. Suppose T is a maximal topen. By Definition 3.2.1(2) T is nonempty, so choose $p \in T$. By Proposition 5.4.3(4) $p_{\between}$ is closed, and using Theorem 4.2.6

$$p \in T = K(p) = \mathit{interior}(p_{\between}) \subseteq p_{\between}.$$

Thus p is weakly regular and by Proposition 5.4.10(1&4) $p_{\between}$ is a least closed neighbourhood of p.

2. It suffices to provide a counterexample. This is Example 5.6.8 below. However, we also provide here a breaking 'proof', which throws light on precisely what Example 5.6.8 is breaking, and illustrates what the difference between semitopology and topology can mean in practical proof.

Suppose $T = \mathit{interior}(C)$ is the nonempty open interior of some minimal closed neighbourhood C: we will try (and fail) to show that this is transitive. By Proposition 3.6.9 it suffices to prove that $p \between p'$ for every $p, p' \in T$.

So suppose $p \in O$ and $p' \in O'$ and $O \not\between O'$. By Definition 5.1.2(1) $p' \notin |O|$, so that $|O| \cap C \subsetneq C$ is a strictly smaller closed set. Also, $O \cap C$ is nonempty because it contains p.

If $(\mathsf{P}, \mathsf{Open})$ is a topology then we are done, because $O \cap T = \mathit{interior}(O \cap C)$ would necessarily be open, contradicting our assumption that C is a minimal closed neighbourhood.

However, if $(\mathsf{P}, \mathsf{Open})$ is a semitopology then this does not necessarily follow: $O \cap T$ need not be open, and we cannot proceed. □

LEMMA 5.6.7. *Consider the semitopology illustrated in Figure 5.2. So:*

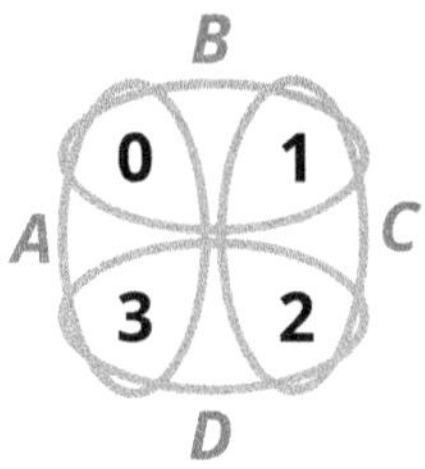

Figure 5.2: An unconflicted, irregular space (Proposition 6.2.1) in which every minimal closed neighbourhood has a non-transitive open interior (Example 5.6.8)

- $\mathsf{P} = \{0,1,2,3\}$.
- Open *is generated by* $\{A,B,C,D\}$ *where:*

$$A = \{3,0\}, \quad B = \{0,1\}, \quad C = \{1,2\}, \quad \textit{and} \quad D = \{2,3\}.$$

Then for every $p \in \mathsf{P}$ *we have:*

1. *p is intertwined only with itself.*
2. $K(p) = \varnothing$.

Proof. Part 1 is by routine calculations from Definition 3.6.1(2). Part 2 follows, noting that $interior(\{p\}) = \varnothing$ for every $p \in \mathsf{P}$. □

Example 5.6.8. The semitopology illustrated in Figure 5.2, and specified in Lemma 5.6.7, contains sets that are minimal amongst closed sets with a nonempty interior, yet that interior is not topen:

- A, B, C, and D are clopen, because C is the complement of A and D is the complement of B, so they are their own interior.
- A is a minimal closed neighbourhood (which is also open, being A itself), because
  - $A = \{3,0\}$ is closed because it is the complement of C, and it is its own interior, and
  - its two nonempty subsets $\{3\}$ and $\{0\}$ are closed (being the complement of $B \cup C$ and $C \cup D$ respectively) but they have empty open interior because $\{3\}$ and $\{0\}$ are not open.
- A is not transitive because 3 and 0 are not intertwined: $3 \in D$ and $0 \in B$ and $B \cap D = \varnothing$.
- Similarly B, C, and D are minimal closed neighbourhoods, which are also open, and they are not transitive.

We further note that:

1. $|0| = \{0\}$, because its complement is equal to $C \cup D$ (Definition 5.1.2; Lemma 5.1.9). Similarly for every other point in P.
2. $0_\between = \{0\}$, as noted in Lemma 5.6.7. Similarly for every other point in P.

3. $K(0) = interior(0_\between) = \varnothing$ as noted in Lemma 5.6.7, so that 0 is not regular (Definition 4.1.4(1)), and 0 is not even weakly regular or quasiregular. Similarly for every other point in P.
4. 0 has *two* minimal closed neighbourhoods: A and B. Similarly for every other point in P.

This illustrates that $p_\between \subsetneq C$ is possible, where C is a minimal closed neighbourhood of p.

REMARK 5.6.9. The results and discussions above tell us something interesting above and beyond the specific mathematical facts which they express.

They demonstrate that points being intertwined (the $p \between p'$ from Definition 3.6.1) is a distinct *semitopological* notion. A reader familiar with topology might be tempted to identify maximal topens with interiors of minimal closed neighbourhood (so that in view of Proposition 3.6.9, being intertwined would be topologically characterised just as two points being in the interior of the same minimal closed neighbourhood).

This works in topologies, but we see from Example 5.6.8 that in semitopologies being intertwined has its own distinct identity.

We conclude with one more example, showing how an (apparently?) slight change to a semitopology can make a big difference to its intertwinedness:

EXAMPLE 5.6.10.

1. $\mathbb{Q}^2$ with open sets generated by any covering collection of pairwise non-parallel **rational lines** — meaning a set of solutions to a linear equation $a.x+b.y = c$ for a, b, and c integers — is a semitopology.
   This consists of a single (maximal) topen: lines are pairwise non-parallel, so any two lines intersect and (looking to Proposition 3.6.9) all points are intertwined. There is only one closed set with a nonempty open interior, which is the whole space.
2. $\mathbb{Q}^2$ with open sets generated by all (possibly parallel) rational lines, is a semitopology. It has no topen sets and (looking to Proposition 3.6.9) no two distinct points are intertwined. For any line l, its complement $\mathbb{Q}^2 \setminus l$ is a closed set, given by the union of all the lines parallel to l. Thus every closed set is also an open set, and vice versa, and every line l is an example of a minimal closed neighbourhood (itself), whose interior is not a topen.

5.7 More on minimal closed neighbourhoods

We make good use of closed neighbourhoods, and in particular minimal closed neighbourhoods, in Subsection 5.6 and elsewhere. We take a moment to give a pleasing alternative characterisation of this useful concept.

5.7.1 Regular open/closed sets

REMARK 5.7.1. The terminology 'regular open/closed set' is from the topological literature. It is not directly related to terminology 'regular point' from Definition 4.1.4(3), which comes from semitopologies. However, it turns out that a mathematical connection does exist between these two notions. We outline some theory of regular open/closed sets, and then demonstrate the connections to what we have seen in our semitopological world.

DEFINITION 5.7.2. Suppose $(\mathsf{P}, \mathsf{Open})$ is a semitopology. Recall some standard terminology from topology [Wil70, Exercise 3D, page 29]:

1. We call an open set $O \in \mathsf{Open}$ a **regular open set** when $O = interior(|O|)$.
2. We call a closed set $C \in \mathsf{Closed}$ a **regular closed set** when $C = |interior(C)|$.
3. Write Open_{reg} and Closed_{reg} for the sets of regular open and regular closed sets respectively.

LEMMA 5.7.3. *Suppose* $(\mathsf{P}, \mathsf{Open})$ *is a semitopology and* $O \in \mathsf{Open}$ *and* $C \in \mathsf{Closed}$. *Then:*

1. $interior(C)$ *is a regular open set.*
2. $|O|$ *is a regular closed set.*

Proof. Direct from Definition 5.7.2 and Corollary 5.2.2. □

COROLLARY 5.7.4. *Suppose* $(\mathsf{P}, \mathsf{Open})$ *is a semitopology and* $p \in \mathsf{P}$. *Then* $K(p) \in \mathsf{Open}_{reg}$.

Proof. We just combine Lemma 5.7.3(1) with Proposition 5.4.3(4). □

COROLLARY 5.7.5. *Suppose* $(\mathsf{P}, \mathsf{Open})$ *is a semitopology and* $O \in \mathsf{Open}$. *Then* $interior(|O|)$ *is a regular open set.*

Proof. By Lemma 5.1.6 $|O|$ is closed, and by Lemma 5.7.3 $interior(|O|)$ is regular open. □

The regular open and the regular closed sets are the same thing, up to an easy and natural bijection:

COROLLARY 5.7.6. *Suppose* $(\mathsf{P}, \mathsf{Open})$ *is a semitopology. Then*

- *the topological closure map* $|\text{-}|$ *and*
- *the topological interior map* $interior(\text{-})$

define a bijection of posets between Open_{reg} *and* Closed_{reg} *ordered by subset inclusion.*

Proof. By Lemma 5.7.3, $|\text{-}|$ and $interior(\text{-})$ map between Open_{reg} to Closed_{reg}. Now we note that the regularity property from Definition 5.7.2, which states that $interior(|O|) = O$ when $O \in \mathsf{Open}_{reg}$ and $|interior(C)| = C$ when $C \in \mathsf{Closed}_{reg}$, expresses precisely that these maps are inverse.

They are maps of posets by Corollary 4.1.3 and Lemma 5.1.3(2). □

LEMMA 5.7.7. *Suppose* $(\mathsf{P}, \mathsf{Open})$ *is a semitopology and* $O \in \mathsf{Open}$ *and* $C \in \mathsf{Closed}$. *Then:*

1. O *is a regular open set if and only if* $\mathsf{P} \setminus O$ *is a regular closed set if and only if* $|O|$ *is a regular closed set.*
2. C *is a regular closed set if and only if* $\mathsf{P} \setminus C$ *is a regular open set if and only if* $interior(C)$ *is a regular open set.*

Proof. By routine calculations from the definitions using parts 3 and 4 of Lemma 5.2.1. □

5.7.2 Intersections of regular open sets

An easy observation about open sets will be useful:

LEMMA 5.7.8. *Suppose* $(\mathsf{P}, \mathsf{Open})$ *is a semitopology and* $O, O' \in \mathsf{Open}$. *Then the following are equivalent:*

1. $O \between O'$.
2. $O \between interior(|O'|)$.
3. $interior(|O|) \between interior(|O'|)$.

Proof. First we prove the equivalence of parts 1 and 2:

1. Suppose $O \between O'$. By Lemma 5.2.1(1) $O \between interior(|O'|)$.

2. Suppose there is some $p \in O \cap interior(|O'|)$. Then O is an open neighbourhood of p and $p \in |O'|$, so by Definition 5.1.2(1) $O \between O'$ as required.[3]

Equivalence of parts 1 and 3 then follows easily by two applications of the equivalence of parts 1 and 2. □

REMARK 5.7.9. Lemma 5.7.8 is true in topologies as well, but it is not prominent in the literature. Two standard reference works [Eng89, Wil70] do not seem to mention it. It appears as equation 10 in Theorem 1.37 of [Kop89], and as a lemma in π-base[4] (thanks to the mathematics StackExchange community for the pointers). We mention this to note an interesting contrast: this result is as true in topologies as it is in semitopologies, but somehow, it *matters* more in the latter than the former.

COROLLARY 5.7.10. *Suppose* $(\mathsf{P}, \mathsf{Open})$ *is a semitopology and* $p, p' \in \mathsf{P}$. *Then the following conditions are equivalent:*

1. *p and p' have a nonintersecting pair of open neighbourhoods.*
2. *p and p' have a nonintersecting pair of regular open neighbourhoods.*

Proof. Part 2 clearly implies part 1, since a regular open set is an open set. Part 1 implies part 2 using Lemma 5.7.8 and Corollary 5.7.5. □

REMARK 5.7.11. In Definition 3.6.1(1) we defined $p \between p'$ in terms of open neighbourhoods of p and p' as follows:

$$\forall O, O' \in \mathsf{Open}.(p \in O \land p' \in O') \Longrightarrow O \between O'.$$

In the light of Corollary 5.7.10, we could just as well have defined it just in terms of regular open neighbourhoods:

$$\forall O, O' \in \mathsf{Open}_{reg}.(p \in O \land p' \in O') \Longrightarrow O \between O'.$$

Mathematically, for what we have needed so far, this latter characterisation is not needed. However, it is easy to think of scenarios in which it might be useful. In particular, *computationally* it could make sense to restrict to the regular open sets, simply because there are fewer of them.

[3] Lemma 6.5.3 packages this argument up nicely with some slick notation, which we have not yet set up.

[4] See https://topology.pi-base.org/theorems/T000420 (permalink: https://web.archive.org/web/20240108192930/https://topology.pi-base.org/theorems/T000420).

5.7.3 Minimal nonempty regular closed sets are precisely the minimal closed neighbourhoods

LEMMA 5.7.12. *Suppose* $(\mathsf{P}, \mathsf{Open})$ *is a semitopology and* $C \in \mathsf{Closed}$. *Then:*

1. *If* C *is a minimal closed neighbourhood (a closed set with a nonempty open interior), then* C *is a nonempty regular closed set (Definition 5.7.2).*
2. *If* C *is a nonempty regular closed set then* C *is a closed neighbourhood (Definition 5.4.1).*

Proof. We consider each part in turn:

1. *Suppose* C *is a minimal closed neighbourhood.*

   Write $O' = interior(C)$ and $C' = |O'| = |interior(C)|$. Because C is a closed neighbourhood, by Definition 5.4.1 $O' \neq \varnothing$. By Lemma 5.1.6 $C' \in \mathsf{Closed}$. Using Corollary 5.2.2 $interior(C') = interior(|interior(C)|) = interior(C) = O' \neq \varnothing$, so that C' is a closed neighbourhood, and by minimality $C' = C$. But then $C = |interior(C)|$ so C is regular, as required.

2. *Suppose* C *is a nonempty regular closed set,* so that $\varnothing \neq C = |interior(C)|$.

   It follows that $interior(C) \neq \varnothing$ and this means precisely that C is a closed neighbourhood. □

In Theorem 5.6.2 we characterised the point p being regular in terms of minimal closed neighbourhoods. We can now characterise the minimal closed neighbourhoods in terms of something topologically familiar:

PROPOSITION 5.7.13. *Suppose* $(\mathsf{P}, \mathsf{Open})$ *is a semitopology and* $C \in \mathsf{Closed}$. *Then the following are equivalent:*

1. C *is a minimal nonempty regular closed set.*
2. C *is a minimal closed neighbourhood.*

Proof. We prove two implications:

- *Suppose* C *is a minimal closed neighbourhood.*

  By Lemma 5.7.12(1) C is a nonempty regular closed set. Furthermore by Lemma 5.7.12(2) if $C' \subseteq C$ is any other nonempty regular closed set contained in C, then it is a closed neighbourhood, and by minimality it is equal to C. Thus, C is minimal.

- *Suppose* C *is a minimal nonempty regular closed set.*

  By Lemma 5.7.12(2) C is a closed neighbourhood. Furthermore by Lemma 5.7.12(1) if $C' \subseteq C$ is any other closed neighbourhood then it is a nonempty regular closed set, and by minimality it is equal to C. □

5.8 How are $p_{\between}$ and $|p|$ related?

REMARK 5.8.1. Recall the definitions of $p_{\between}$ and $|p|$:

- The set $|p|$ is the *closure* of p.
  By Definition 5.1.2 this is the set of p' such that every open neighbourhood $O' \ni p'$ intersects with $\{p\}$. By Definition 5.1.5 $|p|$ is closed.
- The set $p_{\between}$ is the set of points *intertwined* with p.
  By Definition 3.6.1(2) this is the set of p' such that every open neighbourhood $O' \ni p'$ intersects with every open neighbourhood $O \ni p$. By Proposition 5.4.3(4) $p_{\between}$ is closed.

So we see that $|p|$ and $p_{\between}$ give us two canonical ways of generating a closed set from a point $p \in \mathsf{P}$. This invites a question:

How are $p_{\between}$ and $|p|$ related?

Lemma 5.8.2 rephrases Remark 5.8.1 more precisely by looking at it through sets complements. We will use it in Lemma 16.4.4(2):

LEMMA 5.8.2. *Suppose* $(\mathsf{P}, \mathsf{Open})$ *is a semitopology and* $p \in \mathsf{P}$. *Then:*

1. $\mathsf{P} \setminus |p| = \bigcup\{O \in \mathsf{Open} \mid p \notin O\} \in \mathsf{Open}$.
2. $\mathsf{P} \setminus p_{\between} = \bigcup\{O' \in \mathsf{Open} \mid \exists O \in \mathsf{Open}. p \in O \wedge O' \not\between O\} \in \mathsf{Open}$.
3. $\mathsf{P} \setminus p_{\between} = \bigcup\{O \in \mathsf{Open} \mid p \notin |O|\} \in \mathsf{Open}$.

In words, we can say: $\mathsf{P} \setminus |p|$ *is the union of the open sets such that p avoids them, and* $\mathsf{P} \setminus p_{\between}$ *is the union of the open sets such that p avoids their closures.*

Proof.

1. Immediate from Definitions 3.6.1 and 5.1.2.[5] Openness is from Definition 1.2.2(2).
2. By a routine argument direct from Definition 3.6.1. Openness is from Definition 1.2.2(2).
3. Rephrasing part 2 of this result using Definition 5.1.2(1). □

PROPOSITION 5.8.3. *Suppose* $(\mathsf{P}, \mathsf{Open})$ *is a semitopology and* $p \in \mathsf{P}$. *Then:*

1. $|p| \subseteq p_{\between}$.
2. *The subset inclusion may be strict; that is,* $|p| \subsetneq p_{\between}$ *is possible — even if p is regular (Definition 4.1.4(3)).*
3. *If* $interior(|p|) \neq \varnothing$ *(so* $|p|$ *has a nonempty interior) then* $|p| = p_{\between}$.

Proof.

1. We reason as follows:

$$\begin{aligned}
|p| &= |\{p\}| && \text{Definition 5.1.2(2)}\\
&= \bigcap\{C \in \mathsf{Closed} \mid p \in C\} && \text{Corollary 5.1.11(3)}\\
&\subseteq \bigcap\{C \in \mathsf{Closed} \mid p \in interior(C)\} && \text{Fact of intersections}\\
&= p_{\between} && \text{Proposition 5.4.3(3)}
\end{aligned}$$

[5] A longer proof via Corollary 5.1.11(3) and Lemma 5.1.9 is also possible.

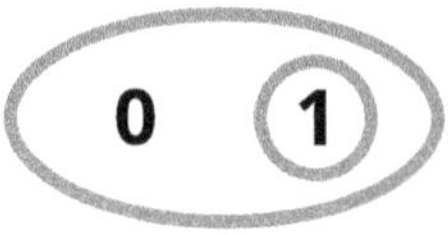

Figure 5.3: The Sierpiński space Sk (Example 5.8.4)

2. Example 5.8.4 below shows that $|p| \subsetneq p_{\between}$ is possible for p regular.

3. Write $O = \mathit{interior}(|p|)$. By standard topological reasoning, $|p|$ is the complement of the union of the open sets that do not contain p, and $O = \mathit{interior}(|p|)$ is the greatest open set such that $\forall O' \in \mathsf{Open}. O \between O' \Longrightarrow p \in O'$. We assumed that O is nonempty, so $O \between O$, thus $p \in O$.

   Then by part 1 of this result $p \in O \subseteq |p| \subseteq p_{\between}$, and by Proposition 5.5.3(2) $p_{\between} = |O|$. Using more standard topological reasoning (since $O \neq \varnothing$) $|O| = |p|$, and the result follows. □

EXAMPLE 5.8.4. Define Sk the **Sierpiński space** [Wil70, Example 3.2(e)] by $\mathsf{P} = \{0, 1\}$ and $\mathsf{Open} = \{\varnothing, \{1\}, \{0, 1\}\}$, as illustrated in Figure 5.3. Then:

- $|0| = \{0\}$ (because $\{1\}$ is open), but
- $0_{\between} = \{0, 1\}$ (because every open neighbourhood of 0 intersects with every open neighbourhood of 1).

Thus we see that $|0| = \{0\} \subsetneq \{0, 1\} = 0_{\between}$, and 0 is regular since $0 \in \mathit{interior}(0_{\between}) = \{0, 1\} \in \mathsf{Topen}$. (The Sierpiński space is also a topology and is a known space. We come back to it in Section 18.)

REMARK 5.8.5. We have one loose end left. We know from Theorem 5.6.2(3) that $p_{\between}$ is a minimal closed neighbourhood (closed set with nonempty open interior) when p is regular. We also know from Proposition 5.8.3 that $|p| \subsetneq p_{\between}$ is possible, even if p is regular.

So a closed *neighbourhood* in between $|p|$ and $p_{\between}$ is impossible by minimality, but can there be any closed *sets* (not necessarily having a nonempty open interior) in between $|p|$ and $p_{\between}$?

Somewhat counterintuitively perhaps, this is possible:

LEMMA 5.8.6. *Suppose* $(\mathsf{P}, \mathsf{Open})$ *is a semitopology and* $p \in \mathsf{P}$. *Then it is possible for there to exist a closed set* $C \subseteq \mathsf{P}$ *with* $|p| \subsetneq C \subsetneq p_{\between}$, *even if* p *is regular.*

Proof. It suffices to provide an example. Consider $\mathbb{N}$ with the semitopology whose open sets are generated by

- final segments $n_{\geq} = \{n' \in \mathbb{N} \mid n' \geq n\}$ for $n \in \mathbb{N}$ (cf. Example 3.7.9(2)), and
- $\{0, 1, 2, 3, 4, 5, 6, 7, 8, 9\}$.

The reader can check that $|0| = \{0\}$ and $0_{\between} = \{0, 1, 2, 3, 4, 5, 6, 7, 8, 9\}$. However, there are also eight closed sets $\{0, 1\}$, $\{0, 1, 2\}$, ..., $\{0, 1, 2, 3, \dots, 8\}$ in between $|0|$ and $0_{\between}$. □

We will study $p_{\between}$ further but to make more progress we need the notion of a(n un)conflicted point. This is an important idea in its own right:

(Un)conflicted points: transitivity of $\between$ 6

6.1 The basic definition

In Lemma 3.6.4 we asked whether the 'is intertwined with' relation $\between$ from Definition 3.6.1(1) is transitive — answer: not necessarily.

Transitivity of $\between$ is a natural condition. We now have enough machinery to study it in more detail, and this will help us gain a deeper understanding of the properties of not-necessarily-regular points.

DEFINITION 6.1.1. Suppose $(\mathsf{P}, \mathsf{Open})$ is a semitopology.

1. Call p a **conflicted point** when there exist p' and p'' such that $p' \between p$ and $p \between p''$ yet $\neg(p' \between p'')$.
2. If $p' \between p \between p''$ implies $p' \between p''$ always, then call p an **unconflicted point**.
3. Continuing Definition 4.1.4(7), if $P \subseteq \mathsf{P}$ and every $p \in P$ is conflicted/unconflicted, then we may call P a **conflicted/unconflicted set** respectively.

EXAMPLE 6.1.2. We consider some examples:

1. In Figure 3.1 top-left diagram, 0 and 2 are unconflicted and intertwined with themselves, and 1 is conflicted (being intertwined with 0, 1, and 2).
   If the reader wants to know what a conflicted point looks like: it looks like 1.
2. In Figure 3.1 top-right diagram, 0 and 2 are unconflicted and intertwined with themselves, and 1 is conflicted (being intertwined with 0, 1, and 2).
3. In Figure 3.1 lower-left diagram, 0 and 1 are unconflicted and intertwined with themselves, and 3 and 4 are unconflicted and intertwined with themselves, and 2 is conflicted (being intertwined with 0, 1, 2, 3, and 4).
4. In Figure 3.1 lower-right diagram, all points are unconflicted, and 0 and 2 are intertwined just with themselves, and 1 and $*$ are intertwined with one another.
5. In Figure 5.2, all points are unconflicted and intertwined only with themselves.

So p is conflicted when it witnesses a counterexample to $\between$ being transitive. We start with an easy lemma (we will use this later, but we mention it now for Remark 6.1.4):

LEMMA 6.1.3. *Suppose* $(\mathsf{P}, \mathsf{Open})$ *is a semitopology and* $p \in \mathsf{P}$. *Then the following are equivalent:*

1. *p is unconflicted.*
2. *If $q \in \mathsf{P}$ and $p \in q_\between$ then $p_\between \subseteq q_\between$.*
3. *$p_\between \subseteq p'_\between$ for every $p' \in p_\between$.*
4. *$p_\between$ is least in the set $\{p'_\between \mid p \between p'\}$ ordered by subset inclusion.*

Proof. The proof is just by pushing definitions around in a cycle of implications.

- *Part 1 implies part 2.*

  Suppose p is unconflicted. Consider $q \in \mathsf{P}$ such that $p \in q_\between$, and consider any $p' \in p_\between$. Unpacking definitions we have that $p' \between p \between q$ and so $p' \between q$, thus $p' \in q_\between$ as required.

- *Part 2 implies part 3.*

  From the fact that $p' \in p_\between$ if and only if $p' \between p$ if and only if $p \in p'_\between$.

- *Part 3 implies part 4.*

  Part 4 just rephrases part 3.

- *Part 4 implies part 1.*

  Suppose $p_\between$ is $\subseteq$-least in $\{p'_\between \mid p \between p'\}$ and suppose $p'' \between p \between p'$. Then $p'' \in p_\between \subseteq p'_\between$, so $p'' \between p'$ as required. □

REMARK 6.1.4. Lemma 6.1.3 is just an exercise in reformulating definitions, but part 4 of the result helps us to contrast the property of being unconflicted, with structurally similar characterisations of *weak regularity* and of *regularity* in Proposition 5.4.10 and Theorem 5.6.2 respectively. For the reader's convenience we collect them here — all sets below are ordered by subset inclusion:

1. p is unconflicted when $p_\between$ *is least in* $\{p'_\between \mid p \between p'\}$.

2. p is weakly regular when $p_\between$ *is least amongst closed neighbourhoods of* p.

   See Proposition 5.4.10 and recall from Definition 5.4.1 that a closed neighbourhood of p is a closed set C such that $p \in interior(C)$.

3. p is regular when $p_\between$ *is a closed neighbourhood of* p *and minimal amongst all closed neighbourhoods.*

   See Theorem 5.6.2 and recall that a closed neighbourhood is any closed set with a nonempty interior (not necessarily containing p).

We know from Lemma 4.1.6(1) that regular implies weakly regular. We now consider how these properties relate to being unconflicted.

6.2 Regular = weakly regular + unconflicted

PROPOSITION 6.2.1. *Suppose* $(\mathsf{P}, \mathsf{Open})$ *is a semitopology and* $p \in \mathsf{P}$. *Then:*

1. *If* p *is regular then it is unconflicted.*
   Equivalently by the contrapositive: if p *is conflicted then it is not regular.*
2. p *may be unconflicted and neither quasiregular, weakly regular, nor regular.*
3. *There exists a semitopological space such that*
   - *every point is unconflicted (so* $\between$ *is a transitive relation) yet*
   - *every point has empty community, so that the space is irregular, not weakly regular, and not quasiregular.*[1]

Proof. We consider each part in turn:

1. So consider $q \between p \between q'$. We must show that $q \between q'$, so consider open neighbourhoods $Q \ni q$ and $Q' \ni q'$. By assumption p is regular, so unpacking Definition 4.1.4(3) $p \in K(p) \in \mathsf{Topen}$. From

$$q \between p \between q' \quad \text{if follows that} \quad Q \between K(p) \between Q',$$

   and by transitivity of $K(p)$ (Definition 3.2.1(1)) we have $Q \between Q'$ as required.

2. Consider the semitopology illustrated in Figure 5.2. By Lemma 5.6.7 the point 0 is trivially unconflicted (because it is intertwined only with itself), but it is also neither quasiregular, weakly regular, nor regular, because its community is the empty set. See also Example 6.3.3.

3. As for the previous part, noting that the same holds of points 1, 2, and 3 in Figure 5.2. □

We can combine Proposition 6.2.1 with a previous result Lemma 4.1.6 to get a precise and attractive relation between being

- regular (Definition 4.1.4(3)),
- weakly regular (Definition 4.1.4(4)), and
- unconflicted (Definition 6.1.1),

as follows:

THEOREM 6.2.2. *Suppose* $(\mathsf{P}, \mathsf{Open})$ *is a semitopology and* $p \in \mathsf{P}$. *Then the following are equivalent:*

- p *is regular.*
- p *is weakly regular and unconflicted.*

More succinctly we can write: regular = weakly regular + unconflicted.[2]

Proof. We prove two implications:

[1] See also Proposition 6.4.11.

[2] See also a similar result Theorem 6.5.8, and a discussion in Remark 6.5.9.

- If p is regular then it is weakly regular by Lemma 4.1.6 and unconflicted by Proposition 6.2.1(1).
- Suppose p is weakly regular and unconflicted. By Definition 4.1.4(4) $p \in K(p)$ and by Lemma 3.6.8 it would suffice to show that $q \between q'$ for any $q, q' \in K(p)$.

  So consider $q, q' \in K(p)$. Now by Definition 4.1.4(1) $K(p) = \mathit{interior}(p_\between)$ so in particular $q, q' \in p_\between$. Thus $q \between p \between q'$, and since p is unconflicted $q \between q'$ as required. □

We can use Theorem 6.2.2 to derive simple global well-behavedness conditions on spaces, as follows:

COROLLARY 6.2.3. *Suppose* $(\mathsf{P}, \mathsf{Open})$ *is a semitopology. Then:*

1. *If the* $\between$ *relation is transitive (i.e. if every point is unconflicted) then a point is regular if and only if it is weakly regular.*
2. *If every point is weakly regular (i.e. if* $p \in K(p)$ *always) then a point is regular if and only if it is unconflicted.*

Proof. Immediate from Theorem 6.2.2. □

6.3 The boundary of $p_\between$

In this short Subsection we ask what points on the topological boundary of $p_\between$ can look like:

NOTATION 6.3.1. Suppose $(\mathsf{P}, \mathsf{Open})$ is a semitopology and $P \subseteq \mathsf{P}$.

1. As standard, we define

$$\mathit{boundary}(P) = P \setminus \mathit{interior}(P)$$

   and we call this the **boundary of** P.
2. In the case that $P = p_\between$ for $p \in \mathsf{P}$ then

$$\mathit{boundary}(p_\between) = p_\between \setminus \mathit{interior}(p_\between) = p_\between \setminus K(p).$$

Points in the boundary of $p_\between$ are *not* regular points:

PROPOSITION 6.3.2. *Suppose* $(\mathsf{P}, \mathsf{Open})$ *is a semitopology and* $p, q \in \mathsf{P}$ *and* $q \in p_\between$. *Then:*

1. *If* q *is regular then* $q \in K(p) = \mathit{interior}(p_\between)$.
2. *If* q *is regular then* $q \notin \mathit{boundary}(p_\between)$.
3. *If* $q \in \mathit{boundary}(p_\between)$ *then* q *is either conflicted or not weakly regular (or both).*

Proof. We consider each part in turn:

1. Suppose q is regular. By Theorem 6.2.2 q is unconflicted, so that by Lemma 6.1.3(3) $q_\between \subseteq p_\between$; and also q is weakly regular, so that $q \in K(q) \in \mathsf{Open}$ and $K(q) \subseteq q_\between \subseteq p_\between$. Thus $K(q)$ is an open neighbourhood of q that is contained in $p_\between$ and thus $q \in \mathit{interior}(p_\between)$ as required.

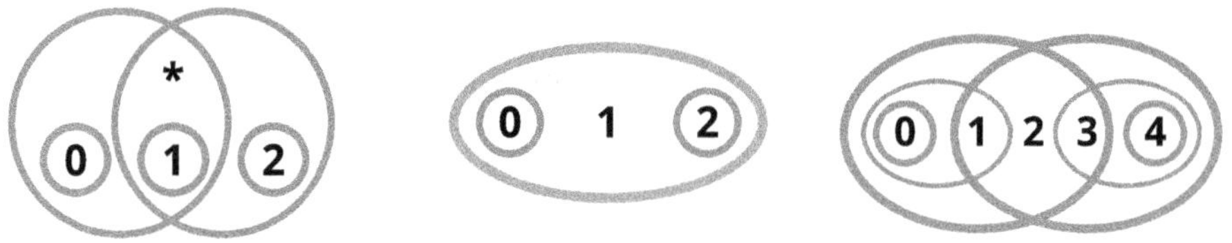

Figure 6.1: Examples of boundary points (Example 6.3.3).

2. This just repeats part 2 of this result, recalling from Notation 6.3.1 that $q \in boundary(p_{\between})$ if and only if $q \notin interior(p_{\between})$.

3. This is just the contrapositive of part 2, combined with Theorem 6.2.2. □

EXAMPLE 6.3.3. Proposition 6.3.2(3) tells us that points on the topological boundary of $p_{\between}$ are either conflicted, or not weakly regular, or perhaps both. It remains to show that all options are possible. It suffices to provide examples:

1. In Figure 6.1 (left-hand diagram) the point $*$ is on the boundary of $1_{\between} = \{*, 1\}$. It is unconflicted (being intertwined just with itself and 1), and not weakly regular (since $* \notin K(*) = \{1\}$).
2. In Figure 6.1 (middle diagram) the point 1 is on the boundary of $0_{\between} = \{0, 1\}$. It is conflicted (since $0 \between 1 \between 2$ yet $0 \not\between 2$) and it is weakly regular (since $1 \in K(1) = \{0, 1, 2\}$).[3]
3. In Figure 6.1 (right-hand diagram) the point 2 is conflicted (since $1 \between 2 \between 3$ yet $1 \not\between 3$) and it is not weakly regular, or even quasiregular (since $K(2) = interior(\{1, 2, 3\}) = \varnothing$).

Example 3 above illustrates a boundary point that does two things — be conflicted *and* be non-weakly-regular — even though examples 1 and 2 already provide examples of boundary points that each do one of these (but not the other). It would also be nice to be able to build an example that does two (bad) things by composing two smaller examples that do one (bad) thing each — e.g. by suitably composing examples 1 and 2 above. In fact this is easy to do using products of semitopologies, but we need a little more machinery for that; see Corollary 7.2.6.

We consider the special case of *regular* spaces (we will pick this thread up again in Subsection 6.4.3 after we have built more machinery):

COROLLARY 6.3.4. *Suppose* $(\mathsf{P}, \mathsf{Open})$ *is a semitopology and* $p \in \mathsf{P}$. *Then:*

1. *If the set* $p_{\between}$ *is regular, then* $boundary(p_{\between}) = \varnothing$ *and* $p_{\between}$ *is clopen (closed and open) and transitive.*
2. *If* P *is a regular space (so every point in it is regular) then* P *partitions into clopen transitive components given by* $\{p_{\between} \mid p \in \mathsf{P}\}$.

[3]This semitopology is also in Figure 3.1. We reproduce it here for the reader's convenience so that the examples are side-by-side.

This particular semitopology is very important for other reasons: it is isomorphic to **3** (Definition 18.1.2); and, it is a counterexample for the plausible-seeming-yet-false non-result that $\exists p \in \mathsf{P}.\mathsf{P} = p_{\between}$ implies that P is intertwined: $\mathsf{P} = 1_{\between}$ yet $0 \not\between 2$. The version of this non-result that *does* hold is Lemma 5.3.5, which is also used in Remark 21.1.7 in a discussion of fast intertwinedness checking.

Proof.

1. By Proposition 6.3.2 $p_{\between} = \mathit{interior}(p_{\between})$, so by Lemma 4.1.2 $p_{\between}$ is open. By Proposition 5.4.3(4) $p_{\between}$ is closed. By Definition 4.1.4(3) $p \in K(p) = \mathit{interior}(p_{\between}) \in \mathsf{Topen}$. It follows that $p_{\between}$ is (topen and therefore) transitive.

2. By part 1 of this result each $p_{\between}$ is a clopen transitive set. Using Theorem 6.2.2 every point is unconflicted and it follows that if $p_{\between} \between p'_{\between}$ then $p_{\between} = p'_{\between}$. □

6.4 The intertwined preorder

6.4.1 Definition and properties

Remark 6.4.1. Recall the *specialisation preorder* on points from topology, defined by

$$p \leq p' \quad \text{when} \quad |p| \subseteq |p'|.$$

In words: we order points p by subset inclusion on their closure $|p|$.

This can also be defined on semitopologies of course, but we will also find a similar preorder interesting, which is defined using $p_{\between}$ instead of $|p|$ (Definition 6.4.2). Recall that:

- $|p|$ is a closed set and is equal to the intersection of all the closed sets containing p, and
- $p_{\between}$ is also a closed set (Proposition 5.4.3(4)) and it is the intersection of all the closed neighbourhoods of p (closed sets with an interior that contains p; see Definition 5.4.1 and Proposition 5.4.3(3)).

Definition 6.4.2. Suppose $(\mathsf{P}, \mathsf{Open})$ is a semitopology.

1. Define the **intertwined preorder** on points $p, p' \in \mathsf{P}$ by:

   $$p \leq_{\between} p' \quad \text{when} \quad p_{\between} \subseteq p'_{\between}.$$

   As standard, we may write $p' \geq_{\between} p$ when $p \leq_{\between} p'$ (pronounced 'p' is intertwined-less / intertwined-greater than p').

   Calling this the 'intertwined preorder' does not refer to the ordering being intertwined in any sense; it just means that we order on $p_{\between}$ (which is read 'intertwined-p').

2. Call $(\mathsf{P}, \mathsf{Open})$ an **$\between$-complete semitopology** (read '**intertwined-complete**') when for every subset $P \subseteq \mathsf{P}$ that is totally ordered by $\leq_{\between}$, there exists some $p \in \mathsf{P}$ such that $p_{\between} \subseteq \bigcap_i \{p_{\between} \mid p \in P\}$.

Remark 6.4.3. Being $\between$-complete (Definition 6.4.2(2)) is a plausible well-behavedness condition, because important classes of semitopologies are $\between$-complete:

1. Finite semitopologies, since a descending chain of subsets of a finite set is terminating. Note that real systems are finite, so assuming that a semitopology is $\between$-complete is justifiable just on these practical grounds.
2. The strongly chain-complete semitopologies which we consider later in Definition 9.1.2 are $\between$-complete; see Lemma 9.3.1.

For now, it suffices to just work with what we need for this subsection, which is being $\between$-complete.

REMARK 6.4.4. There is also the **community preorder** defined such that $p \leq_K p'$ when $K(p) \subseteq K(p')$, which is related to $p \leq p'$ via the fact that by definition $K(p) = \mathit{interior}(p_\between)$, so that $\leq_K$ is a coarser relation (meaning: it relates more points). There is an argument that this would sit more nicely with the condition $q \in K(p)$ in Lemma 6.4.5, but ordering on $K(p)$ would relate all points with empty community, e.g. all of the points in Figure 5.2, and would slightly obfuscate the parallel with the specialisation preorder. This strikes us as unintuitive, so we prefer to preorder on $p_\between$.

LEMMA 6.4.5. *Suppose* $(\mathsf{P}, \mathsf{Open})$ *is a semitopology and* $p, q \in \mathsf{P}$. *Then:*

1. *If* $q \in K(p)$ *then* $q \leq_\between p$ *(meaning that* $q_\between \subseteq p_\between$*).*
2. *If* $q \in K(p)$ *then* $K(q) \subseteq K(p)$.

Proof. We consider each part in turn:

1. Suppose $q \in K(p)$ and recall from Lemma 4.2.3(1) that $K(p) \in \mathsf{Open}$, which means that $|K(p)|$ is a closed neighbourhood of q. We use Proposition 5.4.3(2) and Lemma 5.4.11:[4]

$$q_\between \overset{P5.4.3(2)}{\subseteq} |K(p)| \overset{L5.4.11}{\subseteq} p_\between .$$

2. Suppose $q \in K(p)$. By part 1 of this result and Definition 6.4.2 $q_\between \subseteq p_\between$. It is a fact that then $\mathit{interior}(q_\between) \subseteq \mathit{interior}(p_\between)$. By Definition 4.1.4(1) therefore $K(q) \subseteq K(p)$ as required. □

In the rest of this Subsection we develop corollaries of Lemma 6.4.5 (and compare this with Proposition 4.3.2):

COROLLARY 6.4.6. *Suppose* $(\mathsf{P}, \mathsf{Open})$ *is a semitopology and* $q, q' \in \mathsf{P}$. *Then:*

1. *If* $K(q) \between K(q')$ *then* $q \between q'$.
2. *If* q *and* q' *are weakly regular (so that* $q \in K(q)$ *and* $q' \in K(q')$*) then*

$$q \between q' \quad \text{if and only if} \quad K(q) \between K(q').$$

Proof. We consider each part in turn:

1. Suppose $r \in K(q) \cap K(q')$. Then $r_\between \subseteq q_\between \cap q'_\between$ using Lemma 6.4.5(1). But $q \in r_\between$, so $q \in q'_\between$, and thus $q \between q'$.

2. If q and q' are weakly regular and $q \between q'$ then $K(q) \between K(q')$ follows from Definition 3.6.1(1). The result follows from this and from part 1 of this result. □

[4] If the reader prefers a proof by concrete calculations, it runs as follows: Suppose $p' \in K(p)$, so that in particular $p' \between p$. We wish to prove that $p'_\between \subseteq p_\between$. So consider $p'' \between p'$; we will show that $p'' \between p$, i.e. that every pair of open neighbourhoods of p'' and p must intersect. Consider a pair of open neighbourhoods $p'' \in O'' \in \mathsf{Open}$ and $p \in O \in \mathsf{Open}$. We note that $O'' \between K(p)$, because $p' \in K(p) \in \mathsf{Open}$ and $p'' \between p'$. Choose $q \in K(p) \cap O''$. Now $q \between p$ and $q \in O''$ and $p \in O$, and we conclude that $O'' \between O$ as required.

Theorem 6.4.7 is somewhat reminiscent of the *hairy ball theorem:*[5]

THEOREM 6.4.7. *Suppose* $(\mathsf{P}, \mathsf{Open})$ *is an* $\between$*-complete quasiregular semitopology.*[6] *Then:*

1. *For every* $p \in \mathsf{P}$ *there exists some regular* $q \in K(p)$.
2. P *contains a regular point.*

(See also Proposition 9.3.7, which gives similar result for weak regularity.)

Proof. We consider each part in turn:

1. Consider the subset $\{p' \in \mathsf{P} \mid p' \leq_\between p\} \subseteq \mathsf{P}$ ordered by $\leq_\between$. Using Zorn's lemma (on $\geq_\between$), this contains a $\leq_\between$-minimal element q'. By assumption of quasiregularity $K(q') \neq \varnothing$, so choose $q \in K(q')$. By Lemma 6.4.5(1) $q_\between \subseteq q'_\between$ and by $\leq_\between$-minimality $q_\between = q'_\between$ and it follows that $q \in K(q)$. Thus q is weakly regular. Applying similar reasoning to $p' \in K(q)$ we deduce that $p'_\between = q_\between$, and thus $K(p') = K(q)$, for every $p' \in K(q)$, and so by Corollary 5.4.13 q is regular.

2. Choose any $p \in \mathsf{P}$, and use part 1 of this result. □

REMARK 6.4.8. We care about the existence of regular points as these are the ones that are well-behaved with respect to our semitopological model. A semitopology with a regular point is one that — in some idealised mathematical sense — is capable of some collaboration somewhere to take some action.

So Theorem 6.4.7 can be read as a guarantee that, provided the semitopology is $\between$-complete and quasiregular, there exists somebody, somewhere, who can make sense of their local network and progress to act. This a mathematical guarantee and not an engineering one, much as is the hairy ball theorem of which the result reminds us.

6.4.2 Application to quasiregular conflicted spaces

In Proposition 6.2.1(3) we saw an example of an unconflicted irregular space (illustrated in Figure 5.2): this is a space in which every point is unconflicted but not weakly regular. In this subsection we consider a dual case, of a conflicted quasiregular space: a space in which every point is conflicted yet quasiregular.

One question is: does such a creature even exist? The answer is:

- no, in the finite case (Corollary 6.4.10); and
- yes, in the infinite case (Proposition 6.4.11).

PROPOSITION 6.4.9. *Suppose* $(\mathsf{P}, \mathsf{Open})$ *is a finite quasiregular semitopology (so* P *is finite and every* $p \in \mathsf{P}$ *is quasiregular) — in particular this holds if the semitopology is weakly regular. Then:*

1. *For every* $p \in \mathsf{P}$ *there exist some regular* $q \in K(p)$.

[5] This famous result states that every tangent vector field on a sphere of even dimension — this being the surface of a ball of odd dimension — must vanish at at least one point. Intuitively, if we consider a 'hairy ball' in three-dimensional space and we try to comb its hairs so they all lie smoothly flat (with no discontinuities in direction), then at least one of the hairs is pointing straight up (i.e. its projection onto the ball is zero). A nice combinatorial proof is in [JT04].

[6] Definition 4.1.4(5): a semitopology that is $\between$-complete and whose every point has a nonempty community.

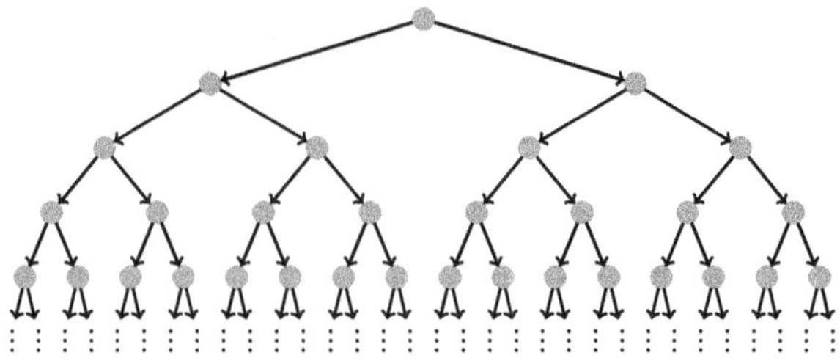

Figure 6.2: A weakly regular, conflicted space (Proposition 6.4.11); the opens are the down-closed sets

2. P *contains a regular point.*

In words we can say: every finite quasiregular semitopology contains a regular point.

Proof. From Theorem 6.4.7, since 'is finite' implies 'is $\between$-complete'.[7] □

COROLLARY 6.4.10. *There exists no finite quasiregular conflicted semitopology (i.e. a semitopology with finitely many points, each of which is quasiregular but conflicted).*

Proof. Suppose $(\mathsf{P}, \mathsf{Open})$ is finite and quasiregular. By Proposition 6.4.9 it contains a regular $q \in \mathsf{P}$ and by Proposition 6.2.1(1) q is unconflicted. □

Corollary 6.4.10 applies to finite semitopologies because these are necessarily $\between$-complete. The infinite case is different, as we shall now observe:

PROPOSITION 6.4.11. *There exists an infinite quasiregular — indeed it is also weakly regular — conflicted semitopology* $(\mathsf{P}, \mathsf{Open})$.

In more detail:

- *every* $p \in \mathsf{P}$ *is weakly regular (so* $p \in K(p) \in \mathsf{Open}$*; see Definition 4.1.4(4)) yet*
- *every* $p \in \mathsf{P}$ *is conflicted (so* $\between$ *is not transitive at* p*; Definition 6.1.1(1)).*

Furthermore: P *is a topology*[8] *and contains no topen sets.*

Proof. Take $\mathsf{P} = [01]^*$ to be the set of words (possibly empty finite lists) from 0 and 1. For $w, w' \in \mathsf{P}$ write $w \leq w'$ when w is an initial segment of w' and define

$$w_{\geq} = \{w' \mid w \leq w'\} \quad \text{and} \quad w_{\leq} = \{w' \mid w' \leq w\}.$$

Let open sets be generated as (possibly empty) unions of the $w_{\geq}$. This space is illustrated in Figure 6.2; open sets are down-closed subsets.

The reader can check that $\neg(w0 \between w1)$, because $w0_{\geq} \cap w1_{\geq} = \varnothing$, and that $w \between w'$ when $w \leq w'$ or $w' \leq w$. It follows from the above that

$$w_{\between} = w_{\geq} \cup w_{\leq} \quad \text{and} \quad K(w) = \mathit{interior}(w_{\between}) = w_{\geq},$$

[7]The proof of Theorem 6.4.7 uses Zorn's lemma. A longer, direct proof of Proposition 6.4.9 is also possible, by explicit induction on size of sets.

[8]Forward reference: it is also a witness semitopology. See Lemma 8.3.4.

and since $w \in w_{\geq}$ every w is weakly regular. Yet every w is also conflicted, because $w0 \between w \between w1$ yet $\neg(w0 \between w1)$.

This example is a topology, because an intersection of down-closed sets is still down-closed. It escapes the constraints of Theorem 6.4.7 by not being $\between$-complete. It contains no topen sets because if it did contain some topen T then by Theorem 4.2.6(1&5) there would exist a regular $p \in T$ in P. □

6.4.3 (Un)conflicted points and boundaries of closed sets

Recall from Definition 5.4.1 that a closed neighbourhood is a closed set with a nonempty interior, and recall that $p_{\between}$ — the set of points intertwined with p from Definition 3.6.1 — is characterised using closed neighbourhoods in Proposition 5.8.3, as the intersection of all closed neighbourhoods that have p in their interior.

This leads to the question of whether the theory of $p_{\between}$ might *be* a theory of closed neighbourhoods. The answer seems to be no: $p_{\between}$ has its own distinct character, as the results and counterexamples below will briefly illustrate.

For instance: in view of Proposition 5.8.3 characterising $p_{\between}$ as an intersection of closed neighbourhoods of p, might it be the case that for C a closed neighbourhood, $C = \bigcup\{p_{\between} \mid p \in \mathit{interior}(C)\}$. In words: is a closed neighbourhood C the union of the points intertwined with its interior? This turns out to be only half true:

Lemma 6.4.12. *Suppose* $(\mathsf{P}, \mathsf{Open})$ *is a semitopology and* $C \in \mathsf{Closed}$ *is a closed neighbourhood. Then:*

1. $\bigcup\{p_{\between} \mid p \in \mathit{interior}(C)\} \subseteq C$.
2. *This subset inclusion may be strict: it is possible for* $p \in \mathsf{P}$ *to be on the boundary of a closed neighbourhood* C*, but not intertwined with any point in that neighbourhood's interior. This is true even if* P *is a regular space (meaning that every* $p \in \mathsf{P}$ *is regular).*

Proof. We consider each part in turn:

1. If $p \in \mathit{interior}(C)$ then $p_{\between} \subseteq C$ by Proposition 5.4.3(3).

2. We provide a counterexample, as illustrated in Figure 6.3 (left-hand diagram):

   - $\mathsf{P} = \{*, 1, 2\}$.
   - Open sets are generated by $\{1\}$, $\{2\}$, and $\{*, 2\}$.
   - We set $p = *$ and $C = \{1, *\}$.

   Then the reader can check that $\mathit{interior}(C) = \{1\}$ $*_{\between} = \{*, 2\}$ and $* \not\between 2$ and every point in P is regular. □

Definition 6.4.13. Suppose $(\mathsf{P}, \mathsf{Open})$ is a semitopology and $P, P' \subseteq \mathsf{P}$. Then define

$$\mathit{kiss}(P, P') = \mathit{boundary}(P) \cap \mathit{boundary}(P')$$

and call this the **kissing set of P and P'**.

Lemma 6.4.14. *Suppose* $(\mathsf{P}, \mathsf{Open})$ *is a semitopology. Then the following are equivalent:*

(a) Regular boundary point of closed neighbourhood that is not intertwined with its interior (Lemma 6.4.12(2))

(b) Regular point in kissing set of closed neighbourhoods, not intertwined with interiors (Corollary 6.4.15(2))

Figure 6.3: Two counterexamples

- *p is conflicted.*
- *There exist $q, q' \in \mathsf{P}$ such that $q \not\between q'$ and $p \in kiss(q_{\between}, q'_{\between})$.*
- *There exist $q, q' \in \mathsf{P}$ such that $q \not\between q'$ and $p \in q_{\between} \cap q'_{\between}$.*

Proof. We prove a cycle of implications:

- *Suppose p is conflicted.*

  Then there exist $q, q' \in \mathsf{P}$ such that $q \between p \between q'$ yet $q \not\between q'$. Rephrasing this, we obtain that $p \in q_{\between} \cap q'_{\between}$.

  We need to check that $p \notin K(q)$ and $p \notin K(q')$. We prove $p \notin K(q)$ by contradiction ($p \notin K(q')$ follows by identical reasoning). Suppose $p \in K(q)$. Then by Lemma 6.4.5(1) $p_{\between} \subseteq q_{\between}$. But $q' \in p_{\between}$, so $q' \in q_{\between}$, so $q' \between q$, contradicting our assumption.

- *Suppose $q \not\between q'$ and $p \in boundary(q_{\between}) \cap boundary(q'_{\between})$.*

  Then certainly $p \in q_{\between} \cap q'_{\between}$.

- *Suppose $q \not\between q'$ and $p \in q_{\between} \cap q'_{\between}$.*

  Then $q \between p \between q'$ and $q \not\between q'$, which is precisely what it means to be conflicted. □

We can look at Definition 6.1.1 and Lemma 6.4.14 and conjecture that a point p is conflicted if and only if it is in the kissing set of a pair of distinct closed sets. Again, this is half true:

COROLLARY 6.4.15. *Suppose $(\mathsf{P}, \mathsf{Open})$ is a semitopology and $p \in \mathsf{P}$. Then:*

1. *If p is conflicted then there exist a pair of closed sets such that $p \in kiss(C, C')$.*
2. *The reverse implication need not hold: it is possible for p to be in the kissing set of a pair of closed sets C and C', yet p is unconflicted. This is even possible if the space is regular (meaning that every point in the space is regular, including p) and C and C' are closed neighbourhoods.*

Proof. We consider each part in turn:

1. If p is conflicted then we use Lemma 6.4.14 and Proposition 5.4.3(4).

2. We provide a counterexample, as illustrated in Figure 6.3 (right-hand diagram):

   - $\mathsf{P} = \{*, 1, 2, 3\}$.
   - Open sets are generated by $\{1\}$, $\{2\}$, $\{3\}$, and $\{*, 2\}$.
   - We set $p = *$ and $C = \{*, 1\}$ and $C' = \{*, 3\}$.

   Note that $*$ is regular (being intertwined with itself and 2), and C and C' are closed neighbourhoods that kiss at $*$, and 1, 2, and 3 are also regular. □

6.5 Regular = quasiregular + hypertransitive

REMARK 6.5.1. In Theorem 6.2.2 we characterised regularity in terms of weak regularity and being unconflicted. Regularity and weak regularity are two of the regularity properties considered in Definition 4.1.4, but there is also a third: *quasiregularity*. This raises the question whether there might be some other property X such that regular = quasiregular + X?[9]

Yes there is, and we develop it in this Subsection, culminating with Theorem 6.5.8.

6.5.1 Hypertransitivity

NOTATION 6.5.2. Suppose $(\mathsf{P}, \mathsf{Open})$ is a semitopology and $O' \in \mathsf{Open}$ and $\mathcal{O} \subseteq \mathsf{Open}$.

1. Write $O' \between \mathcal{O}$, or equivalently $\mathcal{O} \between O'$, when $O' \between O$ for every $O \in \mathcal{O}$. In symbols:

$$O' \between \mathcal{O} \quad \text{when} \quad \forall O \in \mathcal{O}. O' \between O.$$

2. As a special case of part 1 above taking $\mathcal{O} = nbhd(p)$ (Definition 5.4.4), if $p \in \mathsf{P}$ then write $O' \between nbhd(p)$, or equivalently $nbhd(p) \between O'$, when $O' \between O$ for every $O \in \mathsf{Open}$ such that $p \in O$.

LEMMA 6.5.3. *Suppose* $(\mathsf{P}, \mathsf{Open})$ *is a semitopology and* $p \in \mathsf{P}$ *and* $O' \in \mathsf{Open}$. *Then*

$$p \in |O'| \quad \text{if and only if} \quad O' \between nbhd(p).$$

See also Corollary 9.5.5.

Proof. This just rephrases Definition 5.1.2(1). □

DEFINITION 6.5.4. Suppose $(\mathsf{P}, \mathsf{Open})$ is a semitopology. Call $p \in \mathsf{P}$ a **hypertransitive point** when for every $O', O'' \in \mathsf{Open}$,

$$O' \between nbhd(p) \between O'' \quad \text{implies} \quad O' \between O''.$$

Call $(\mathsf{P}, \mathsf{Open})$ a **hypertransitive semitopology** when every $p \in \mathsf{P}$ is hypertransitive.

LEMMA 6.5.5. *Suppose* $(\mathsf{P}, \mathsf{Open})$ *is a semitopology and* $p \in \mathsf{P}$. *Then the following are equivalent:*

[9] By Lemma 4.1.6(2) being weakly regular is a stronger condition than being quasiregular, thus we would expect X to be stronger than being unconflicted. And indeed this will be so: see Lemma 6.5.6(2).

1. *p is hypertransitive.*
2. *For every pair of open sets $O', O'' \in \mathsf{Open}$, $p \in |O'| \cap |O''|$ implies $O' \between O''$.*
3. *For every pair of* regular *open sets $O', O'' \in \mathsf{Open}_{reg}$, $p \in |O'| \cap |O''|$ implies $O' \between O''$ (cf. Remark 5.7.11).*

Proof. For the equivalence of parts 1 and 2 we reason as follows:

- Suppose p is hypertransitive and suppose $p \in |O'|$ and $p \in |O''|$. By Lemma 6.5.3 it follows that $O' \between nbhd(p) \between O''$. By hypertransitivity, $O' \between O''$ as required.
- Suppose for every $O, O' \in \mathsf{Open}$, $p \in |O| \cap |O'|$ implies $O' \between O''$, and suppose $O' \between nbhd(p) \between O''$. By Lemma 6.5.3 $p \in |O| \cap |O'|$ and therefore $O' \between O''$.

For the equivalence of parts 2 and 3 we reason as follows:

- Part 2 implies part 3 follows since every open regular set is also an open set.
- To show part 3 implies part 2, suppose for every pair of regular opens $O', O'' \in \mathsf{Open}_{reg}$, $p \in |O'| \cap |O''|$ implies $O' \between O''$, and suppose $O', O'' \in \mathsf{Open}$ are two open sets that are not necessarily regular, and suppose $p \in |O'| \cap |O''|$. We must show that $O' \between O''$. Write $P' = interior(|O'|)$ and $P'' = interior(|O''|)$ and note by Lemmas 5.7.3 and 5.1.6 that P' and P'' are regular open sets and $|P'| = |O'|$ and $|P''| = |O''|$. Then $|P'| \between |P''|$, so $P' \between P''$, and $O' \between O''$ follows from Lemma 5.7.8 □

6.5.2 The equivalence

LEMMA 6.5.6. *Suppose $(\mathsf{P}, \mathsf{Open})$ is a semitopology and $p \in \mathsf{P}$. Then:*

1. *If p is regular then it is hypertransitive.*
2. *If p is hypertransitive then it is unconflicted.*
3. *The reverse implication need not hold: it is possible for p to be unconflicted but not hypertransitive.*
4. *It is possible for p to be hypertransitive (and unconflicted), but not quasiregular (and thus not weakly regular or regular).*

Proof. We consider each part:

1. Suppose p is regular and $O, O' \in \mathsf{Open}$ and $O \between nbhd(p) \between O'$. By Definition 4.1.4(3) (since p is regular) $K(p)$ is a topen (= open and transitive) neighbourhood of p. Therefore by transitivity $O \between O'$ as required.

2. Suppose p is hypertransitive and suppose $p', p'' \in \mathsf{P}$ and $p' \between p \between p''$. Now consider $p' \in O' \in \mathsf{Open}$ and $p'' \in O'' \in \mathsf{Open}$. By our intertwinedness assumptions we have that $O' \between nbhd(p) \between O''$. But p is hypertransitive, so $O' \between O''$ as required.

3. It suffices to provide a counterexample. Consider the bottom right semitopology in Figure 3.1, and take $p = *$ and $O' = \{1\}$ and $O'' = \{0, 2\}$. Note that:

   - $*$ is unconflicted, since it is intertwined only with itself and 1.
   - O' and O' intersect every open neighbourhood of $*$, but $O' \not\between O''$, so $*$ is not hypertransitive.

4. It suffices to provide an example. Consider the semitopology illustrated in Figure 3.1, top-right diagram; so $\mathsf{P} = \{0,1,2\}$ and $\mathsf{Open} = \{\varnothing, \{0\}, \{2\}, \{1,2\}, \{0,1\}, \{0,1,2\}\}$. The reader can check that $p = 1$ is hypertransitive, but $1_{\between} = \{1\}$ and $K(1) = \varnothing$ so p is not quasiregular. □

(Yet) another characterisation of being quasiregular will be helpful:

LEMMA 6.5.7. *Suppose* $(\mathsf{P}, \mathsf{Open})$ *is a semitopology and* $p \in \mathsf{P}$. *Then the following conditions are equivalent:*

1. *p is quasiregular (meaning by Definition 4.1.4(5) that $K(p) \neq \varnothing$).*
2. *$K(p) \between nbhd(p)$ (meaning by Notation 6.5.2(2) that $K(p) \between O$ for every $O \in nbhd(p)$).*
3. *$p \in |K(p)|$.*

Proof. Equivalence of parts 2 and 3 is immediate from Lemma 6.5.3.

For equivalence of parts 1 and 2, we prove two implications:

- Suppose p is quasiregular, meaning by Definition 4.1.4(5) that $K(p) \neq \varnothing$. Pick some $p' \in K(p)$ (it does not matter which). It follows by construction in Definitions 3.6.1(2) and 4.1.4(1) and Lemma 4.1.2 that $p' \between p$, so that $p' \in K(p)$. Using Definition 3.6.1(1) it follows that $K(p) \between O$ for every $O \in nbhd(p)$, as required.
- Suppose $K(p) \between nbhd(p)$. Then in particular $K(p) \between \mathsf{P}$ (because $p \in \mathsf{P} \in \mathsf{Open}$), and by Notation 3.1.1(1) it follows that $K(p) \neq \varnothing$. □

Compare and contrast Theorem 6.5.8 with Theorem 6.2.2:

THEOREM 6.5.8. *Suppose* $(\mathsf{P}, \mathsf{Open})$ *is a semitopology and* $p \in \mathsf{P}$. *Then the following are equivalent:*

1. *p is regular.*
2. *p is quasiregular and hypertransitive.*

Proof. We consider two implications:

- *Suppose p is regular.*

  Then p is quasiregular by Lemma 4.1.6(1&2), and hypertransitive by Lemma 6.5.6(1).

- *Suppose p is quasiregular and hypertransitive.*

  By Lemma 6.5.6(2) p is unconflicted. If we can prove that p is weakly regular (meaning by Definition 4.1.4(4) that $p \in K(p)$), then by Theorem 6.2.2 it would follow that p is regular as required. Thus, it would suffice to show that $p \in K(p)$, thus that there is an open neighbourhood of points with which p is intertwined.

  Write $O'' = interior(\mathsf{P} \setminus K(p))$. We have two subcases to consider:

  - *Suppose $nbhd(p) \between O''$.*
    By Lemma 6.5.7 (since p is quasiregular) we have that $K(p) \between nbhd(p)$. Thus $K(p) \between nbhd(p) \between O''$, and by hypertransitivity of p it follows that $K(p) \between O''$. But this contradicts the construction of O'' as being a subset of $\mathsf{P} \setminus K(p)$, so this case is impossible.

- *Suppose* $nbhd(p) \not\between O''$. Then there exists some $O \in nbhd(p)$ such that $O \not\between O''$, and it follows that $O \subseteq K(p)$ so that $p \in K(p)$ as required.

Thus p is weakly regular, as required. □

REMARK 6.5.9. So we have obtained two nice characterisations of regularity of points from Definition 4.1.4(3):

1. Regular = weakly regular + unconflicted, by Theorem 6.2.2.
2. Regular = quasiregular + hypertransitive, by Theorem 6.5.8.

It remains an open problem to check whether there is some natural property X' such that regular = indirectly regular + X' (see Definition 9.3.2). Regularity properties are important to us and characterising them is a theme for us: see in particular Theorems 16.5.4 and 16.8.7, and the discussions in Remarks 16.5.5, 16.8.8(4) and 22.6.18.

The product semitopology 7

Products of semitopologies can be defined just as for topologies. We do this in Definition 7.1.2, then study how semitopological properties — like being a (maximal) topen or being a regular point — interact with products.

7.1 Basic definitions and results (shared with topologies)

DEFINITION 7.1.1. Suppose P_1 and P_2 are sets and suppose $P_1 \subseteq \mathsf{P}_1$ and $P_2 \subseteq \mathsf{P}_2$. Then:

1. Call the set
$$P_1 \times P_2 = \{(p_1, p_2) \mid p_1 \in P_1,\ p_2 \in P_2\}$$
a **square**, and
2. call P_1 and P_2 the **sides of the square**.

DEFINITION 7.1.2 (Product semitopology). Suppose $(\mathsf{P}_1, \mathsf{Open}_1)$ and $(\mathsf{P}_2, \mathsf{Open}_2)$ are semitopologies.

1. As for topologies, define the **product semitopology** $\mathsf{P}_1 \times \mathsf{P}_2$ such that:
   - The set of points is the Cartesian product $\mathsf{P}_1 \times \mathsf{P}_2$.
   - Open sets are (possibly empty, possibly infinite) unions of squares $O_1 \times O_2$ for $O_1 \in \mathsf{Open}_1$ and $O_2 \in \mathsf{Open}_2$. By abuse of notation we may write this set $\mathsf{Open}_1 \times \mathsf{Open}_2$.
2. Define the **first projection** $\pi_1 : \mathsf{P}_1 \times \mathsf{P}_2 \to \mathsf{P}_1$ and the **second projection** $\pi_2 : \mathsf{P}_1 \times \mathsf{P}_2 \to \mathsf{P}_2$ as usual such that $\pi_1(p_1, p_2) = p_1$ and $\pi_2(p_1, p_2) = p_2$.
3. For this Subsection, if X is a set and f is a function on X then we define the **pointwise application** $f(X)$ by
$$f(X) = \{f(x) \mid x \in X\}.$$
In particular we will use this notation for pointwise application of π_1 and π_2 to subsets $P \subseteq \mathsf{P}_1 \times \mathsf{P}_2$.

LEMMA 7.1.3. *Suppose* $(\mathsf{P}_1, \mathsf{Open}_1)$ *and* $(\mathsf{P}_2, \mathsf{Open}_2)$ *are semitopologies. Then the first and second projections* π_1 *and* π_2 *from Definition 7.1.2 are*

1. *continuous (inverse image of open set is open / inverse image of closed set is closed), and*
2. *open (pointwise image of open set is open).*

Proof. By routine calculations, as for topologies; see for example [Eng89], page 79, just before Example 2.3.10. □

Lemma 7.1.4 below is a special case of a general result from topology [Eng89, Lemma 2.3.3, page 78] that (in our terminology from Definition 7.1.1) the closure of a square is the square of the closure of its sides. We do need to check that this still works for semitopologies, and it does:

LEMMA 7.1.4. *Suppose* $(\mathsf{P}_1, \mathsf{Open}_1)$ *and* $(\mathsf{P}_2, \mathsf{Open}_2)$ *are semitopologies and* $p_1 \in \mathsf{P}_1$ *and* $p_2 \in \mathsf{P}_2$. *Then*

$$|(p_1, p_2)| = |p_1| \times |p_2|.$$

Proof. The closure of a set is the complement of the largest open set disjoint from it.[1] By construction in Definition 7.1.2, open sets in the product topology are unions of squares of opens, and the result now just follows noting that for $O_1 \in \mathsf{Open}_1$ and $O_2 \in \mathsf{Open}_2$, $(p_1, p_2) \in O_1 \times O_2$ if and only if $p_1 \in O_1$ and $p_2 \in O_2$. □

7.2 Componentwise composition of semitopological properties

We prove a sequence of results checking how properties such as being intertwined, regular, weakly, regular, and conflicted relate between a product space and the component spaces. Most notably perhaps, we show that 'being intertwined', 'being regular', 'being weakly regular', and 'being conflicted' hold componentwise — i.e. the results have the form

"$(\mathsf{P}_1, \mathsf{P}_2)$ has property ϕ if and only if P_1 and P_2 have ϕ".

We will then use this to generate examples with complex behaviour that is obtained by composing the behaviour of their (simpler) components: see in particular Corollary 7.2.6 and Theorem 7.3.4.

LEMMA 7.2.1 (Intersecting squares is componentwise). *Suppose* $(\mathsf{P}_1, \mathsf{Open}_1)$ *and* $(\mathsf{P}_2, \mathsf{Open}_2)$ *are semitopologies and suppose* $O, O' \in \mathsf{Open}_1 \times \mathsf{Open}_2$ *are squares. Then*

$$O \between O' \quad \textit{if and only if} \quad \pi_1(O) \between \pi_1(O') \ \wedge \ \pi_2(O) \between \pi_2(O').$$

Proof. By routine sets calculations, noting that since O and O' are squares by definition $O = \pi_1(O) \times \pi_2(O)$ and $O' = \pi_1(O') \times \pi_2(O')$. □

PROPOSITION 7.2.2 (Intertwined is componentwise). *Suppose* $(\mathsf{P}_1, \mathsf{Open}_1)$ *and* $(\mathsf{P}_2, \mathsf{Open}_2)$ *are semitopologies. Then:*

1. $(p_1, p_2) \between (p'_1, p'_2)$ *if and only if* $p_1 \between p'_1 \ \wedge \ p_2 \between p'_2$.
2. *As an immediate corollary,* $(p_1, p_2)_{\between} = p_{1\between} \times p_{2\between}$.

Proof. For part 1 of this result we prove two implications:

[1] That is: the complement of the interior of the complement.

- *Suppose* $p_1 \between p'_1$ *and* $p_2 \between p'_2$.

  Consider two open neighbourhoods $O \ni (p_1, p_2)$ and $O' \ni (p'_1, p'_2)$. We wish to show that $O \between O'$.

  Without loss of generality we may assume that O and O' are squares, since: opens are unions of squares so we just choose squares in O and O' that contain (p_1, p_2) and (p'_1, p'_2) respectively. Thus, $O = O_1 \times O_2$ and $O' = O'_1 \times O'_2$.

  Now $p_1 \in O_1$ and $p'_1 \in O'_1$ and $p_1 \between p'_1$, thus $O_1 \between O'_1$. Similarly for p_2 and p'_2. We use Lemma 7.2.1.

- *Suppose* $(p_1, p_2) \between (p'_1, p'_2)$.

  Then in particular all square open neighbourhoods intersect, and by Lemma 7.2.1 so must their sides.

Part 2 just rephrases part 1 of this result using Definition 3.6.1(2). □

Corollary 7.2.3 ((Maximal) topen is componentwise). *Suppose* $(\mathsf{P}_1, \mathsf{Open}_1)$ *and* $(\mathsf{P}_2, \mathsf{Open}_2)$ *are semitopologies and* $T \in \mathsf{Open}_1 \times \mathsf{Open}_2$ *is a square. Then for each of 'a topen' / 'a maximal topen' below, the following are equivalent:*

- $T \in \mathsf{Open}_1 \times \mathsf{Open}_2$ *is a topen / a maximal topen in* $\mathsf{P}_1 \times \mathsf{P}_2$.
- *The sides* $\pi_1(T)$ *and* $\pi_2(T)$ *of* T *are topens / maximal topens in* $\mathsf{P}_1 \times \mathsf{P}_2$.

Proof. First, we consider the versions without 'maximal':

1. *Suppose* $T \in \mathsf{Open}_1 \times \mathsf{Open}_2$ *is a topen in* $\mathsf{P}_1 \times \mathsf{P}_2$.

   By Lemma 7.1.3(2) its sides $\pi_1(T)$ and $\pi_2(T)$ are open. Now consider $p_1, p'_1 \in \pi_1(T)$ and choose any $p_2 \in \pi_2(T)$. We know $(p_1, p_2) \between (p'_1, p_2)$ must hold, because both points are in T and by Proposition 3.6.9 all points in T are intertwined. By Proposition 7.2.2(1) it follows that $p_1 \between p'_1$. Since p_1 and p'_1 were arbitrary in $\pi_1(T)$ it follows using Proposition 3.6.9 again that $\pi_1(T)$ is topen.

   The reasoning for $\pi_2(T)$ is precisely similar.

2. *Suppose* $T_1 \in \mathsf{Open}_1$ *and* $T_2 \in \mathsf{Open}_2$ *are topen in* $\mathsf{P}_1 \times \mathsf{P}_2$.

   By construction in Definition 7.1.2 the square $T_1 \times T_2$ is open, and it follows using Proposition 7.2.2(1) and Proposition 3.6.9 that $T_1 \times T_2$ is topen.

We now consider maximality:

1. *Suppose* $T \in \mathsf{Open}_1 \times \mathsf{Open}_2$ *is a maximal topen in* $\mathsf{P}_1 \times \mathsf{P}_2$.

   By our reasoning above its sides are topens, but if those sides were not maximal topens — so at least one of them is included in a strictly larger topen — then, again using our reasoning above, we could use obtain a larger topen square in $\mathsf{Open}_1 \times \mathsf{Open}_2$, contradicting maximality of T.

2. *Suppose* $T_1 \in \mathsf{Open}_1$ *and* $T_2 \in \mathsf{Open}_2$ *are maximal topens in* P_1 *and* P_2.

   By our reasoning above the square $T_1 \times T_2$ is a topen. If it were not a maximal topen — so it is included in some strictly larger topen T — then by our reasoning above $\pi_1(T)$ and $\pi_2(T)$ are also topens and one of them would have to be larger than T_1 or T_2, contradicting their maximality. □

COROLLARY 7.2.4 (Regular is componentwise). *Suppose* $(\mathsf{P}_1, \mathsf{Open}_1)$ *and* $(\mathsf{P}_2, \mathsf{Open}_2)$ *are semitopologies and* $p_1 \in \mathsf{P}_1$ *and* $p_2 \in \mathsf{P}_2$. *Then the following are equivalent:*

- (p_1, p_2) *is regular in* $\mathsf{P}_1 \times \mathsf{P}_2$.
- p_1 *is regular in* P_1 *and* p_2 *is regular in* P_2.

Proof. Suppose (p_1, p_2) is regular. By Theorem 4.2.6(1&5) it has a topen neighbourhood T. Using Corollary 7.2.3 $\pi_1(T)$ and $\pi_2(T)$ are topen neighbourhoods of p_1 and p_2 respectively. By Theorem 4.2.6(1&5) p_1 and p_2 are regular.

If conversely p_1 and p_2 are regular then we just reverse the reasoning of the previous paragraph. □

Proposition 7.2.5 does for 'is conflicted' and 'is weakly regular' what Corollary 7.2.4 does for 'is regular'. With the machinery we now have, the argument is straightforward:

PROPOSITION 7.2.5 (Unconflicted & weakly regular is componentwise). *Suppose* $(\mathsf{P}_1, \mathsf{Open}_1)$ *and* $(\mathsf{P}_2, \mathsf{Open}_2)$ *are semitopologies and suppose* $p_1 \in \mathsf{P}_1$ *and* $p_2 \in \mathsf{P}_2$. *Then:*

1. (p_1, p_2) *is unconflicted in* $\mathsf{P}_1 \times \mathsf{P}_2$ *if and only if* p_1 *is unconflicted in* P_1 *and* p_2 *is unconflicted in* P_2.
2. (p_1, p_2) *is weakly regular in* $\mathsf{P}_1 \times \mathsf{P}_2$ *if and only if* p_1 *is weakly regular in* P_1 *and* p_2 *is weakly regular in* P_2.

Proof. For part 1 we prove two implications:

- *Suppose* (p_1, p_2) *is unconflicted.* We will show that p_1 is unconflicted (the case of p_2 is precisely similar).

  Consider $p', p'' \in \mathsf{P}_1$ and suppose $p' \between p_1 \between p''$. Using Proposition 7.2.2(1) $(p', p_2) \between (p_1, p_2) \between (p'', p_2)$, by transitivity (since we assumed (p_1, p_2) is unconflicted) $(p', p_2) \between (p'', p_2)$, and using Proposition 7.2.2(1) we conclude that $p' \between p''$ as required.

  Suppose p_1 *and* p_2 *are unconflicted.* We will assume $(p'_1, p'_2) \between (p_1, p_2) \between (p''_1, p''_2)$ and prove $(p'_1, p'_2) \between (p''_1, p''_2)$.

  Using Proposition 7.2.2(1) $p'_1 \between p_1 \between p''_1$ and by transitivity (since we assumed p_1 is unconflicted) we have $p'_1 \between p''_1$. Similarly $p'_2 \between p''_2$, and using Proposition 7.2.2(1) $(p'_1, p'_2) \between (p''_1, p''_2)$ as required.

Part 2 follows by routine reasoning just combining part 1 of this result and Corollary 7.2.4 with Theorem 6.2.2. □

We now have the machinery that we need to make good on a promise made at the end of Example 6.3.3:

COROLLARY 7.2.6. *There exists a semitopology* $(\mathsf{P}, \mathsf{Open})$ *and points* $p, q \in \mathsf{P}$ *such that*

- *q is on the boundary of $p_{\between}$ and*
- *q is conflicted and not weakly regular.*

Proof. We already know this from Example 6.3.3(3), as illustrated in the right-hand diagram in Figure 6.1, but now we can give a more principled construction: we let $(\mathsf{P}_1, \mathsf{Open}_1)$ and $(\mathsf{P}_2, \mathsf{Open}_2)$ be examples 1 and 2 from Example 6.3.3, as illustrated in Figure 6.1 (left-hand and middle diagram).

The point $* \in \mathsf{P}_1$ is on the boundary of $1_{\between}$ and it is unconflicted and not weakly regular. The point $1 \in \mathsf{P}_2$ is on the boundary of $0_{\between}$ and it is conflicted and weakly regular. It follows from Proposition 7.2.5 that $(*, 1)$ is conflicted and not weakly regular.

By Proposition 7.2.2(2) $(1,0)_{\between} = 1_{\between} \times 0_{\between}$, and by some routine topological calculation we see that $(*, 1)$ is on the boundary of this set. □

7.3 Minimal closed neighbourhoods, and a counterexample

We continue the development of Subsection 7.2 and the example in Corollary 7.2.6 with some slightly more technical results, leading up to another example.

LEMMA 7.3.1. *Suppose that:*

- $(\mathsf{P}_1, \mathsf{Open}_1)$ *and* $(\mathsf{P}_2, \mathsf{Open}_2)$ *are semitopologies.*
- *C is a square (Definition 7.1.1) in $\mathsf{P}_1 \times \mathsf{P}_2$.*

Then

- *if C is a minimal closed neighbourhood in $\mathsf{P}_1 \times \mathsf{P}_2$,*
- *then the sides of C, $C_1 = \pi_1(C)$ and $C_2 = \pi_2(C)$, are minimal closed neighbourhoods in P_1 and P_2 respectively.*

Proof. Suppose C is a square minimal closed neighbourhood, and consider $C_1' \subseteq C_1$ a closed neighbourhood in P_1. We will show that $C_1' = C_1$ (the argument for P_2 is no different). Using Lemma 7.1.3, $C_1' \times C_2$ is a closed neighbourhood in $\mathsf{P}_1 \times \mathsf{P}_2$. By routine sets calculations and minimality we have that $C_1' \times C_2 = C_1 \times C_2$, and it follows that $C_1' = C_1$. □

COROLLARY 7.3.2. *Suppose that:*

- $(\mathsf{P}_1, \mathsf{Open}_1)$ *and* $(\mathsf{P}_2, \mathsf{Open}_2)$ *are semitopologies.*
- $p_2 \in \mathsf{P}_2$.
- *$p_{2\between}$ is not a minimal closed neighbourhood of p_2.*

Then for every $p_1 \in \mathsf{P}_1$ and for every C a minimal closed neighbourhood of (p_1, p_2), we have that $(p_1, p_2)_{\between} \subsetneq C$.

Proof. By Proposition 7.2.2(2) $(p_1, p_2)_{\between} = p_{1\between} \times p_{2\between}$ and by Proposition 5.4.3(3) $(p_1, p_2)_{\between} \subseteq C$.

If $C = (p_1, p_2)_{\between} = p_{1\between} \times p_{2\between}$ then by Lemma 7.3.1 its side $p_{2\between}$ is a minimal closed neighbourhood of p_2, but we assumed this is not the case. Thus, $(p_1, p_2)_{\between} \subsetneq C$ as required. □

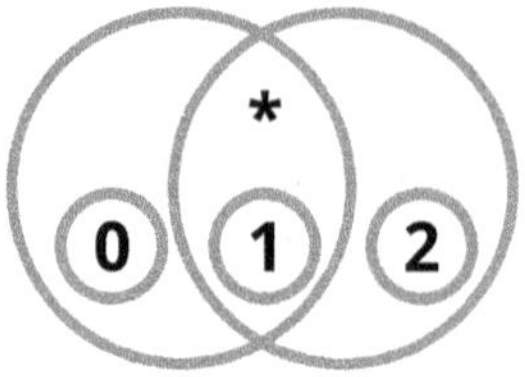

Figure 7.1: Example 7.3.5: $|*| \subsetneq *_{\between} \subsetneq \{0,1,*\}$

REMARK 7.3.3. Recall that Proposition 5.8.3(1) shows that $|p| \subseteq p_{\between}$, and Example 5.8.4 shows that this inclusion may be strict by giving a semitopology in which $|p| \subsetneq p_{\between}$. Recall also that it follows from Proposition 5.4.3(3) that $p_{\between} \subseteq C$ for any C a (minimal) closed neighbourhood of p, and Example 5.6.8 shows that this inclusion may be strict by giving a semitopology in which $p_{\between} \subsetneq C$ for C a minimal closed neighbourhood of p. What we have not done so far is show that both inclusions may be strict *for a single p*: we can now apply what we have shown about the product semitopology in this Subsection, to 'glue' our examples together:

THEOREM 7.3.4. *There exists a semitopology* $(\mathsf{P}, \mathsf{Open})$ *and a* $p \in \mathsf{P}$ *and a minimal closed neighbourhood* C *of* p *such that the inclusions below are strict:*

$$|p| \subsetneq p_{\between} \subsetneq C.$$

Proof. Let $(\mathsf{P}_1, \mathsf{Open}_1)$ be the semitopology from Example 5.8.4, and $(\mathsf{P}_2, \mathsf{Open}_2)$ be that from Example 5.6.8. We set:

- $(\mathsf{P}, \mathsf{Open}) = (\mathsf{P}_1, \mathsf{Open}_1) \times (\mathsf{P}_2, \mathsf{Open}_2)$, the product semitopology.
- $p_1 = 1 \in \mathsf{P}_1$, for which $|1| \subsetneq 1_{\between} = \{0,1\}$, and
- $p_2 = (0,0)$, which has a minimal closed neighbourhood $A = \{(0,0),(1,0)\}$ which is not equal to $p_{2\between} = (0,0)_{\between} = \{(0,0)\}$, and
- $C = \{0,1\} \times \{(0,0),(1,0)\}$.

We show that $|(p_1,p_2)| \subsetneq (p_1,p_2)_{\between}$, as follows:

$$\begin{aligned} |(p_1,p_2)| &= |p_1| \times |p_2| && \text{Lemma 7.1.4} \\ &\subsetneq p_{1\between} \times |p_2| && |1| \subsetneq 1_{\between} \\ &\subseteq p_{1\between} \times p_{2\between} && \text{Proposition 5.8.3(1).} \end{aligned}$$

Furthermore, by Corollary 7.3.2 $(p_1,p_2)_{\between} \subsetneq C$, because $p_{2\between} \subsetneq A$. □

EXAMPLE 7.3.5. We now give a smaller, but less compositional, example for Theorem 7.3.4. Set

- $\mathsf{P} = \{0,1,2,*\}$ and
- let Open be generated by $\{0\}$, $\{1\}$, $\{2\}$ (so $\{0,1,2\}$ has the discrete semitopology) and by $\{0,1,*\}$, and $\{1,2,*\}$,

as illustrated in Figure 7.1 (we used this same example in Figure 6.1, left-hand diagram, and it was also one of our first examples of topens in Figure 3.1). Then:

- $|*| = \{*\}$, because $\{0,1,2\}$ is open.
- $*_{\between} = \{1,*\}$, since $\{1,2,*\}$ is disjoint from $\{0\}$ and $\{0,1,*\}$ is disjoint from $\{2\}$.
- $\{0,1,*\}$ and $\{1,2,*\}$ are distinct minimal closed neighbourhoods of $*$, with open interiors $\{0,1\}$ and $\{1,2\}$ respectively.

The witnesses semitopology 8

8.1 Discussion

REMARK 8.1.1. In this Section, we turn to the problem of computing with semitopologies. We want two things from our maths:

- that it will deliver algorithms; and also
- that these algorithms should be **local**, by which we mean *executable by points knowing only information near (local) to them, by communicating with local peers.*

In particular, a local algorithm should not assume that points can globally synchronise or agree.[1]

We now note that our notion of 'open neighbourhood of a point' from semitopologies is not *a priori* particularly local. The simplest illustration is perhaps to note that $(\mathsf{P}, \mathsf{Open}) = (\mathbb{N}, \{\varnothing, \mathbb{N}\})$ expresses that points coordinate on whether they all agree, but the lack of locality shows up in the mathematics in other, perhaps unexpected ways, because we can encode nontrivial information in the structure of open sets. Consider the following example of a semitopology with (by design) poor algorithmic behaviour:

EXAMPLE 8.1.2. Let the **uncomputable semitopology** have

- $\mathsf{P} = \mathbb{N}$ and
- open sets generated as unions of *uncomputable subsets* of $\mathbb{N}$.

(Call a subset $U \subseteq \mathbb{N}$ *uncomputable* when there is no algorithm that inputs $n \in \mathbb{N}$ and returns 'true' if $n \in U$ and 'false' if $n \notin U$.) This is a semitopology. It is not a topology, since the intersection of two uncomputable subsets need not be uncomputable. By construction, no algorithm can compute its open sets.

[1] Indeed, to do this would be to assume a solution to the problem that semitopologies were created to study.

REMARK 8.1.3. Example 8.1.2 just comes from the fact that the definition of semitopologies involves a subset of the (powerset of) $\mathbb{N}$. This is not unusual, and the existence of such uncomputable subsets is well-known [Chu36, Theorem XVIII, page 360].

What we should do now is determine and study algorithmically tractable semitopologies. So: what is an appropriate and useful definition?

In this Section will identify a class of algorithmically tractable semitopologies, and furthermore this in the strong sense that the definition is clean, makes a novel connection to declarative programming, and from it we extract distributed and local algorithms in the sense discussed above. To do this, we will define witnessed sets (Definition 8.2.2) and show that they determine computationally tractable semitopologies in a sense made formal by results including

- Propositions 8.4.5 and 8.4.13 (which show that algorithms exist to compute open and closed sets) and
- the remarkable Theorem 9.4.1 (which shows intuitively that witness semitopologies behave locally like finite sets, even if they are globally infinite).

The impatient reader can jump to Remarks 8.4.6 and 8.4.14, where we describe these algorithms. They are described at a high level, but what matters is that they exist, and what is nice about them is that they correspond to natural (semi)topological operations.

8.2 The witness function and semitopology

NOTATION 8.2.1. We extend Notation 1.2.1. Suppose P is a set.

1. Call a nonempty subset of P a **witness-set**, and write $pow_{\neq\varnothing}(\mathsf{P})$ for the set of witness-sets (nonempty subsets) of P.
2. Write $\mathit{fin}(\mathsf{P})$ for the finite powerset of P (the set of finite subsets of P).
3. Write $\mathit{fin}_{\neq\varnothing}(\mathsf{P})$ for the finite powerset of P (the set of finite subsets of P).
4. Write
$$\mathcal{W}(\mathsf{P}) = \mathit{fin}_{\neq\varnothing}(pow_{\neq\varnothing}(\mathsf{P}))$$
(finite sets of witness-sets of P), and call $\mathcal{W}(\mathsf{P})$ the **witnessing universe of P**.

DEFINITION 8.2.2. Suppose P is a set. Then:

1. A **witness function** on P is a function
$$wf : \mathsf{P} \to \mathcal{W}(\mathsf{P}) = \mathit{fin}_{\neq\varnothing}(pow_{\neq\varnothing}(\mathsf{P})).$$
Intuitively, a witness function assigns to each $p \in \mathsf{P}$ finitely many witness-sets. We call each $w \in wf(p)$ a **witness-set** for p.
2. A **witnessed set** is a pair (P, wf) of a set and a witness function on that set.

REMARK 8.2.3.

1. A witness function wf gives rise to a relation $wf \subseteq \mathsf{P} \times pow_{\neq\varnothing}(\mathsf{P})$ by taking
$$p\ wf\ w \quad \text{when} \quad w \in wf(p).$$

2. If we read the relation from right to left then for each $w \in wf(p)$ we can read w as an abstract notion of 'potential set of witness for the beliefs of p'.

3. The nonemptiness conditions implies that every p is witnessed by some nonempty $\varnothing \neq w \in wf(p)$ — even if w is just equal to $\{p\}$.

DEFINITION 8.2.4. Suppose $wf : \mathsf{P} \to \mathcal{W}(\mathsf{P})$ is a witness function on a finite set P (Definition 8.2.2), and suppose $p \in \mathsf{P}$ and $P \subseteq \mathsf{P}$.

1. Define $p \,!_{wf}\, P$, or synonymously $P \,!_{wf}\, p$, by:

$$p \,!_{wf}\, P \quad \text{when} \quad \forall w \in wf(p). w \between P$$

   and say that P **blocks** p, and call P a **blocking set for** p.
   In words: P blocks p when it intersects with all of p's witness-sets.
2. Define $p \,?_{wf}\, P$, or synonymously $P \,?_{wf}\, p$, by

$$p \,?_{wf}\, P \quad \text{when} \quad \exists w \in wf(p). w \subseteq P$$

   and say that P **enables** p, and call P an **enabling set for** p.
   In words: P enables p when it contains at least one of p's witness-sets.

DEFINITION 8.2.5. Suppose $wf : \mathsf{P} \to \mathcal{W}(\mathsf{P})$ is a witness function on P (Definition 8.2.2).

1. Call $O \subseteq \mathsf{P}$ a (wf-)**open set** when

$$\forall p \in \mathsf{P}. p \in O \Longrightarrow p \,?_{wf}\, O$$

   In words, O is open when it enables its own elements.[2]
2. Call $C \subseteq \mathsf{P}$ a (wf-)**closed set** when

$$\forall p \in \mathsf{P}. p \,!_{wf}\, C \Longrightarrow p \in C.$$

   In words, C is closed when it contains every element that it blocks.
3. Let the **witness semitopology** $\mathsf{Open}(wf)$ on P be the set of wf-open sets. In symbols:

$$\begin{aligned} \mathsf{Open}(wf) &= \{O \subseteq \mathsf{P} \mid O \text{ is } wf\text{-open}\} \\ &= \{O \subseteq \mathsf{P} \mid \forall p \in \mathsf{P}. p \in O \Longrightarrow p \,?_{wf}\, O\} \\ &= \{O \subseteq \mathsf{P} \mid \forall p \in O. \exists w \in wf(p). w \subseteq O\}. \end{aligned}$$

   We also define $\mathsf{Closed}(wf)$ by:

$$\begin{aligned} \mathsf{Closed}(wf) &= \{C \subseteq \mathsf{P} \mid C \text{ is } wf\text{-closed}\} \\ &= \{C \subseteq \mathsf{P} \mid \forall p \in \mathsf{P}. p \,!_{wf}\, P \Longrightarrow p \in C\}. \end{aligned}$$

By Lemma 5.1.9, being open and being closed are dual. We make the elementary observation that $?_{wf}$ and $!_{wf}$, and $\mathsf{Open}(wf)$ and $\mathsf{Closed}(wf)$, do indeed match up as they should:

[2]Note that if $p \in O$ then O need not contain *every* enabling witness-set of p. In Definition 8.2.4(2) $p \,?_{wf}\, O$ is existential, that O contains *some* witness-set of p.

LEMMA 8.2.6. *Suppose* $wf : \mathsf{P} \to \mathcal{W}(\mathsf{P})$ *is a witness function on* P *and suppose* $p \in \mathsf{P}$ *and* $P \subseteq \mathsf{P}$. *Then*

1. $p \mathrel{!}_{wf} P$ *if and only if* $p \mathrel{\not{?}}_{wf} \mathsf{P} \setminus P$.
2. $p \mathrel{?}_{wf} P$ *if and only if* $p \mathrel{\not{!}}_{wf} \mathsf{P} \setminus P$.
3. $P \in \mathsf{Open}(wf)$ *if and only if* $\mathsf{P} \setminus P \in \mathsf{Closed}(wf)$, *and* $P \in \mathsf{Closed}(wf)$ *if and only if* $\mathsf{P} \setminus P \in \mathsf{Open}(wf)$.

Proof. By routine calculations from Definition 8.2.5. □

LEMMA 8.2.7. *Suppose* $wf : \mathsf{P} \to \mathcal{W}(\mathsf{P})$ *is a witness function on a finite set* P, *and suppose* $\mathcal{O} \subseteq pow(\mathsf{P})$. *Then if every* $O \in \mathcal{O}$ *is open in the sense of Definition 8.2.4(2), then* $\bigcup \mathcal{O}$ *is also open.*

Proof. Suppose $p \in \bigcup \mathcal{O}$. Then $p \in O$ for some $O \in \mathcal{O}$. By openness of O, p is enabled by some $w \in wf(p)$ such that $w \subseteq O$. But then also $w \subseteq \bigcup \mathcal{O}$, so we are done. □

COROLLARY 8.2.8. *Suppose* $wf : \mathsf{P} \to \mathcal{W}(\mathsf{P})$ *is a witness function on a finite set* P. *Then* $\mathsf{Open}(wf)$ *from Definition 8.2.5 makes* P *into a semitopology in the sense of Definition 1.2.2.*

Proof. Unpacking conditions 1 and 2 of Definition 1.2.2, we must check that $\varnothing$ and P are open — which is routine — and that an arbitrary union of open sets is open — which is Lemma 8.2.7. So we are done. □

REMARK 8.2.9. There is design freedom, whether we want to include (or exclude) $p \in w \in wf(p)$: Definition 8.2.2 makes no commitment either way. Lemma 8.2.10 is an easy observation that expresses a precise mathematical sense in which this choice does not really matter; so we can choose whatever is most convenient for a particular case. We will use Lemma 8.2.10 later, to prove Lemma 8.3.5.

LEMMA 8.2.10. *Suppose* (P, wf) *is a witnessed set. Let* wf' *and* wf'' *be defined by*[3]

$$
\begin{aligned}
wf'(p) &= \{w \cup \{p\} \mid w \in wf(p)\} \\
wf''(p) &= \{w \setminus \{p\} \mid w \in wf(p) \land w \neq \{p\}\} \cup \{w \mid w \in wf(p) \land w = \{p\}\}
\end{aligned}
$$

Then (P, wf') *and* (P, wf'') *are also witnessed sets, and they generate the same witness semitopology as does* (P, wf).

Proof. By a routine calculation. □

[3] The case-split in wf'' is required just because witness function in Definition 8.2.2(1) must return a finite set of *nonempty* sets.

8.3 Examples

REMARK 8.3.1.

- Sometimes, proving the existence of a witness function wf to generate a given semitopology $(\mathsf{P}, \mathsf{Open})$ as a witness semitopology (Definition 8.2.5) is fairly straightforward. Lemma 8.3.2 gives a natural example of this.
- Sometimes, the existence of a witness function is less evident. Lemma 8.3.3 illustrates one example of a non-obvious witness function for a semitopology, and Lemma 8.3.5 conversely illustrates an apparently not dissimilar semitopology, but for which no witness function exists.

LEMMA 8.3.2. *Suppose* $(\mathsf{P}, \mathsf{Open})$ *is a* finite *semitopology (so* P *is finite, and so is* Open*). Then* $(\mathsf{P}, \mathsf{Open})$ *can be generated as witness semitopology. Thus: every finite semitopology is also a witness semitopology for a witnessed set.*[4]

Proof. Set $wf(p) = \{O \in \mathsf{Open} \mid p \in O\}$. The reader can check that this satisfies the finiteness conditions on a witness function in Definition 8.2.2; it remains to show that $\mathsf{Open}(wf) = \mathsf{Open}$. If $X \in \mathsf{Open}(wf)$ then by Definition 8.2.5(1) X is a union of open sets, and thus $X \in \mathsf{Open}$. Conversely, if $O \in \mathsf{Open}$ then $O \in \mathsf{Open}(wf)$ because each $p \in O$ is witnessed by O. □

LEMMA 8.3.3. *Consider the* all-but-one semitopology *on* $\mathbb{Z}$ *from Example 2.1.4(5c):*

- $\mathsf{P} = \mathbb{Z}$ *and*
- $\mathsf{Open} = \{\varnothing, \mathbb{Z}\} \cup \{\mathbb{Z} \setminus \{i\} \mid i \in \mathbb{Z}\}$.

Then a witness function for this semitopology is:

$$wf(i) = \{\{i\text{-}1, i{+}1\}, \mathbb{Z} \setminus \{i{+}1\}, \mathbb{Z} \setminus \{i\text{-}1\}\}$$

Proof. We prove that $\mathsf{Open} = \mathsf{Open}(wf)$ by checking two subset inclusions.

- *We check that if* $O \in \mathsf{Open}$ *then* $O \in \mathsf{Open}(wf)$*:*

  If $O = \varnothing$ or $O = \mathbb{Z}$ then there is nothing to prove. So suppose $O = \mathbb{Z} \setminus \{i\}$.

  We must show that every $j \in O$ is witnessed by some element $w_j \in wf(j)$. This is routine:

  - For $j \notin \{i\text{-}1, i{+}1\}$ we use witness-set $\{j\text{-}1, j{+}1\}$;
  - for $j = i\text{-}1$ we use witness-set $\mathbb{Z} \setminus \{j{+}1\}$; and
  - for $j = i{+}1$ use witness-set $\mathbb{Z} \setminus \{j\text{-}1\}$.

- *We check that if* $O \in \mathsf{Open}(wf)$ *then* $O \in \mathsf{Open}$.

  If $O = \varnothing$ or $O = \mathbb{Z}$ then there is nothing to prove. So suppose $O \notin \{\varnothing, \mathbb{Z}\}$.

  Then there exists an $i \in \mathbb{Z}$ such that $i \in O$ and $\{i\text{-}1, i{+}1\} \not\subseteq O$. We assumed $O \in \mathsf{Open}(wf)$, so one of the following must hold:

[4]The reader might consider Lemma 8.3.2 to be a satisfactory answer to the open problem we describe later in Remark 9.5.13, since all semitopologies realisable in the real world are finite. We are not so sure — even if all you care about is physically realisable semitopologies — for reasons outlined in Remark 9.4.7.

- $\{i\text{-}1, i{+}1\} \subseteq O$, which we assumed is not the case, or
- $i{+}1 \notin O$ and $\mathbb{Z}\setminus\{i{+}1\} \subseteq O$, so we are done because, with $O \neq \mathbb{Z}$, it must be that $O = \mathbb{Z}\setminus\{i{+}1\}$, or
- $i\text{-}1 \notin O$ and $\mathbb{Z}\setminus\{i\text{-}1\} \subseteq O$, and again we are done. □

LEMMA 8.3.4. *A witness function for the semitopology used in Proposition 6.4.11, as illustrated in Figure 6.2, is*

$$wf(w) = \{\{w0, w1\}\}.$$

Proof. Setting $wf(w) = \{\{w0, w1\}\}$ just expresses that if $w \in O$ then $w0, w1 \in O$, i.e. that O is down-closed — for 'down' as illustrated in Figure 6.2. □

Lemma 8.3.5 will provide a key counterexample later in Lemma 9.4.2:

LEMMA 8.3.5. *Consider the* more-than-one semitopology *on $\mathbb{N}$ from Example 2.1.4(5d): so $X = \mathbb{N}$ and opens have the form $\varnothing$ or any set of cardinality more than one (i.e. containing at least two elements). There is no witness function for this semitopology.*

Proof. Suppose some such witness function wf exists. Using Lemma 8.2.10 we may assume without loss of generality that $n \in w$ for every $w \in wf(n)$, for every $n \in \mathbb{N}$ (that is, $n \in \bigcap wf(n)$ always). Furthermore because no singletons are open, we know that $\{n\} \notin wf(n)$ for every $n \in \mathbb{N}$.

Now consider two distinct $n \neq n' \in \mathbb{N}$. We know that $\{n, n'\}$ is open, so it follows that one of the following must hold:

1. *Suppose $\{n, n'\} \in wf(n)$ and $\{n, n'\} \notin wf(n')$.*
   This is impossible because $\{n'\} \notin wf(n')$ and $wf(n')$ is not empty, so $\{n, n'\}$ could not be open.
2. *Suppose $\{n, n'\} \in wf(n')$ and $\{n, n'\} \notin wf(n)$.*
   This is also impossible because $\{n\} \notin wf(n)$ and $wf(n)$ is not empty, so $\{n, n'\}$ could not be open.
3. It follows that $\{n, n'\} \in wf(n)$ and $\{n, n'\} \in wf(n')$.

It follows that $\{n, n'\} \in wf(n)$ for *every* n' other than n. But this contradicts finiteness of $wf(n)$. □

8.4 Computing open and closed sets in witness semitopologies

8.4.1 Computing open sets: X is open when $X \prec X$

DEFINITION 8.4.1. Suppose that (P, wf) is a witnessed set (Definition 8.2.2) and $X, X' \subseteq \mathsf{P}$. Define the **witness ordering** $X \prec X'$ by

$$X \prec X' \quad \text{when} \quad X \subseteq X' \land \forall p \in X. \exists w \in wf(p). w \subseteq X'.$$

If $X \prec X$ then call X a **$\prec$-fixedpoint**.

REMARK 8.4.2. Intuitively, $X \prec X'$ when X' extends X with (at least) one witness-set for every element $p \in X$.

LEMMA 8.4.3. *Suppose* (P, wf) *is a witnessed set, and recall the witness ordering* $\prec$ *from Definition 8.4.1. Then:*

1. *If* $X \prec X'$ *then* $X \subseteq X'$, *or in symbols:* $\prec \subseteq \subseteq$.
2. $\prec$ *is a transitive (*$X \prec X' \prec X''$ *implies* $X \prec X''$*) and antisymmetric (*$X \prec X'$ *and* $X' \prec X$ *implies* $X = X'$*) relation on* $pow(\mathsf{P})$.

Proof. By routine calculations from Definition 8.4.1. □

LEMMA 8.4.4. *Suppose* (P, wf) *is a witnessed set. Then the following are equivalent:*

- O *is open in the witness semitopology (Definition 8.2.5).*
- O *is a* $\prec$*-fixedpoint (Definition 8.4.1).*

In symbols:

$$\mathsf{Open}(wf) = \{X \subseteq \mathsf{P} \mid X \prec X\}.$$

Proof. Being a $\prec$-fixedpoint in Definition 8.4.1 — every point in O is witnessed by a subset of O — simply reformulates the openness condition from Definition 8.2.5. □

PROPOSITION 8.4.5. *Suppose* (P, wf) *is a witnessed set and suppose* $\mathcal{X} = (X_0 \prec X_1 \prec \dots)$ *is a countably ascending* $\prec$*-chain. Write* $\bigcup \mathcal{X}$ *for the union* $\bigcup_i X_i$ *of the elements in* $\mathcal{X}$. *Then:*

1. $\bigcup \mathcal{X}$ *is a* $\prec$*-limit for* $\mathcal{X}$: $\forall i. X_i \prec \bigcup \mathcal{X}$.
2. $\bigcup \mathcal{X}$ *is a* $\prec$*-fixedpoint and so (by Lemma 8.4.4) is open:* $\bigcup \mathcal{X} \prec \bigcup \mathcal{X} \in \mathsf{Open}(wf)$.

Proof.

1. We must show that if $p \in X_i$ then $w \subseteq \bigcup \mathcal{X}$ for some $w \in wf(p)$. But this is automatic from the fact that $X_i \prec X_{i+1} \subseteq \bigcup \mathcal{X}$.

2. From part 1 noting that if $p \in \bigcup \mathcal{X}$ then $p \in X_i$ for some i. □

REMARK 8.4.6 (Computing open sets). Proposition 8.4.5 and Lemma 8.4.4 above are not complicated[5] and they say something important: in the *witness* semitopology, open sets can be computed with a simple loop that accumulates a set of points; and for each point in the set so far, add some choice of witness-set of that point to the set (if one is not already present); repeat until we reach a fixed point; then return the result.

In more detail, to compute an open set in the witness semitopology:

1. Nondeterministically choose an initial R_0 — in particular, to compute an open neighbourhood of $p \in \mathsf{P}$ we can set $R_0 = \{p\}$.
2. Given R_i, for each $p \in R_i$ nondeterministically pick some witness-set $w(p) \in wf(p)$ and set $R_{i+1} = R_i \cup \bigcup_{p \in R_i} w(p)$.
3. If $R_{i+1} = R_i$ then terminate with result R_i; otherwise loop back to 2.

This algorithm is nondeterministic and could run forever if P is infinite, but it is an algorithm and it is local in the sense of Remark 8.1.1. We continue this thread in Remarks 8.4.14 and 8.5.1.

[5]This is a feature and did not happen by accident: it required design effort.

8.4.2 Computing closed sets using limit points: $|P| = lim(P)$

DEFINITION 8.4.7. Suppose P is a set and $\mathcal{W}$ is a set (or a sequence) of sets. Define $P \between \mathcal{W}$ by

$$P \between \mathcal{W} \quad \text{when} \quad \forall W \in \mathcal{W}. P \between W.$$

In words: $P \between \mathcal{W}$ when P intersects with every $W \in \mathcal{W}$.

DEFINITION 8.4.8. Suppose (P, wf) is a witnessed set and $P \subseteq \mathsf{P}$. Define $lim_w(P)$ by

$$lim_w(P) = P \cup \{p \in \mathsf{P} \mid P \between wf(p)\}.$$

In words: $lim_w(P)$ is the set of points p whose every witness-set contains a P-element.

We iterate this:

$$\begin{aligned} lim_0(P) &= P \\ lim_{i+1}(P) &= lim_w(lim_i(P)) \\ lim(P) &= \bigcup_{n \geq 0} lim_n(P) \end{aligned}$$

We call $lim(P)$ the set of **limit points of** P.

REMARK 8.4.9. In Definition 8.2.2(1) we insisted that $wf(p)$ is nonempty for every point p. This avoids a degenerate situation in the definition of $lim_w(P)$ in Definition 8.4.8 above in which the condition $P \between wf(p)$ is vacuously satisfied by a p with empty $wf(p)$ (i.e. by a p with no witness sets). Definition 8.2.2(1) excludes this by insisting that p has to have at least one witness, even if it is just $wf(p) = \{\{p\}\}$.

LEMMA 8.4.10. *Suppose* (P, wf) *is a witnessed set and* $P \subseteq \mathsf{P}$. *Then*

$$P \subseteq lim(P).$$

Proof. It is a fact of Definition 8.4.8 that $P = lim_0(P) \subseteq lim_1(P) \subseteq lim(P)$. □

LEMMA 8.4.11. *Suppose* (P, wf) *is a witnessed set and* $p \in \mathsf{P}$ *and* $P \subseteq \mathsf{P}$. *Then:*

1. *If* $lim(P) \between wf(p)$ *(Definition 8.4.7) then* $p \in lim(P)$.
2. *By the contrapositive and expanding Definition 8.4.7,*

$$p \in \mathsf{P} \setminus lim(P) \quad \textit{implies} \quad \exists w \in wf(p). w \cap lim(P) = \varnothing.$$

Proof. Suppose $lim(P) \between wf(p)$. Unpacking Definitions 8.4.7 and 8.4.8 it follows that for every $w \in wf(p)$ there exists $n_w \geq 0$ such that $lim_{n_w}(P) \between w$. Now by Definition 8.2.2(1) $wf(p)$ — the set of witness-sets to p — is finite, and it follows that for some/any n greater than the maximum of all the n_w, we have $lim_n(P) \between wf(p)$. Thus $p \in lim_w(lim_n(P)) \subseteq lim(P)$ as required. □

LEMMA 8.4.12. *Suppose* (P, wf) *is a witnessed set and* $p \in \mathsf{P}$ *and* $P \subseteq \mathsf{P}$ *and* $O \in \mathsf{Open}(wf)$. *Then:*

1. *If* $O \between lim_w(P)$ *then* $O \between P$.
2. *If* $O \between lim(P)$ *then* $O \between P$.
3. *As a corollary, if* $O \cap P = \varnothing$ *then* $O \cap lim(P) = \varnothing$.

Proof.

1. Consider $p \in \mathbf{P}$ such that $p \in O$ and $p \in lim_w(P)$. By assumption there exists $w \in wf(p)$ such that $w \subseteq O$. Also by assumption $w \between P$. It follows that $O \between P$ as required.

2. If $O \between lim(P)$ then $O \between lim_n(P)$ for some finite $n \geq 0$. By a routine induction using part 1 of this result, it follows that $O \between P$ as required.

3. This is just the contrapositive of part 2 of this result, noting that $O \between P$ when $O \cap P = \varnothing$ by Notation 3.1.1, and similarly for $O \between lim(P)$. $\qquad\square$

PROPOSITION 8.4.13. *Suppose* $(\mathbf{P}, wf)$ *is a witnessed set and suppose* $P \subseteq \mathbf{P}$. *Then:*

$$lim(P) = |P|.$$

In words: the set of limit points of P from Definition 8.4.8 is equal to the topological closure of P from Definition 5.1.2.

Proof. We prove two implications:

- *Suppose* $p \notin |P|$.

  Then there exists some $p \in O \in \mathsf{Open}(wf)$ such that $O \cap P = \varnothing$. Thus by Lemma 8.4.12(3) also $O \cap lim(P) = \varnothing$.

- *Suppose* $p \notin lim(P)$.

  By Definition 5.1.2 we need to exhibit an $p \in O \in \mathsf{Open}(wf)$ that is disjoint from P, and since $P \subseteq lim(P)$ by Lemma 8.4.10, it would suffice to exhibit $p \in O \in \mathsf{Open}(wf)$ that is disjoint from $lim(P)$. We set
  $$O = \mathbf{P} \setminus lim(P).$$

  Lemma 8.4.11(2) expresses precisely that this is an open set in the witness semitopology, and by construction it is disjoint from $lim(P)$. $\qquad\square$

REMARK 8.4.14 (Computing closed sets). As in Remark 8.4.6 we see that in the *witness* semitopology, closed sets can be computed with a simple loop that accumulates a set of points so far: and for each point in the space, if all of its witness-sets intersect with the set of points so far, add that point to the set so far; repeat until we reach a fixed point; return the result.

In more detail, to compute a closed set in the witness semitopology:

1. Nondeterministically choose an initial P_0 — in particular, to compute a closed set containing $p \in \mathbf{P}$ we can set $P_0 = \{p\}$.
2. Given P_i, for every $p \in \mathbf{P}$ check if $w \between P_i$ for every witness-set $w \in wf(p)$ and collect these p into a set B_i. Set $P_{i+1} = P_i \cup B_i$.
3. If $P_{i+1} = P_i$ then terminate with result P_i; otherwise loop back to 2.

This algorithm could run forever if $\mathbf{P}$ is infinite, but it is an algorithm and it is local in the sense of Remark 8.1.1. Note that quantification over every point is local in the sense of Remark 8.1.1, in spite of the quantification over all $p \in \mathbf{P}$ in step 2 above: participants would listen for queries from peers on the channel "I am trying to compute an open set; here is my R_i; do you want to join it?".

REMARK 8.4.15 (Summing up). It might at first appear that working with semitopologies would require some form of prior coordination: e.g. for participants to at least have common knowledge of their shared, minimal open neighbourhoods. For, consider a new participant p joining a system based on semitopology: how is p supposed to know which are the open sets?

Surprisingly, we have seen that *witness semitopologies can be built without any coordination*. Each participant just unilaterally chooses a set of witness-sets. As discussed in Remarks 8.4.6 and 8.4.14, and even in an infinite semitopology, a participant can compute open and closed sets — they do not have to, but they can if they wish to spend the bandwidth — by exploring witness-sets using nondeterministic algorithms.

We make no claims to efficiency (we have not even set up machinery to measure what that would mean) but what matters is that for witness semitopologies such procedures exist, in contrast e.g. to the uncomputable semitopology from Example 8.1.2.

In the next subsection we offer an interpretation of witness functions that in some sense explains why this should be so, and gives a new intuition of why witness semitopologies are amenable to decentralised computation in the style that we require.

8.5 Declarative content of witness semitopologies

8.5.1 Witnessed sets and Horn clause theories

REMARK 8.5.1. Recall that a *sequential space* is one in which the sets closed under convergent sequences, are precisely the closed sets. Proposition 8.4.13 ($lim(P) = |P|$) looks, just a bit, like a sequential space closure result. Looking more closely, we see that the similarity comes from the fact that the definition uses an ω-iteration that is, just a little, reminiscent of a converging ω-sequence of points. Perhaps surprisingly, we can make this resemblance into something much more precise, as follows:

DEFINITION 8.5.2. Suppose (P, wf) is a finite witnessed set (so P is a finite set).

1. Let the **derived logic** $\mathsf{Prop}(\mathsf{P}, wf)$ be a propositional syntax with connectives $\bot$, $\top$, $\vee$, $\wedge$, and $\Rightarrow$ over a set of **atomic proposition symbols** $\bar{\mathsf{P}} = \{\bar{p} \mid p \in \mathsf{P}\}$.

   Note that $\bar{p}$ is just a symbol in our formal syntax; there is one such for each point $p \in \mathsf{P}$.

2. For each $p \in \mathsf{P}$ define an **axiom** $\bar{wf}(p)$ by[6]

   $$\bar{wf}(p) = \left(\wedge_{w \in wf(p)} \vee_{q \in w} \bar{q}\right) \Rightarrow \bar{p}$$

   and collect these axioms into a set

   $$\mathsf{Ax}(\mathsf{P}, wf) = \{\bar{wf}(p) \mid p \in \mathsf{P}\}.$$

3. A **sequent** $\Phi \vDash \Psi$ is a pair of finite sets of propositions in the syntax of $\mathsf{Prop}(\mathsf{P}, wf)$.

4. Call $\Phi \vDash \Psi$ a **derivable sequent** when $\Phi, \mathsf{Ax}(\mathsf{P}, wf) \vdash \Psi$ is derivable in propositional logic.

[6]Below, $\wedge$ and $\vee$ denote a finite list of $\wedge$ and $\vee$ connectives. We use this instead of $\bigwedge$ and $\bigvee$ to emphasise that this is formal syntax in $\mathsf{Prop}(\mathsf{P}, wf)$.

5. If $S \subseteq \mathsf{P}$ write $\bar{S} = \{\bar{p} \mid p \in S\}$. Call $\bar{S}$ a **model** or **answer set** for $\mathsf{Ax}(\mathsf{P}, wf)$ when

$$\forall p \in \mathsf{P}.(\bar{S} \vDash \bar{p} \Longrightarrow \bar{p} \in \bar{S}).$$

PROPOSITION 8.5.3 (Declarative interpretation). *Suppose* (P, wf) *is a finite witnessed set and* $C \subseteq \mathsf{P}$. *Then the following are equivalent:*

- *C is closed in the witness semitopology (Definition 8.2.5).*
- *$\bar{C}$ is a model (Definition 8.5.2(5)).*

Proof. By Definitions 8.2.5(2) and 8.5.2(2), the condition in Definition 8.5.2(5) for $\bar{C}$ to be a model precisely expresses the property that C is closed. □

COROLLARY 8.5.4. *Every finite semitopology can be exhibited as the set of (set complements of) models of a propositional Horn clause theory.*

Proof. Lemma 8.3.2 shows how to exhibit a finite semitopology as the witness semitopology of a witnessed set, and Proposition 8.5.3 shows how to interpret that witnessed set as a Horn clause theory in a propositional logic. □

REMARK 8.5.5. An axiom $\bar{wf}(p)$ consists of a propositional goal implied by a conjunction of disjunctions of (unnegated) propositional goals. This fits the Horn clause syntax from Section 3 of [MNPS91], and it can be translated into a more restricted Prolog-like syntax if required, just by expanding the disjuncts into multiple clauses using the $(\mathbf{\vee L})$ rule.[7]

Thus closed sets — and so also open sets, which are their complements — can be computed from the axioms $\mathsf{Ax}(\mathsf{P}, wf)$ by asking a suitable propositional solver to compute *models*. Answer Set Programming (ASP) tool is one such tool [Lif08, Lif19]. Thus:

- We can view the algorithm for computing closed sets described in Remark 8.4.14 as 'just' (see next Remark) an ASP solver for the Horn clause theory $\mathsf{Ax}(\mathsf{P}, wf)$ in the logic $\mathsf{Prop}(\mathsf{P}, wf)$.
- Conversely, we can view this Subsection as observing that the set of all solutions to a finite Horn clause theory has a semitopological structure, via witnessed sets.

REMARK 8.5.6. Proposition 8.5.3 is not a 'proof' that we should, or even could, actually use an ASP solver to do this.

Proposition 8.5.3 assumes complete and up-to-date information on the witness function. Mathematically this is fine, just as writing 'consider an uncomputable subset of $\mathbb{N}$' is mathematically fine — we can prove that this exists. As a *computational* statement about possible implementations, this is more problematic, because a point of working with distributed systems is that we do not suppose that a participant could ever collect a global snapshot of the network state; and if they somehow did, it could become out-of-date; and in any case, in the presence of failing or adversarial participants it could be inaccurate. So just because there *is* a network state at some point in time, does not mean we have access to it.

Even mathematically, Proposition 8.5.3 is not the full story of (witness) semitopologies:

[7] An example makes the point: $((\bar{q} \vee \bar{q}') \wedge \bar{q}'') \Rightarrow \bar{p}$ is equivalent to two simpler clauses $(\bar{q} \wedge \bar{q}'') \Rightarrow \bar{p}$ and $(\bar{q}' \wedge \bar{q}'') \Rightarrow \bar{p}$; for more details see [MNPS91].

- it concerns *finite* semitopologies, whereas we are also interested in *infinite* ones (see Remark 9.4.7); and
- the questions we ask in the mathematics — especially the second-order ones such as "Are these two points intertwined?" or "Find a maximal topen neighbourhood of this point, or confirm that none such exists." — have not been considered in declarative programming, so far as we know.

So it is important to appreciate that while Proposition 8.5.3 characterises closed and open sets in a witness semitopology in terms of solutions to Horn clause theories, and so helps us to understand what these sets really are at a mathematical level, this is not in and of itself automatically useful to actually turning such a semitopology into working network — for that, we need algorithms like that described in Remark 8.4.14 — nor is it a full mathematical account of all the facts of interest about semitopologies.

One practical use case where the correspondence with declarative programming might be immediately useful would be a monitoring tool, especially one testing mathematical properties to detect leading indicators of network malfunction. Thus, for a network that is operating well and not changing too quickly, it would be feasible to traverse the network collecting information, and then use something like an ASP solver as part of a monitoring tool to compute the closed and open sets and so monitor properties such as the current intertwinedness of the network. This is fine, so long as the reader is clear that a (centralised) network monitor that works in good conditions is not the same thing as the robust decentralised network itself.

8.5.2 Witnessed sets and topologies

If finite semitopologies can be thought of as sets of solutions to Horn clause theories via witnessed sets, as outlined in Corollary 8.5.4, what do finite topologies correspond to? We will find answers just by unrolling definitions and doing some simple reasoning, but the results are perhaps illuminating and a little bit surprising:

DEFINITION 8.5.7.

1. Call a semitopology $(\mathsf{P}, \mathsf{Open})$ a **deterministic** when each point p has a unique least open neighbourhood $p \in M_p \in \mathsf{Open}$.
2. Call a Horn clause theory (in the sense used in Definition 8.5.2(2)) **deterministic** when for each propositional atom $\bar{p} \in \bar{\mathsf{P}}$ there exists at most one axiom in which $\bar{p}$ appears in its head.[8]
3. Call a witnessed set (P, wf) (Definition 8.2.2(1)) **deterministic** when for each point p, $wf(p)$ is a singleton set; thus $wf(p) = \{W_p\}$.[9] In words: wf is deterministic when *every point has precisely one (possibly empty) witness-set.*

REMARK 8.5.8. Recall the algorithms for computing open and closed sets from witness functions from Remarks 8.4.6 and 8.4.14. When wf is deterministic, the algorithms simplify: there is precisely one witness-set to each point, and this removes the *nondeterminism* from the algorithms and they become deterministic — as our choice of name in Definition 8.5.7 suggests.

---

[8] The *head* of the axiom is its final propositional atom, written to the right-hand side of the $\Rightarrow$ in Definition 8.5.2(2).

[9] W_p (the witness-set to p) is not necessarily equal to M_p (the least open set containing p). The witness function *generates* a witness semitopology, but is not necessarily *equal* to it.

LEMMA 8.5.9. *Suppose* $(\mathsf{P}, \mathsf{Open})$ *is a finite semitopology. Then the following are equivalent:*

1. $(\mathsf{P}, \mathsf{Open})$ *is a topology (intersections of open sets are open).*
2. $(\mathsf{P}, \mathsf{Open})$ *is a deterministic semitopology (every* $p \in \mathsf{P}$ *has a unique least open neighbourhood* $M_p \in \mathsf{Open}$*).*

Proof. Suppose $(\mathsf{P}, \mathsf{Open})$ is a topology and consider some $p \in \mathsf{P}$. We must find a least open neighbourhood $p \in M_p$. We just set $M_p = \bigcap\{O \in \mathsf{Open} \mid p \in O\}$. Open sets in topologies are closed under finite unions, so M_p is an open neighbourhood of p, and by construction it is least.

Suppose $(\mathsf{P}, \mathsf{Open})$ is deterministic and consider $O, O' \in \mathsf{Open}$. We must show that $O \cap O'$ is open. We just note that $O \cap O' = \bigcup\{M_p \mid p \in O \cap O'\}$. This is a union of open sets and so an open set, and by construction it contains $O \cap O'$. But also it is contained in $O \cap O'$, since if $p \in O$ then $M_p \subseteq O$, and similarly for O'. □

REMARK 8.5.10. Returning to the terminology *deterministic* in Definition 8.5.7 above: when we are doing resolution in the Horn clause theory, and when we are building an open set using the algorithm in Remark 8.4.6, there is only ever one witness/clause for each point. Thus resolution never has to backtrack; and building the open set never has to make any choices.

LEMMA 8.5.11. *Suppose* $(\mathsf{P}, \mathsf{Open})$ *is a finite semitopology. Then the following are equivalent:*

- $(\mathsf{P}, \mathsf{Open})$ *is a topology.*
- $(\mathsf{P}, \mathsf{Open}) = (\mathsf{P}, \mathsf{Open}(wf))$ *for some deterministic witness function* wf *on* P.

Proof. Suppose $(\mathsf{P}, \mathsf{Open})$ is a topology. We just modify the construction from Lemma 8.3.2 and set $wf(p) = \{M_p\}$. The reader can check that $\mathsf{Open} = \mathsf{Open}(wf)$.

Conversely, suppose $\mathsf{Open} = \mathsf{Open}(wf)$ for deterministic wf, and write $wf(p) = \{W_p\}$. Now consider $O, O' \in \mathsf{Open}$; we need to show that $O \cap O' \in \mathsf{Open}$. By the construction of the witness semitopology in Definition 8.2.5(3) it would suffice to show that if $p \in O \cap O'$ then $W_p \subseteq O \cap O'$. But this is immediate, since $O, O' \in \mathsf{Open}(wf)$ so that if $p \in O$ then $W_p \subseteq O$, and similarly for O'. □

PROPOSITION 8.5.12. *Suppose* $(\mathsf{P}, \mathsf{Open})$ *is a finite semitopology. Then the following are equivalent:*

- $(\mathsf{P}, \mathsf{Open})$ *is a topology.*
- $(\mathsf{P}, \mathsf{Open})$ *is a deterministic semitopology.*
- $(\mathsf{P}, \mathsf{Open})$ *is the witness semitopology of a deterministic witness function.*

Proof. We combine Lemmas 8.5.9 and 8.5.11. □

REMARK 8.5.13. The definitions and proofs in this Subsection are quite easy, but they capture a nice intuition which is not immediately obvious from just looking at the definitions:

- Finite semitopologies correspond to computation with nondeterminism and backtracking.
- Finite topologies correspond to computation that does not require backtracking.

Proposition 8.5.12 makes this intuition formal up to a point, but it is not the full story. What is missing is that a semitopology may have more than one presentation as the witness semitopology of a witnessed set.[10] In particular, it is possible to create a non-deterministic witness function that generates a topology; intuitively, just because there might be a choice of witness-set, does not mean that the choice makes any difference to the final result. Put another way: *determinism* ensures that backtracking is impossible, but nondeterminism not necessarily imply that it is *required*.[11]

We speculate that Proposition 8.5.12 could be strengthened to show that topologies correspond to Horn clause theories that (may not be deterministic in the sense of Definition 8.5.7, but that) do not require backtracking. We leave this for future work.

---

[10]Let's spell that out: it is possible for $\mathsf{Open} = \mathsf{Open}(wf) = \mathsf{Open}(wf')$ for distinct wf and wf'.

[11]Think of reducing a simply-typed λ-calculus term; there are many reduction paths, but they all lead to the same normal form.

(Strongly) chain-complete semitopologies 9

9.1 Definition and discussion

REMARK 9.1.1. Just as for topologies, in semitopologies it is not true in general that the intersection of a descending chain of open sets is open.

Consider $\mathbb{N}$ with the semitopology generated by $O \subseteq \mathbb{N}$ such that $\{0\} \subsetneq O$. Then $(\{0\} \cup i_{\geq} \mid i \geq 1)$ where $i_{\geq} = \{i' \mid i' \geq i\}$ is a descending chain of open sets, but its intersection $\{0\}$ is not open.

For the special case of *witness* semitopologies, we can say something considerably stronger, as we shall see in Definition 9.1.2 and Theorem 9.4.1.

Recall from Definition 3.4.4 the notion of an ascending/descending *chain of sets*:

DEFINITION 9.1.2.

1. Call a semitopology **chain-complete** when for every descending chain of open sets $\mathcal{O} \subseteq \mathsf{Open}$ (Definition 3.4.4), its intersection $\bigcap \mathcal{O}$ is open.
2. Call a semitopology **strongly chain-complete** when for every nonempty descending chain of nonempty open sets $\mathcal{O} \subseteq \mathsf{Open}_{\neq\varnothing}$, its intersection $\bigcap \mathcal{O}$ is open and nonempty.[1]

REMARK 9.1.3 (Chain-completeness in context). We make a few general observations about Definition 9.1.2 in the context of topology:

1. The strong chain-completeness condition (every descending chain of nonempty open sets is nonempty and open) is reminiscent of, though different from, a standard *compactness* condition on metric spaces, that every descending chain of nonempty closed sets should be nonempty and closed.

2. Call a topological space **Alexandrov** when its open sets are closed under arbitrary (and not just finite) intersections.

[1] We insist the chain is nonempty to exclude the pathological case of an empty chain over the semitopology $(\varnothing, \{\varnothing\})$.

In the case that a semitopology (P, Open) is a topology (so open sets are closed under finite intersections), and assuming that open sets can be well-ordered, being chain-complete is equivalent to being Alexandrov. Clearly, an Alexandrov space is chain-complete; and conversely if we have an infinite collection of open sets in a chain-complete topology then (assuming that this collection can be well-ordered) we obtain their intersection by a transfinite induction taking limits of infinite descending chains of intersections.

The Alexandrov condition is unnatural in semitopologies in the sense that we do not assume even that finite intersections exist, so there is no finite-intersections condition to strengthen to the infinite case. However, the chain-completeness condition *is* natural in semitopologies, so in the light of the previous paragraph we could argue that chain-completeness is to semitopologies as being Alexandrov is to topologies.

This is an intuitive observation, not a mathematical one, but it may help to guide the reader's intuitions.

3. Strong chain-completeness has a stronger flavour of finiteness than chain-completeness.

   For example: a strongly chain-complete space can contain only finitely many disjoint open sets — since otherwise it would be easy to form an infinite descending chain of open sets with an empty intersection — so, in the light of the topen partitioning result in Theorem 3.5.4, we see that the topen partition of a strongly chain-complete semitopology is *actually finite*.

REMARK 9.1.4. Definition 9.1.2 abstracts two useful properties of two important classes of semitopologies:

1. Every finite semitopology is strongly chain-complete, because a strictly descending chain of finite sets is finite.[2]
2. Every witness semitopology is chain-complete; we will prove this shortly, in Theorem 9.4.1.

More discussion of these points is in Remark 9.4.6. The main mathematical/technical properties that come out of a semitopology being chain-complete and strongly chain-complete are respectively:

- Lemma 9.5.3 and Corollary 9.5.4 (existence of open covers), and
- Lemma 9.5.8 and Corollary 9.5.9 (existence of open atoms) respectively.

However, before we come to that, we will set up some machinery and check some useful properties.

[2] For the record, it is easy to come up with other conditions. For instance, an even stronger condition is that a descending chain of open sets strictly above some $O \in \mathsf{Open}$ has an open intersection that is also strictly above O (we recover the strong chain-completeness condition just by restricting O to be equal to $\varnothing$). This is a very reasonable thing to say: it is in a footnote and not the main text just because we have not (yet) found a direct use for it. In contrast, strong chain-completeness turns out to be natural and useful, so we focus on that.

9.2 Elementary properties of the definition

LEMMA 9.2.1. *Suppose* $(\mathsf{P}, \mathsf{Open})$ *is a semitopology. Then:*

1. *If* $(\mathsf{P}, \mathsf{Open})$ *is strongly chain-complete, then it is chain-complete.*
2. *The reverse implication need not hold: it is possible for a semitopology to be chain-complete but not strongly chain-complete.*
3. *Not every semitopology is chain-complete.*

Proof. We consider each part in turn:

1. Consider a descending chain of open sets $\mathcal{O}$. If one of the elements in $\mathcal{O}$ is empty then $\bigcap \mathcal{O} = \varnothing$ and $\varnothing \in \mathsf{Open}$ so we are done. If all of the elements in $\mathcal{O}$ are nonempty then by chain-completeness $\bigcap \mathcal{O}$ is nonempty and open, and thus in particular it is open.

2. A counterexample is $(\mathbb{N}, pow(\mathbb{N}))$ (the discrete semitopology on the infinite set of natural numbers). Then $i_{\geq} = \{i' \mid i' \geq i\}$ for $i \geq 0$ is a descending chain of nonempty open sets whose intersection $\varnothing$ is open but not nonempty.

3. This just repeats Remark 9.1.1, which gives an easy counterexample. □

EXAMPLE 9.2.2.

1. The *all-but-one* and *more-than-one* semitopologies (see Examples 2.1.4(5c&5d)) are (strongly) chain-complete.

2. The closed interval $[-1, 1]$ with its usual topology is not chain-complete (and not strongly chain-complete): e.g. $\{(-1/i, 1/i) \mid i \geq 1\}$ is a descending chain of open sets but its intersection $\{0\}$ is not open. Similarly for the two semitopologies on $\mathbb{Q}^2$ in Example 5.6.10.

   (Looking ahead just for a moment to Theorem 9.4.1, this tells us that these semitopologies cannot be generated by witness functions.)

LEMMA 9.2.3. *Suppose* $(\mathsf{P}, \mathsf{Open})$ *is a semitopology. Then:*

1. P *is chain-complete if and only if the union of any ascending chain of closed sets, is closed.*
2. P *is strongly chain-complete if and only if the union of any ascending chain of closed sets that are not equal to* P*, is closed and not equal to* P*.*

Proof. Direct from Definition 9.1.2 using Lemma 5.1.9, which notes that closed sets are the complements of open sets (just as for topologies). □

9.3 Consequences of being strongly chain-complete

Being strongly chain-complete is a useful well-behavedness condition. We consider some of its consequences.

9.3.1 Strongly chain-complete implies $\between$-complete

We saw a chain-completeness condition before: $\between$-completeness from Definition 6.4.2(2). As promised in Remark 6.4.3(2), we now note that strongly chain-complete semitopologies are also $\between$-complete:

LEMMA 9.3.1. *Suppose* $(\mathsf{P}, \mathsf{Open})$ *is a quasiregular semitopology. Then if* $(\mathsf{P}, \mathsf{Open})$ *is strongly chain-complete then it is* $\between$*-complete (Definition 6.4.2(2)).*

Proof. Suppose we have a $\geq_{\between}$-descending chain of points $p_1 \geq_{\between} p_2 \geq_{\between} \dots$.

Since P is quasiregular, $K(p_i) \in \mathsf{Open}_{\neq\varnothing}$ for every i. Write $I = \bigcap_i K(p_i)$.

Since P is strongly chain-complete (Definition 9.1.2(2)), $I \in \mathsf{Open}_{\neq\varnothing}$ (we need *strong* chain-completeness here to know that I is not just open but also nonempty). Choose some $p \in I$.

It follows from Lemma 6.4.5(1) that $p \leq_{\between} p_i$ for every i, thus p is a $\leq_{\between}$-lower bound for the chain. □

9.3.2 Indirectly regular points: inherent properties

In Definition 4.1.4 we saw three regularity conditions on points: quasiregular, weakly regular, and regular. We now add a fourth condition to this mix: *indirect regularity*. A point is indirectly regular when it is intertwined with a regular point; intuitively, if regular points are 'nice' then an indirectly regular point is a point that is not necessarily nice itself, but it is intertwined with a point that *is* nice.[3] It is not at all obvious that this should have anything to do with strong chain-completeness, but it does: a punchline of this Subsection will come in Remark 9.3.6, where we note that if a semitopology in strongly chain-complete then indirect regularity slots in particularly nicely with the three regularity conditions from Definition 4.1.4. We now set about building the machinery we need to tell this story:

DEFINITION 9.3.2. Suppose $(\mathsf{P}, \mathsf{Open})$ is a semitopology. Call p an **indirectly regular point** when $p \between q$ for some regular q.

LEMMA 9.3.3. *Suppose* $(\mathsf{P}, \mathsf{Open})$ *is a semitopology and* $p \in \mathsf{P}$. *Then p is indirectly regular if and only if p is in the closure of a topen set.*

Proof. We prove two implications:

- *Suppose p is indirectly regular; so* $p \between q$ *for some regular* $q \in \mathsf{P}$.

  By Definition 4.1.4(3) $q \in K(q) \in \mathsf{Topen}$. By Definition 3.6.1(2) (since $p \between q$) $p \in q_{\between}$, and by Theorem 5.4.12(2) $q_{\between} = |K(q)|$. Thus p is in the closure of the topen set $K(q)$.

- *Suppose* $p \in |T|$ *for some* $T \in \mathsf{Topen}$.

  Choose any $q \in T$. Note by Theorem 4.2.6(5) that q is regular; we will now show that $p \between q$.

  Consider a pair of open neighbourhoods $p \in O \in \mathsf{Open}$ and $q \in O' \in \mathsf{Open}$. Then $O \between T$ (because $p \in O$ and $p \in |T|$), and $T \between O'$ (because $q \in T \cap O'$). By transitivity of T, $O \between O'$. It follows that $p \between q$ as required. □

[3] Like in the movies: where a gangster falls in love with a nice person; the gangster may not stop being a gangster, but they now have a moral compass, if only indirectly.

We will need Lemma 9.3.4 below. We can think of this as a version of Lemma 6.4.5 where we know that one of the points is regular:

LEMMA 9.3.4. *Suppose* $(\mathsf{P}, \mathsf{Open})$ *is a semitopology and* $p, q \in \mathsf{P}$ *and* q *is regular. Then:*

1. *If* $q \between p$ *then* $K(q) \subseteq p_{\between}$.
2. *As a corollary, if* $p \in q_{\between}$ *or* $q \in p_{\between}$, *then* $K(q) \subseteq p_{\between}$.

Proof. The corollary follows because by Definition 3.6.1(2), $p \in q_{\between}$ and $q \in p_{\between}$ are both equivalent to $p \between q$.

So suppose q is regular and $q \in p_{\between}$. By Definition 4.1.4(3) $q \in K(q) \in \mathsf{Topen}$ and by Theorem 6.2.2 q is unconflicted. Consider any other $q' \in K(q)$; unpacking Definition 4.1.4(1) and 3.6.1(2) $p \between q \between q'$ and so by Definition 6.1.1(2) (since q is unconflicted) $p \between q'$. Thus $K(q) \subseteq p_{\between}$ as required. □

COROLLARY 9.3.5. *Suppose* $(\mathsf{P}, \mathsf{Open})$ *is a semitopology and* $p \in \mathsf{P}$. *Then the following are equivalent:*

1. p *is indirectly regular.*
2. $p_{\between}$ *contains a topen set.*

Proof. We prove two implications:

- *Suppose there exists* $T \in \mathsf{Topen}$ *such that* $T \subseteq p_{\between}$.

  Take any $q \in T$. By Theorem 4.2.6(5) q is regular, and by Definition 3.6.1(2) $q \between p$.

- *Suppose* p *is indirectly regular.*

  By Definition 9.3.2 $p \between q$ for some regular $q \in \mathsf{P}$. By Lemma 9.3.4 $K(q) \subseteq p_{\between}$. By Definition 4.1.4(3) $K(q) \in \mathsf{Topen}$. □

9.3.3 Indirectly regular points in the context of other regularity properties

We can now continue the observations made in Remark 4.1.5:

REMARK 9.3.6. This Subsection develops a sequence of results that are interesting in themselves, but also taken together with Lemma 4.1.6 they indicate that in a strongly chain-complete semitopology, our regularity conditions organise into a nice list ordered by increasing strength as follows:

- Being quasiregular (having a nonempty community).
- Being indirectly regular (intertwined with a regular point / being on the boundary of a topen set).
- Being weakly regular (being an element of your community).
- Being regular (being an element of your *topen* community).

As a diagram, in strongly chain-complete semitopologies we get the following *chain of implications*:

$$\text{quasiregular} \implies \text{indirectly regular} \implies \text{weakly regular} \implies \text{regular}.$$

If the semitopology is not strongly chain-complete, then (by Lemma 4.1.6 we still have the other implications, but) being indirectly regular does not fall so neatly in line.

We read this as evidence that the *strongly chain-complete semitopologies* are a particularly natural class of semitopologies for us to study, and they are a useful abstraction of the finite semitopologies (much as e.g. Alexandrov topologies, or compact topologies, capture aspects of finiteness for topologies).

PROPOSITION 9.3.7. *Suppose* $(\mathsf{P}, \mathsf{Open})$ *is a strongly chain-complete semitopology (by Remark 9.1.4 this holds in particular if* P *is finite). Then:*

1. *If* $p \in \mathsf{P}$ *is weakly regular then* p *is indirectly regular.*
2. *The converse implication need not hold: it is possible for* $(\mathsf{P}, \mathsf{Open})$ *to be strongly chain-complete, and even actually* finite, *and* $p \in \mathsf{P}$ *is indirectly regular yet not weakly regular.*
3. *If* $(\mathsf{P}, \mathsf{Open})$ *is not strongly chain-complete then the implication in part 1 might fail: it is possible for* $p \in \mathsf{P}$ *to be weakly regular but not indirectly regular, even if* $(\mathsf{P}, \mathsf{Open})$ *is chain-complete (but not strongly chain-complete).*

Proof. We consider each part in turn:

1. Suppose $p \in \mathsf{P}$ is weakly regular. From Proposition 5.4.10 $p_{\between}$ is a closed neighbourhood (a closed set with a nonempty open interior). Using strong chain-completeness and Zorn's lemma on $\supseteq$, the set of closed neighbourhoods that are subsets of $p_{\between}$ contains a minimal closed neighbourhood $C \subseteq p_{\between}$ (we need *strong* chain-completeness to ensure that C has a *nonempty* open interior). Take $q \in interior(C)$; By Theorem 5.6.2 q is regular.

2. For a counterexample consider point $*$ in Figure 7.1. Then $K(*) = \{1\}$ and 1 is regular, but $* \notin K(*)$ so $*$ is not weakly regular.

   (It *does* follow from existence of a regular $q \in K(p)$ that p is quasiregular, but only because existence of *any* (not necessarily regular) $q \in K(p)$ means precisely that $K(p) \neq \varnothing$.)

3. A counterexample is in Figure 6.2. □

REMARK 9.3.8. Proposition 9.3.7 above is just an easy corollary of Theorem 5.6.2. We can think of this as another version of the 'hairy ball' result that we saw in Theorem 6.4.7, but for the case of a weakly regular point, instead of for a quasiregular space.

Recall that we care about regular points because these are (for our purposes) well-behaved: they have a topen neighbourhood (Theorem 4.2.6), by which fact local consensus is guaranteed where algorithms succeed (Remark 3.2.5). Thus the interest of Proposition 9.3.7 is that it provides certain guarantees of progress; a weakly regular point may not be able to progress (even if algorithms succeed), but it guaranteed to be intertwined with some well-behaved regular point.

LEMMA 9.3.9. *Suppose* $(\mathsf{P}, \mathsf{Open})$ *is a semitopology and suppose* $p \in \mathsf{P}$. *Then:*

1. *If* p *is indirectly regular then* p *is quasiregular.*
2. *The converse implication need not hold: it is possible for* p *to be quasiregular but not indirectly regular.*

Proof. We consider each part in turn:

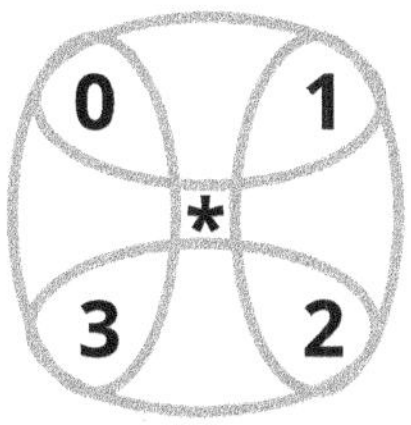

Figure 9.1: Lemma 9.3.9(2): a point $*$ that is quasiregular but not indirectly regular

1. Suppose p is indirectly regular. By Definition 9.3.2 $p \in q_{\between}$ for some regular q. By Lemma 4.1.6(1&2) q is quasiregular, meaning that $\varnothing \neq K(q)$. By Lemma 9.3.4 $K(q) \subseteq K(p)$, so that $K(p)$ is nonempty and p is quasiregular.

2. It suffices to provide a counterexample. Consider the point $*$ in the semitopology illustrated in Figure 9.1:[4]

   - $\mathsf{P} = \{*, 0, 1, 2, 3\}$.
   - Open is generated by $\{\{3,0\}, \{0,1\}, \{1,2\}, \{2,3\}\}$; note that the only open neighbourhood of $*$ is all of P.

   The reader can check that $*_{\between} = \mathsf{P}$, so $K(*) = \mathsf{P} \neq \varnothing$ so $*$ is quasiregular. However, the reader can also check that no point in this space is regular, so $*$ is not intertwined with any regular point. □

This completes the chain of implications from Remark 9.3.6.

9.4 Witness semitopologies are chain-complete

THEOREM 9.4.1. *Suppose that (P, wf) is a witnessed set. Then the witness semitopology $\mathsf{Open}(wf)$ from Definition 8.2.5 is chain-complete.*

Unpacking this we can say:

> *In a witness semitopology, intersections of descending chains of open sets are open, and unions of ascending chains of closed sets are closed.*

Proof. Consider a chain of open sets $\mathcal{O} \subseteq \mathsf{Open}$. There are three cases:

- *Suppose $\bigcap \mathcal{O} = \varnothing$.*

  We note that $\varnothing \in \mathsf{Open}$ (Definition 1.2.2(1)) and we are done.

- *Suppose $\mathcal{O}$ has a least element O.*

  Then $O = \bigcap \mathcal{O}$ and $O \in \mathsf{Open}$ and we are done.

[4] This is an elaboration of the semitopology we have already seen in Figure 5.2).

- *Suppose $\mathcal{O} \neq \varnothing$ and $\mathcal{O}$ has no least element.*

  Then note that $\mathcal{O}$ is infinite. Consider some $p \in \bigcap \mathcal{O}$. By construction of the witness semitopology (Definition 8.2.5) for each $O \in \mathcal{O}$ there exists a witness-set $w_O \in wf(p)$ such that $w_O \subseteq O$. Now by Definition 8.2.2(1) $wf(p)$ is finite, so by the pigeonhole principle, there exists some $w \in wf(p)$ such that $w \subseteq O$ for for every $O \in \mathcal{O}$, and thus $w \subseteq \bigcap \mathcal{O}$.

  Now p in the previous paragraph was arbitrary, so we have shown that if $p \in \bigcap \mathcal{O}$ then also there exists $w \in wf(p)$ such that $w \subseteq \bigcap \mathcal{O}$. It follows by construction of the witness semitopology in Definition 8.2.5 that $\bigcap \mathcal{O}$ is open as required. □

LEMMA 9.4.2. *The reverse implication in Theorem 9.4.1 does not hold: there exists a chain-complete semitopology (indeed, it is also strongly chain-complete) that is not generated as the witness semitopology of a witnessed set.*

Proof. It is a fact that the more-than-one semitopology on $\mathbb{N}$ (having open sets generated by distinct pairs $\{i, i'\} \subseteq \mathsf{P}$; see Example 2.1.4(5d)) is strongly chain-complete, but by Lemma 8.3.5 is is not generated by a witness function. □

REMARK 9.4.3. Elaborating further on Lemma 9.4.2, suppose $(\mathsf{P}, \mathsf{Open})$ is a chain-complete semitopology. Then to every p we can assign a nonempty set $\mathcal{O}_p$ of *covers* (minimal open sets containing p; see Definition 9.5.2).

Can we obtain a witness function just by setting $wf(p) = \mathcal{O}_p$? No: because p need not have finitely many covers, and Definition 8.2.2 insists on a *finite* set of (possibly infinite) nonempty witness-sets.[5]

We could allow an infinite set of witness-sets in Definition 8.2.2, but at a price:

- The proof of Theorem 9.4.1 depends on the pigeonhole principle, which uses finiteness of the set of witness-sets.
- The proof of Lemma 8.4.11 depends on the set of witness-sets being finite, and this is required for Proposition 8.4.13.

REMARK 9.4.4. Theorem 9.4.1 shows that witness semitopologies are chain-complete, but Lemma 9.4.2 demonstrates that this cannot precisely characterise witness semitopologies. Might there be another way?

We might look at Corollary 9.5.4 (open covers exist), cross-reference with Definition 8.2.2(1) (every p has only finitely many witness-sets), and ask if we might characterise witness semitopologies as those topologies that are chain-complete *and* every p has finitely many open covers (Definition 9.5.2(2)).

No: by Lemma 8.3.3, the all-but-one semitopology from Example 2.1.4(5c) is a witness semitopology, and if the underlying set of points is infinite then every point has infinitely many covers. See also Remark 9.5.13(1).

PROPOSITION 9.4.5.

1. *Not every witness semitopology (Definition 8.2.5) is strongly chain-complete (Definition 9.1.2(2)).*

[5] See Example 2.1.4(5c) for an example of a semitopology containing points with infinitely many covers, though interestingly, this *can* be generated by a witness function, as noted in Lemma 8.3.3.

2. *Part 1 holds even if we restrict the witness function* $wf : \mathsf{P} \to \mathit{fin}_{\neq\varnothing}(\mathit{pow}_{\neq\varnothing}(\mathsf{P}))$ *in Definition 8.2.2(1) to return a finite set of finite witness-sets, so that* $wf : \mathsf{P} \to \mathit{fin}_{\neq\varnothing}(\mathit{fin}_{\neq\varnothing}(\mathsf{P}))$.
3. *Every* finite *semitopology (this includes every finite witness semitopology) is strongly chain-complete.*

Proof.

1. It suffices to provide a counterexample. Consider $\mathbb{N}$ with witness function $wf(i) = \{\{i+1\}\}$. This generates a semitopology with open sets generated by $i_{\geq} = \{i' \in \mathbb{N} \mid i' \geq i\}$. Then $(i_{\geq} \mid i \in \mathbb{N})$ is a descending chain of open sets with an open, but empty, intersection.

2. We just use the counterexample in part 1.

3. We noted already in Remark 9.1.4 that if the semitopology is finite then every descending chain of open sets is eventually stationary; so we just take the final element in the chain. □

REMARK 9.4.6. We are particularly interested in the concrete example of finite witnessed semitopologies, since these are the ones that we can actually implement. But we can ask what it is about this class of examples that makes them mathematically well-behaved; what essential algebraic features might we identify here? Proposition 9.4.5 suggests that being strongly chain-complete may be a suitable mathematical abstraction:

- by Proposition 9.4.5 the abstraction is both non-trivial and sound (not every witness semitopology is strongly chain-complete, but every *finite* witness semitopology is), and
- Theorem 9.4.1 asserts that for a (possibly infinite) $(\mathsf{P}, \mathsf{Open})$, any convergence using a descending sequence of open sets has a flavour of being 'locally finite' in the sense of being guaranteed to have a nonempty open intersection.[6]

So strongly chain-complete semitopologies are a plausible abstraction of finite witness semitopologies.[7] The test is now to explore the theory of strongly chain-complete semitopologies. Key results are Corollaries 9.5.4 and 9.5.9, which ensure that in a strongly chain-complete semitopology, open covers and atoms always exist, and we will build from there.

REMARK 9.4.7 (Why infinity?). Following on from Remark 9.4.6, we sometimes get asked, especially by engineers, why we care about infinite models when all practical computer networks are finite.

A simple answer is that we do this for the same reason that Python (and many other programming languages) have a datatype for infinite precision integers. Any given execution will only compute numbers in a finite subset this infinity, but since we may not be able to predict how large this subset is, it is natural to support the notion of an infinite datatype. Note this this

[6]There is also a computational interpretation to (strong) chain-completeness: think of a descending chain of open sets as a computation that computes to narrow down possibilities to smaller and smaller nonempty open sets, then this possibly infinite computation does deliver a final answer that is a (nonempty) open set.

[7]There may be more than one such abstraction; identifying one candidate does not mean there may not be others. For example, both *rings* and *models of first-order arithmetic* are valid abstractions of the notion of 'number'. Which of these mathematical structures we work with, depends on which aspects of the concrete thing we are interested in studying.

holds for data, not just datatypes: e.g. Python accommodates values for π, e, and j even though these are not rational numbers, and for infinite streams and may other 'infinite' objects.[8]

However there is another reason: participants cannot depend on an exhaustive search of the full network ever terminating (nor that even an attempt at this would be cost-effective), so this requires a theory and algorithms that make sense on at least countably infinitely many points.

In fact, arguably the natural cardinality for semitopology is at least *uncountable*, since for a participant on a system with network latency, the system is not just unbounded, but also unenumerable. This is another reason that Theorem 9.4.1 is remarkable. See also Remark 23.3.6.

9.5 Minimal sets: open covers and atoms

9.5.1 Open covers (minimal open neighbourhoods)

First, some useful notation:

NOTATION 9.5.1. Suppose $(\mathsf{P}, \mathsf{Open})$ is a semitopology and $P \subseteq \mathsf{P}$. Write

$$O \gtrdot P \quad \text{and synonymously} \quad P \lessdot O$$

when O is a minimal nonempty open set containing P. In symbols:

$$O \gtrdot P \quad \text{when} \quad O \neq \varnothing \wedge P \subseteq O \wedge \forall O' {\in} \mathsf{Open}_{\neq\varnothing}.(P \subseteq O' \subseteq O \Longrightarrow O' = O).$$

We may combine $\gtrdot$ with other relations for compactness. For example:

- $p{\in}O \gtrdot P$ is shorthand for $p{\in}O \wedge O \gtrdot P$; and
- $P \supseteq O \gtrdot Q$ is shorthand for $O \subseteq P \wedge O \gtrdot Q$.

Definition 9.5.2 collects some (standard) terminology.

DEFINITION 9.5.2. Suppose $(\mathsf{P}, \mathsf{Open})$ is a semitopology and $p{\in}\mathsf{P}$.

1. Call $O{\in}\mathsf{Open}$ an **(open) neighbourhood of** p when $p{\in}O$.

2. Call $O{\in}\mathsf{Open}$ an **(open) cover of** p, write

   $$O \gtrdot p \quad \text{and/or} \quad p \lessdot O,$$

   and say that O **covers** p, when $O \gtrdot \{p\}$ (Notation 9.5.1).

   In words using the terminology of part 1: $O \gtrdot p$ when O is a minimal open neighbourhood of p.

[8] I once struggled to convince a Computer Science undergraduate student that $1/3$ is finite. The blockage was that the student only believed in the `float` datatype, and the decimal expansion of $1/3$ as $0.333\ldots$ is infinite. This deadlock was broken by inviting the student to implement a base-3 float type. The deeper point here is that what we consider 'infinite' may depend on what representation we assume as primitive. We see something similar in model theory, where we may distinguish between *internal* and *external* notions of size in a model. The bottom line is: obsessing about size can become a dead end; we also need to pay attention to what seems elegant and natural, i.e. to what our brains want — and then model *that*.

3. Write $Covers(p)$ for the set of **open covers of** p, and if $O' \in \mathsf{Open}$ then write $Covers_{O'}(p)$ for the open covers of p that are subsets of O'. In symbols:

$$Covers(p) = \{O \in \mathsf{Open} \mid p \lessdot O\} \quad \text{and} \quad Covers_{O'}(p) = \{O \in \mathsf{Open} \mid p \lessdot O \subseteq O'\}.$$

Note of course that $Covers(p) = Covers_{\mathsf{P}}(p)$.

LEMMA 9.5.3. *Suppose* $(\mathsf{P}, \mathsf{Open})$ *is a strongly chain-complete semitopology and suppose* $\varnothing \neq \mathcal{O} \subseteq \mathsf{Open}_{\neq\varnothing}$ *is a nonempty set of nonempty open sets that is* $\subseteq$*-down-closed (meaning that if* $\varnothing \neq O' \subseteq O \in \mathcal{O}$ *then* $O' \in \mathcal{O}$*).*

Then $\mathcal{O}$ *contains a* $\subseteq$*-minimal element.*

Proof. We use Zorn's Lemma [Jec73, Cam78]: By strong chain-completeness, $\mathcal{O}$ ordered by the *superset* relation (the reverse of the subset inclusion relation), contains limits, and so upper bounds, of ascending chains. By Zorn's Lemma, $\mathcal{O}$ contains a $\supseteq$-maximal element. This is the required $\subseteq$-minimal element. □

COROLLARY 9.5.4 (Existence of open covers). *Suppose* $(\mathsf{P}, \mathsf{Open})$ *is a chain-complete semitopology and* $p \in \mathsf{P}$.[9] *Then:*

1. *If* $p \in O' \in \mathsf{Open}$ *then* O' *contains an open cover of* p*. In symbols:*

$$Covers_{O'}(p) \neq \varnothing, \quad \textit{equivalently} \quad \exists O \in \mathsf{Open}.p \lessdot O \subseteq O'.$$

2. *p has at least one open cover. In symbols:*

$$Covers(p) \neq \varnothing, \quad \textit{equivalently} \quad \exists O \in \mathsf{Open}.p \lessdot O.$$

Proof. Direct from Lemma 9.5.3, considering $\{O \in \mathsf{Open} \mid p \in O \subseteq O'\}$ (nonempty because it contains O'), and then setting $O' = \mathsf{P}$. □

We can apply Corollary 9.5.4 in an elementary way to extend Lemma 6.5.3:

COROLLARY 9.5.5. *Suppose* $(\mathsf{P}, \mathsf{Open})$ *is a chain-complete semitopology (in particular this holds if* P *is finite) and suppose* $p \in \mathsf{P}$ *and* $O' \in \mathsf{Open}$*. Then the following are equivalent:*

1. $p \in |O'|$.
2. $O' \between nbhd(p)$.
3. $O' \between Covers(p)$.

Proof. Parts 1 and 2 just repeat Lemma 6.5.3. Equivalence of parts 2 and 3 is routine as follows:

- If $O' \between nbhd(p)$ then $O' \between Covers(p)$, since by construction $Covers(p) \subseteq nbhd(p)$.
- Suppose $O' \between Covers(p)$ and consider some $O \in nbhd(p)$. By Corollary 9.5.4(1) O contains some cover $p \lessdot O'' \subseteq O$, and since $O' \between O''$ also $O' \between O$. Thus $O' \between nbhd(p)$. □

[9] Note that we only require *chain-completeness* here (Definition 9.1.2(1)), not strong chain-completeness (Definition 9.1.2(2)).

REMARK 9.5.6. Recall that our semitopological analysis of consensus is all about continuity and value assignments being locally constant — as per Definitions 2.2.1(3) and 2.1.3 and results like Lemma 2.2.4 — and these discussions are about the open neighbourhoods of p. Thus, to understand consensus at p we need to understand its open neighbourhoods.

Corollary 9.5.4 tells us that in a witness semitopology, we can simplify and just consider the open covers of p. This is because if a continuous function $f : \mathsf{P} \to \mathsf{P}'$ such that $f(p) = p' \in O'$ is continuous at $p \in \mathsf{P}$, then using continuity and Corollary 9.5.4 there exists some open cover $p \lessdot P \subseteq f^{-1}(O')$.

Turning this around, if we want to *create* consensus around p — perhaps as part of a consensus algorithm — it suffices to find some open cover of p, and convince that cover. This fact is all the more powerful because Corollary 9.5.4 does not assume that P is finite: it is a fact of witness semitopologies of any cardinality.

9.5.2 Atoms (minimal nonempty open sets)

DEFINITION 9.5.7. Suppose $(\mathsf{P}, \mathsf{Open})$ is a semitopology.

1. Call $A \in \mathsf{Open}$ an **(open) atom** when A is a minimal nonempty open set.[10] In symbols using Notation 9.5.1 this is:

$$A \gtrdot \varnothing \quad \text{and synonymously} \quad \varnothing \lessdot A.$$

2. If $P \subseteq \mathsf{P}$ then write $Atoms(P)$ for the atoms that are subsets of P. In symbols:

$$Atoms(P) = \{A \in \mathsf{Open} \mid \varnothing \lessdot A \subseteq P\}.$$

LEMMA 9.5.8. *Suppose* $(\mathsf{P}, \mathsf{Open})$ *is a strongly chain-complete semitopology and suppose* $\varnothing \neq \mathcal{O} \subseteq \mathsf{Open}_{\neq\varnothing}$ *is a nonempty set of nonempty open sets that is* $\subseteq$*-down-closed (meaning that if* $\varnothing \neq O' \subseteq O \in \mathcal{O}$ *then* $O' \in \mathcal{O}$*).*

Then $\mathcal{O}$ *contains an atom.*

Proof. Just from Lemma 9.5.3, noting that an atom is precisely a $\subseteq$-minimal nonempty open set. □

COROLLARY 9.5.9 (Existence of atoms). *Suppose* $(\mathsf{P}, \mathsf{Open})$ *is a strongly chain-complete semitopology and* $O \in \mathsf{Open}_{\neq\varnothing}$ *is a nonempty open set. Then* O *contains an atom. In symbols:*

$$Atoms(O) \neq \varnothing.$$

Proof. From Lemma 9.5.8, considering $\{O' \in \mathsf{Open} \mid \varnothing \neq O' \subseteq O\}$ (which is nonempty because it contains O). □

REMARK 9.5.10. A simple observation is that if $(\mathsf{P}, \mathsf{Open})$ is a strongly chain-complete topology — thus, a strongly chain-complete semitopology whose opens are closed under finite intersections — then the atom that exists by Corollary 9.5.9 is unique, simply because if we have atoms A and A' then $A \cap A'$ is less than both and so by minimality must be equal to both. See also Lemma 10.1.5.

[10] An open atom covers every point that it contains, but an open cover for a point p need not be an atom, since it may contain a smaller open set — just not one that contains p. See Example 9.5.12(3).

9.5.3 Discussion

REMARK 9.5.11 (Origin of terminology).

1. The terminology "O covers p" in Definition 9.5.2(2) is adapted from order theory (see e.g. [DP02, §1.14]), where we say that y covers x when $y > x$ and there exists no z such that $y > z > x$.
2. The terminology "A is an atom" in Definition 9.5.7(1) is also adapted from order theory (see e.g. [DP02, §5.2]), where we call x an atom when it is a least element not equal to $\bot$ (i.e. when x covers $\bot$).

EXAMPLE 9.5.12 ((Counter)examples of atoms and open covers).

1. *p can be in multiple distinct atoms (minimal nonempty open sets), and/or open covers (minimal open sets that contain p).*

   For instance, consider $\mathbb{N}$ with the semitopology generated by $1_{\leq} = \{0, 1\}$ and $1_{\geq} = \{1, 2, 3, \ldots\}$. Then $1 \in 1_{\leq}$ and $1 \in 1_{\geq}$, and $1_{\leq}$ and $1_{\geq}$ are distinct minimal open sets (and also open covers of 1).

   A topology would compress this down to nothing: if $\{0, 1\}$ is open and $\{1, 2, 3, \ldots\}$ is open then their intersection $\{1\}$ would be open, and this would be the unique least open set containing 1. Because open sets in semitopologies are not necessarily closed under intersection, semitopologies permit richer structure.

2. *An open cover O of p is a minimal open set that contains p — but O need not be an atom (a minimal nonempty open set).*

   Consider $\mathbb{N}$ with the semitopology generated by $i_{\geq} = \{i' \in \mathbb{N} \mid i' \geq i\}$. Then $\mathit{Covers}(i) = \{i_{\geq}\}$ but (with this semitopology) $\mathit{Atoms}(\mathsf{P}) = \varnothing$; there are no least nonempty open sets.

3. *An atom $A \in \mathit{atoms}(p)$ is a minimal nonempty open set that is a subset of a minimal open set that contains p — but A need not contain p.*

   For instance, consider $\mathbb{N}$ with the semitopology generated by $i_{\leq} = \{i' \in \mathbb{N} \mid i' \leq i\}$. Then $\mathit{atoms}(i) = \{\{0\}\}$ for every i, because with this semitopology $\mathit{Atoms}(\mathbb{N}) = \{\{0\}\}$ and each i is covered by $i_{\leq}$, and $\{0\} \subseteq i_{\leq}$. However, we only have $i \in \{0\}$ when $i = 0$.

REMARK 9.5.13 (Two open problems).

1. *Topological characterisation of witness semitopologies.*

   Following on from Remark 9.4.4, we have seen that witness semitopologies are chain-complete, but that this does not precisely characterise witness semitopologies. A topological characterisation of witness semitopologies, or a proof that such a characterisation is impossible, remains an open problem. To this end, the material in Subsection 8.5 may be relevant, which relates witness semitopologies to a Turing-complete model of computation.

2. *Conditions on witness functions to guarantee (quasi)regularity.*

It remains an open problem to investigate conditions on witness functions to guarantee that every point is quasiregular. In view of Proposition 6.4.9 and Corollary 6.4.10, such conditions would suffice to guarantee the existence of a regular point in the finite case. Regular points are well-behaved, so a system with at least one regular point is a system that is in some sense 'somewhere sensible'.

Kernels: the atoms in a community 10

10.1 Definition and examples

REMARK 10.1.1. We have studied $K(p)$ the community of a point and have seen that is has a rich mathematics. We also know from results (like Theorem 3.2.2) and discussions (like Remark 5.5.1) that to understand consensus in a semitopology, we have to understand its communities.

It is now interesting to look at the atoms in a community (Definition 9.5.7; minimal nonempty open sets). As we shall see later, the atoms in a community dictate its ability to act — see e.g. Corollary 11.6.11, which is reminiscent of *Arrow's theorem* from social choice theory — so that understanding $K(p)$ is, in a sense we will make formal, much the same thing as understanding the atoms in $K(p)$ (see e.g. Proposition 11.3.2).

Kernels are also interesting in and of themselves, so we start in this Section by studying them (culminating, out of several results, with Propositions 10.2.7 and 10.3.2).

DEFINITION 10.1.2. Suppose $(\mathsf{P}, \mathsf{Open})$ is a semitopology and $p \in \mathsf{P}$.

1. Define $ker(p)$ the **kernel of** p to be the union of the atoms in its community. We give equivalent formulations which we may use as convenient:

$$\begin{aligned} ker(p) &= \bigcup\{A \in Atoms(\mathsf{P}) \mid A \subseteq p_{\between}\} \\ &= \bigcup\{A \in Atoms(\mathsf{P}) \mid A \subseteq K(p)\} \\ &= \bigcup\{A \in \mathsf{Open} \mid \varnothing \lessdot A \subseteq K(p)\}. \end{aligned}$$

   Above, $\varnothing \lessdot A$ is just another way of saying that A is an atom (minimal nonempty open set; see Definition 9.5.7), and $A \subseteq p_{\between}$ if and only if $A \subseteq K(p)$ because A is open and $K(p)$ is just the open interior of $p_{\between}$ (Definition 4.1.4(1)).
2. If A is an atom that is a subset of $ker(p)$ (in symbols: $\varnothing \lessdot A \subseteq ker(p)$) then we might call A a **kernel atom of** p.

3. Extend ker to subsets $P \subseteq \mathsf{P}$ by taking a sets union:
$$ker(P) = \bigcup \{ker(p) \mid p \in P\}.$$

We return to and extend Example 4.4.1, and we include details of the kernels:

EXAMPLE 10.1.3.

1. Take P to be $\mathbb{R}$ the real numbers, with its usual topology (which is also a semitopology), as per Example 4.4.1(1). Then:
   - $x_{\between} = \{x\}$ and $K(x) = \varnothing$ for every $x \in \mathbb{R}$.
   - $ker(x) = \varnothing$ for every $x \in \mathbb{R}$.
   - $ker(\mathbb{R}) = \varnothing$.
2. We take, as per Example 4.4.1(2) and as illustrated in Figure 3.1, top-left diagram:
   - $\mathsf{P} = \{0, 1, 2\}$.
   - Open is generated by $\{0\}$ and $\{2\}$.

   Then:
   - $0_{\between} = \{0, 1\}$ and $K(0) = interior(0_{\between}) = \{0\} = ker(0)$.
   - $2_{\between} = \{1, 2\}$ and $K(2) = interior(2_{\between}) = \{2\} = ker(2)$.
   - $1_{\between} = \{0, 1, 2\}$ and $K(1) = \{0, 1, 2\}$ and $ker(1) = \{0, 2\}$.
   - $ker(\mathsf{P}) = \{0, 2\}$.
3. We take, as per Example 4.4.1(3), as illustrated in Figure 4.1, and as reproduced for convenience here in Figure 10.1 (left-hand diagram):
   - $\mathsf{P} = \{0, 1, 2, 3, 4\}$.
   - Open is generated by $\{1, 2\}$, $\{0, 1, 3\}$, $\{0, 2, 4\}$, $\{3\}$, and $\{4\}$.

   Then:
   - $x_{\between} = \{0, 1, 2\}$ and $K(x) = interior(x_{\between}) = \{1, 2\}$ for $x \in \{0, 1, 2\}$.
   - $x_{\between} = \{x\} = K(x)$ for $x \in \{3, 4\}$.
   - $ker(x) = \{1, 2\}$ for $x \in \{0, 1, 2\}$.
   - $ker(x) = \{x\}$ for $x \in \{3, 4\}$.
   - $ker(\mathsf{P}) = \{1, 2, 3, 4\}$.
   - By construction $ker(\mathsf{P}) \subseteq \bigcup Atoms(\mathsf{P})$, but but we see here that the inclusion may be strict, since e.g. $\{0, 1, 3\}$ is an atom in this example but $0 \notin ker(\mathsf{P})$.
4. We add one point to part 3 of this example, -1, which is intertwined with 0, 1, and 2 but is not in a minimal nonempty open set, as illustrated in Figure 10.1 (right-hand diagram):
   - $\mathsf{P} = \{\text{-}1, 0, 1, 2, 3, 4\}$.
   - Open is generated by $\{\text{-}1, 1, 2\}$, $\{1, 2\}$, $\{0, 1, 3\}$, $\{0, 2, 4\}$, $\{3\}$, and $\{4\}$.

   Then:
   - $x_{\between} = \{\text{-}1, 0, 1, 2\}$ and $K(x) = interior(x_{\between}) = \{\text{-}1, 1, 2\}$ for $x \in \{\text{-}1, 0, 1, 2\}$.

Figure 10.1: Illustration of Example 10.1.3(3&4)

- $x_{\between} = \{x\} = K(x)$ for $x \in \{3, 4\}$.
- $ker(x) = \{1, 2\}$ for $x \in \{\text{-}1, 0, 1, 2\}$.
- $ker(x) = \{x\}$ for $x \in \{3, 4\}$.
- $ker(\mathsf{P}) = \{1, 2, 3, 4\}$.
- By construction $ker(p) \subseteq K(p) \subseteq p_{\between}$, but the inclusions may be strict. For instance:

$$ker(0) = ker(\text{-}1) = \{1, 2\} \subsetneq K(0) = K(\text{-}1) = \{\text{-}1, 1, 2\}$$
$$\subsetneq 0_{\between} = \text{-}1_{\between} = \{\text{-}1, 0, 1, 2\}.$$

5. We take $\mathsf{P} = \mathbb{N}$, with the semitopology (also a topology) generated by final subsets $n_{\geq} = \{n' \in \mathbb{N} \mid n' \geq n\}$ for $n \in \mathbb{N}$. Then $n_{\between} = \mathbb{N} = K(n)$ for every $n \in \mathbb{N}$, and $ker(n) = \varnothing$ (because there is no minimal nonempty open set).

We warm up with a couple of simple lemmas:

LEMMA 10.1.4. *Suppose* $(\mathsf{P}, \mathsf{Open})$ *is a semitopology and* $p \in \mathsf{P}$ *is a regular point.*
Then all kernel atoms of p *intersect, or in symbols:*

$$\varnothing \lessdot A, A' \subseteq ker(p) \quad \textit{implies} \quad A \between A'.$$

Proof. By construction in Definition 10.1.2(1) $A, A' \subseteq ker(p) \subseteq K(p)$. By regularity (Definition 4.1.4(3)) $K(p)$ is transitive. Then $A \between K(p) \between A'$ and by transitivity (Definition 3.2.1) it follows that $A \between A'$. □

LEMMA 10.1.5. *Suppose* $(\mathsf{P}, \mathsf{Open})$ *is a* topology *— thus: a semitopology whose open sets are closed under intersections — and* $p \in \mathsf{P}$ *is regular. Then one of the following holds:*

- $ker(p) = \varnothing$.
- $ker(p) = A$ *for some atom* $\varnothing \lessdot A$.

Proof. Suppose $ker(p) \neq \varnothing$, and suppose there exist two atoms $A, A' \subseteq ker(p)$. Then (just as already noted in Remark 9.5.10) $A \cap A'$ is an open set. It is not empty because $A \between A'$ by Lemma 10.1.4. By minimality, $A = A \cap A' = A'$. Thus, being a topology crushes Definition 10.1.2 down to be at most a single atom. □

10.2 Characterisations of the kernel

We open with a non-implication:

LEMMA 10.2.1. *Suppose* $(\mathsf{P}, \mathsf{Open})$ *is a semitopology and* $p \in \mathsf{P}$ *is regular (so* $p \in K(p) \in \mathsf{Topen}$*). Then it is not necessarily the case that* $ker(p) \neq \varnothing$.

Proof. A counterexample is Example 10.1.3(5). In full: we consider $\mathbb{N}$ with the semitopology generated by $n_{\geq} = \{n' \in \mathbb{N} \mid n' \geq n\}$ for $n \in \mathbb{N}$. Then $n_{\between} = \mathbb{N} = K(n)$ for every $n \in \mathbb{N}$, but $ker(n) = \varnothing$ because there is no minimal nonempty open set. □

We can exclude the case noted in the proof of Lemma 10.2.1 by restricting to strongly chain-complete semitopologies.

LEMMA 10.2.2. *Suppose* $(\mathsf{P}, \mathsf{Open})$ *is a strongly chain-complete semitopology — in particular, this holds if* P *is finite by Proposition 9.4.5 — and* $p \in \mathsf{P}$*. Then:*

1. $K(p) = \varnothing$ *if and only if* $ker(p) = \varnothing$*, and equivalently* $K(p) \neq \varnothing$ *if and only if* $ker(p) \neq \varnothing$*.* *In words: p has a nonempty community if and only if it has a nonempty kernel.*
2. *If p is regular then* $ker(p)$ *is a topen subset of* $K(p)$ *(nonempty transitive and open, see Definition 3.2.1(2)).* *(See also Lemma 10.2.5, which proves a stronger version of this property for the kernel atoms of a regular p.)*

Proof.

1. Suppose $\varnothing \neq K(p) = interior(p_{\between})$. Then by Corollary 9.5.9 (since P is strongly chain-complete) $K(p)$ contains at least one atom A, which is a subset of $K(p)$ by construction, and so $A \in ker(p)$.

   Conversely, if there exists an atom $A \in ker(p)$ then (since an atom is by assumption a nonempty set) we have $\varnothing \neq A \subseteq K(p)$.

2. Suppose p is regular. Unpacking Definition 4.1.4(3) this means that $p \in K(p)$. Thus in particular $K(p) \neq \varnothing$, and by part 1 of this result $ker(p) \neq \varnothing$.

   So $ker(p)$ is a nonempty subset of $K(p)$. By Theorem 4.2.6 $K(p)$ is a (maximal) topen, and by Lemma 3.4.2(2) $\varnothing \neq ker(p) \subseteq K(p)$ is topen as required. □

REMARK 10.2.3. Note in Lemma 10.2.2(2) that $ker(p)$ need not be a topen neighbourhood of p, simply because p (even if it is regular) might generate a topen community $K(p)$ but need not necessarily be in an atom in that community. See Example 10.1.3(4) taking $p = 0$ or $p = $ -1, or Lemma 10.2.4(2).

We complement Lemma 10.2.2 with some non-implications:

LEMMA 10.2.4. *Suppose* $(\mathsf{P}, \mathsf{Open})$ *is a semitopology. Then:*

1. $ker(p) \neq \varnothing$ *does not imply that p is regular.*
2. $ker(p)$ *topen does not imply that p is regular.*

Proof.

1. See Example 4.4.1(2): then $ker(1) = \{0, 2\} \neq \varnothing$ but (as noted in Lemma 4.4.2) 1 is not regular.

2. See Example 4.4.1(3): so $\mathsf{P} = \{0, 1, 2, 3, 4\}$ and Open is generated by $\{1, 2\}$, $\{0, 1, 3\}$, $\{0, 2, 4\}$, $\{3\}$, and $\{4\}$ and $ker(0) = K(0) = \{1, 2\}$, and this is (nonempty and) topen, but 0 is not regular since $0 \notin K(0) = \{1, 2\}$.

3. Take $\mathsf{P} = \{0, 1\}$ and set $\mathsf{Open} = \{\varnothing, \{0\}, \{0, 1\}\}$. Then 1 is regular but $1 \notin ker(p) = \{0\}$. □

Lemma 10.2.5. *Suppose* $(\mathsf{P}, \mathsf{Open})$ *is a semitopology and* $p \in \mathsf{P}$ *is a regular point and* $A \subseteq K(p)$. *Then the following are equivalent:*

1. *A is a kernel atom of p* ($\varnothing \lessdot A \subseteq K(p)$).
2. *A is a minimal topen in* $K(p)$.

Proof. We prove two implications:

- Suppose A is a kernel atom of p. By assumption in Definition 10.1.2(2) it is an atom (a minimal nonempty open set) in $K(p)$, and by Lemma 3.4.2(2) it is topen; so it is necessarily a minimal topen.

- Suppose A is a minimal topen in $K(p)$ and suppose $A' \subseteq A$ is any nonempty open set. By Lemma 3.4.2(2) A' is topen, so by minimality $A = A'$. Thus, A is an atom in $K(p)$, and so is a kernel atom of p. □

Remark 10.2.6. The proof of Lemma 10.2.5 above is elementary given our results so far, but it makes a useful observation. Recall from Theorem 4.2.6 that if p is regular (so $p \in K(p) \in \mathsf{Topen}$) then $K(p)$ is a maximal topen, and recall from Definition 10.1.2 that a kernel atom is an atom (i.e. a minimal nonempty open set) in $K(p)$. So we can read Lemma 10.2.5 as follows:

A kernel atom is a minimal topen inside a maximal topen.

Thus for regular p, $ker(p)$ and $K(p)$ are in some sense dual: the community of p is the maximal topen containing p, and the kernel of p is the union of the minimal topens inside that maximal topen.

So Lemma 10.2.5 tells us that for regular p, the kernel atoms of p are the minimal topens that are subsets of the community of p. Proposition 10.2.7 strengthens this to show that in fact, the kernel atoms of regular p are also the minimal topens that even *intersect* with the community of p (the significance of this to consensus is discussed in Remark 5.5.1):

Proposition 10.2.7. *Suppose that:*

- $(\mathsf{P}, \mathsf{Open})$ *is a semitopology.*[1]
- $A \in Atoms(\mathsf{P})$ *is an atom.*
- $p \in \mathsf{P}$ *is a regular point (so by Definition 4.1.4(1)* $p \in K(p) \in \mathsf{Topen}$*).*

[1]We do not seem to need P to be (strongly) chain-complete here. This is simply because we normally use strong chain-completeness to ensure that atoms and open covers exist, but in this result this is assumed.

- $O \in \mathsf{Topen}$ *and* $O \between K(p)$, *so* O *is any topen set that intersects the community of p (at least one such exists, by regularity, namely* $K(p)$ *itself).*[2]

Then the following are all equivalent:

1. $A \subseteq ker(p)$.
   In words: A *is a kernel atom of p.*
2. $A \subseteq K(p)$.
   In words: A *is an atom in the community of p.*
3. A *is topen and* $A \between O$.
   In words: A *is topen and intersects* O.

In particular, if A *is a topen atom*[3] *then we have:*

$$A \subseteq ker(p) \iff A \subseteq K(p) \iff A \between O \iff A \between ker(p) \iff A \between K(p).$$

Proof. We consider a cycle of implications:

- Suppose $A \subseteq ker(p)$. By construction in Definition 10.1.2(1) $ker(p) \subseteq K(p)$, so $A \subseteq K(p)$.

- Suppose $A \subseteq K(p)$. By Definition 9.5.7(1) (since A is an atom) A is nonempty. Then $A \between K(p) \between O$. By regularity $K(p)$ is topen, so by transitivity (Definition 3.2.1) $A \between O$ as required.

- Suppose A is topen and $A \between O$. By assumption $A \between O \between K(p)$ so by transitivity of O, $A \between K(p)$. By Proposition 4.3.1 $A \subseteq K(p)$ and it follows from Definition 10.1.2(2) that $A \subseteq ker(p)$ as required.

The equivalence

$$A \between O \iff A \between ker(p) \iff A \between K(p)$$

then follows routinely from the above, noting the equivalence $A \between O \iff A \subseteq K(p)$ and choosing $O = ker(p)$ or $O = K(p)$. □

Corollary 10.2.8. *Suppose* $(\mathbf{P}, \mathsf{Open})$ *is a semitopology and* $p \in \mathbf{P}$ *and* $O \in \mathsf{Topen}$ *and* $p \in O$ *(so p is regular and* O *is some topen neighbourhood of p). Then*

$$ker(p) = \bigcup\{A \in \mathsf{Topen}(\mathbf{P}) \mid \varnothing < A \between O\}.$$

In words: for any topen neighbourhood O *of p,* $ker(p)$ *is equal to the union of the topen atoms that intersect that neighbourhood.*

Proof. Unpacking Definition 10.1.2(1), $ker(p)$ is the union of atoms $A \subseteq K(p)$. We use Proposition 10.2.7. □

[2] By Proposition 4.3.1, this is equivalent to $O \subseteq K(p)$. We use $O \between K(p)$ because it yields a stronger form of the result.

[3] An atom is a minimal nonempty open set, and a topen is a nonempty open transitive set; so saying 'topen atom' is just the same as saying 'transitive atom'.

Lemma 10.2.9 explicitly characterises the union of all kernels as the union of all transitive atoms, which (given the results above) is what one might expect:

LEMMA 10.2.9. *Suppose* $(\mathsf{P}, \mathsf{Open})$ *is a semitopology. Then:*

1. *If* $A \subseteq \mathsf{P}$ *is a transitive atom then* $A \subseteq ker(p)$ *for every* $p \in A$.
   In words we can say: a transitive atom is a kernel atom for any points that it contains.
2. $ker(\mathsf{P})$ *is the union of the transitive atoms in* P.

Proof.

1. If $p \in A \in \mathsf{Topen}$ then A is a topen neighbourhood for p. By Theorem 4.2.6 $p \in A \subseteq K(p)$. But then by construction A is an atom in $K(p)$ so by Definition 10.1.2(1) $A \subseteq ker(p)$.

2. It follows from Lemma 10.2.5 and Definition 10.1.2(3) that every atom in $ker(\mathsf{P})$ is (topen and so) transitive. Conversely by part 1 of this result every transitive atom is in the kernel of the community of its points. □

10.3 Further properties of kernels

10.3.1 Intersections between the kernel of p and its open neighbourhoods

Lemma 10.3.1 is quite easy to prove by following definitions and applying transitivity properties, but it makes a useful point:

LEMMA 10.3.1. *Suppose that:*

- $(\mathsf{P}, \mathsf{Open})$ *is a semitopology.*
- $p \in \mathsf{P}$ *is a regular point.*
- A *is a kernel atom for* p. *In symbols:* $\varnothing \lessdot A \subseteq ker(p)$.

Then

$$\forall O \in \mathsf{Open}. p \in O \Longrightarrow A \between O.$$

In words:

Every open neighbourhood of a regular p intersects every kernel atom of p.[4]

Proof. By our assumption that p is regular we have $p \in K(p) \in \mathsf{Topen}$ (Definition 4.1.4(3)), and we assumed $p \in O$, so $O \between K(p)$.

Also by assumption $A \between K(p)$, since $A \subseteq ker(p) \subseteq K(p)$ by Definition 10.1.2(2&1). Thus $O \between K(p) \between A$. Now $K(p)$ is topen, thus it is transitive (Definition 3.2.1(1)) and so $O \between A$ as required. □

PROPOSITION 10.3.2. $(\mathsf{P}, \mathsf{Open})$ *is a semitopology and p is regular and* $p \in O \in \mathsf{Open}$. *Then:*

[4]This property is a bit subtle, because it is not necessarily the case that $p \in ker(p)$ (cf. Remark 10.2.3). So a kernel atom $\varnothing \lessdot A \subseteq ker(p)$ is is not itself necessarily a neighbourhood of p, but it still has a property of 'oversight' over p in the sense that it intersects with every quorum (open neighbourhood) that p has.

1. *The kernel of p is a subset of the union of the atoms intersecting O. In symbols:*

$$ker(p) \subseteq \bigcup\{A \in \mathsf{Open} \mid \varnothing \lessdot A \between O\} = \bigcup\{A \in Atoms(\mathbf{P}) \mid A \between O\}.$$

2. *The inclusion may be strict, even if O is an open cover of p (in symbols: $O \gtrdot p$).*
3. *The inclusion may be strict, even if $O \gtrdot p$ is a topen (transitive open) cover of p.*
4. *If O is a topen cover of p, then the kernel of p is precisely equal to the union of the transitive atoms intersecting O. In symbols:*

$$ker(p) = \bigcup\{A \in \mathsf{Topen}(\mathbf{P}) \mid \varnothing \lessdot A \between O\}.$$

Proof.

1. For the inclusion we just combine Lemma 10.3.1 with Definition 10.1.2(1).
2. To see how the inclusion may be strict, see Example 10.3.3(1).
3. To see how the inclusion may be strict, even for transitive O, see Example 10.3.3(2).
4. This just repeats Corollary 10.2.8. □

EXAMPLE 10.3.3.

1. Take $\mathbf{P} = \{0,1,2\}$ and let opens be generated by $\{0,1\}$ and $\{0,2\}$ and $\{1,2\}$ and $\{2\}$, as illustrated in Figure 10.2 (left-hand diagram).

   Set $p = 1$ and $O = \{1,2\}$. Then we can calculate that:

   - 0, 1, and 2 are all regular.
   - The community and kernel of 1 and 0 are equal to $\{0,1\}$ — 2 is not intertwined with 0 or 1 because $\{2\} \cap \{0,1\} = \varnothing$.
   - The community and kernel of 2 are equal to $\{2\}$.
   - $\{1,2\}$ is an open cover of 1.
   - The union of the atoms that intersect with $\{1,2\}$ is the whole space $\{0,1,2\}$.

   Thus $ker(1) = \{0,1\} \subsetneq \bigcup\{A \in Atoms(\mathbf{P}) \mid A \between \{1,2\}\} = \{0,1\} \cup \{2\} = \{0,1,2\}$.

2. Take $\mathbf{P} = \{0,1,2,3\}$ and let opens be generated by $\{0,1\}$ and $\{1,2\}$ and $\{2,3\}$, as illustrated in Figure 10.2 (right-hand diagram).

   Set $p = 1$ and $O = \{0,1\}$. Then we can calculate that:

   - $\mathbf{P}$ splits into two disjoint topen sets: $\{0,1\}$ and $\{2,3\}$. So O is topen.
   - The community and kernel of 0 and 1 are equal to $\{0,1\}$ — 2 is not intertwined with 0 or 1 because $\{2,3\} \cap \{0,1\} = \varnothing$. So $\{1,2\}$ is an atom, but it is not transitive.
   - The community and kernel of 2 and 3 are equal to $\{2,3\}$.
   - $\{0,1\}$ is an open cover of 1.

   Thus $ker(1) = \{0,1\} \subsetneq \bigcup\{A \in Atoms(\mathbf{P}) \mid A \between \{0,1\}\} = \{0,1\} \cup \{1,2\} = \{0,1,2\}$.

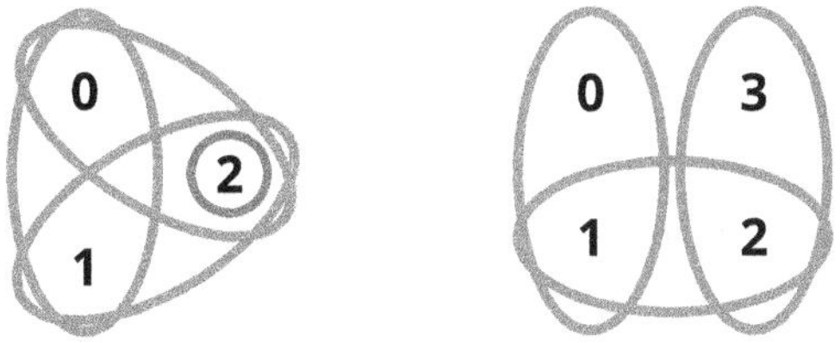

Figure 10.2: The semitopologies in Example 10.3.3

Remark 10.3.4 (Algorithmic content of Proposition 10.3.2). Proposition 10.3.2 reduces the problem of computing kernels to the problem of identifying transitive sets.[5] Once we have this, an algorithm for computing $ker(p)$ for regular p follows:

- Compute a transitive open neighbourhood O of p — for example using the algorithm outlined in Remark 8.4.6 to compute open neighbourhoods of p, and testing until we find one that is transitive. At least one transitive cover of p exists, by our assumption that p is regular.
- For each $p' \in O$, compute all the atoms that contain p' — for example by computing the open neighbourhoods of p' and checking which are atoms, and are transitive.

By Proposition 10.3.2(4), this collection of transitive atoms that intersect with O, will return the kernel atoms of p.

We conclude by noting a non-result:

Lemma 10.3.5. *Suppose* $(\mathsf{P}, \mathsf{Open})$ *is a semitopology and* $p \in \mathsf{P}$ *is regular. Recall from Theorem 5.6.2 that* $K(p)$ *the community of* p *is the greatest transitive open neighbourhood of* p*, so that any* transitive *open neighbourhood of* p *is contained in the community of* p*.*

However, there may still exist a non-transitive open cover of p *that is not contained in the community of* p*.*

Proof. It suffices to provide a counterexample, and as it happens we have just considered one. Consider Example 10.3.3(2), as illustrated in Figure 10.2 (right-hand diagram). Then $1_{\between} = K(1) = \{0, 1\}$ and $\{1, 2\}$ is an open cover of 1 and $\{1, 2\} \not\subseteq \{0, 1\}$. □

10.3.2 Idempotence properties of the kernel and community

Remark 10.3.6. In Definitions 4.1.4(2) and 10.1.2(3) we extend the notions of community and kernel of a set of points, using sets union. This allows us to take the community of a community $K(K(p))$, then kernel of a kernel $ker(ker(p))$, and so forth. Does doing this add any information? We would hope not — but we need to check.

In this Subsection we take check this for regular points, and see that they display good behaviour (e.g.: the community of the community is just the community, and so forth). The proofs also illuminate how regularity condition ensures good behaviour.

[5]We considered *that* question in results including Proposition 5.4.10 and Theorem 5.6.2.

LEMMA 10.3.7. *Suppose* $(\mathsf{P}, \mathsf{Open})$ *is a semitopology and suppose* $p \in \mathsf{P}$ *is regular. Then*

$$K(K(p)) = K(p).$$

Proof. We prove two subset inclusions:

- Suppose $q \in K(K(p))$, so unpacking Definition 4.1.4(2) there exists $p' \in K(p)$ such that $q \in K(p')$. By Corollary 4.2.8 (since p is regular) $K(p') = K(p)$, so $q \in K(p)$.

  q was arbitrary, and it follows that $K(K(p)) \subseteq K(p)$.

- Suppose $q \in K(p)$. Then by Corollary 4.2.8 (since p is regular) $K(q) = K(p) \in \mathsf{Topen}$, so in particular $q \in K(q)$.

  q was arbitrary, and it follows that $K(p) \subseteq K(K(p))$. □

COROLLARY 10.3.8. *Suppose* $(\mathsf{P}, \mathsf{Open})$ *is a semitopology and* $p \in \mathsf{P}$ *is regular. Suppose further that* $ker(p) \neq \varnothing$ *(if* P *is strongly chain-complete or finite then by Lemma 10.2.2(2) and Proposition 9.4.5* $ker(p) \neq \varnothing$ *is guaranteed). Then*

$$K(p) = K(ker(p)).$$

Proof. Suppose $q \in K(p)$ and pick any $k \in ker(p) \subseteq K(p)$. Then $k \in K(p)$ so by Corollary 4.2.8 $K(p) = K(k)$ so $q \in K(k)$. Thus $K(p) \subseteq K(ker(p))$.

Furthermore $K(ker(p)) \subseteq K(K(p))$ is a structural fact of Definition 4.1.4(2) and the fact, noted above, that $ker(p) \subseteq K(p)$.

We finish with Lemma 10.3.7:

$$K(p) \subseteq K(ker(p)) \subseteq K(K(p)) \overset{L10.3.7}{=} K(p).$$ □

LEMMA 10.3.9. *Suppose* $(\mathsf{P}, \mathsf{Open})$ *is a semitopology and suppose* $p \in \mathsf{P}$ *is regular. Then*

$$ker(K(p)) = ker(p).$$

Proof. Unpacking Definition 10.1.2, $ker(K(p)) = \bigcup\{ker(p') \mid p' \in K(p)\}$ and for each $p' \in K(p)$ we have $ker(p') = \bigcup\{A \subseteq \mathsf{P} \mid \varnothing \lessdot A \subseteq K(p')\}$. By Corollary 4.2.8, $K(p) = K(p')$ for every $p' \in K(p)$, and threading this equality through the definitions above, we obtain the result. □

LEMMA 10.3.10. *Suppose* $(\mathsf{P}, \mathsf{Open})$ *is a semitopology and suppose* $p \in \mathsf{P}$ *is regular. Then*

$$ker(p) = ker(ker(p)).$$

Proof. If $ker(p) = \varnothing$ then the result is immediate. So suppose $ker(p) \neq \varnothing$. We show two subset inclusions.

- To prove $ker(p) \subseteq ker(ker(p))$ we can reason as follows:

$$\begin{aligned} ker(ker(p)) &\subseteq ker(K(p)) && ker(p) \subseteq K(p), \text{Def. 10.1.2(3)} \\ &= ker(p) && \text{Lemma 10.3.9} \end{aligned}$$

- To prove $ker(ker(p)) \subseteq ker(p)$ we note that a kernel is a union of atoms in Definition 10.1.2(1), and we reason as follows, for an atom $\varnothing \lessdot A$ (which exists because $ker(p)$ is a union of atoms and we assumed $ker(p) \neq \varnothing$):

$$\begin{aligned} A \subseteq ker(p) &\iff A \subseteq K(p) && \text{Definition 10.1.2(1)} \\ &\iff A \subseteq K(ker(p)) && K(p) = K(ker(p)) = K(K(p)) \\ &\iff A \subseteq ker(ker(p)) && \text{Definition 10.1.2(1).} \end{aligned}$$

Above, $K(p) = K(ker(p)) = K(K(p))$ follows from Lemma 10.3.7 and Corollary 10.3.8 (since we assumed $ker(p) \neq \varnothing$). □

Dense subsets & continuous extensions 11

11.1 Definition and basic properties

REMARK 11.1.1. Suppose $(\mathsf{P}, \mathsf{Open})$ is a semitopology and suppose $\varnothing \neq D \subseteq P \in \mathsf{Open}$ (D need not be open). The following four standard definitions of what it means for D to be *dense* in P are equivalent in topology:

1. Every nonempty open subset of P intersects D.
2. The interior of $P \setminus D$ is empty.
3. Every open subset that intersects P, intersects D.
4. $|D| = |P|$.

We shall see that in semitopologies, these definitions split into two groups.

DEFINITION 11.1.2. Suppose $(\mathsf{P}, \mathsf{Open})$ is a semitopology and suppose $\varnothing \neq D \subseteq P \in \mathsf{Open}$ (D need not be open). Then:

1. Call D **weakly dense in** P when

   $$\forall O \in \mathsf{Open}. \varnothing \neq O \subseteq P \Longrightarrow D \between O.$$

   In words:

   D is weakly dense in P when every nonempty open subset of P intersects D.

2. Call D **strongly dense in** P when

   $$\forall O \in \mathsf{Open}. P \between O \Longrightarrow D \between O.$$

   In words:

   D is strongly dense in P when every open set that intersects P, intersects D.[1]

[1] We do not need to explicitly state that O is nonempty because if O is empty then $O \between P$ is false.

3. If D is strongly dense in P and $interior(D) \neq \varnothing$ then, following Definition 5.4.1(1), we may call D a **strongly dense neighbourhood in** P.

In a topology, the two notions of being *dense* described in Definition 11.1.2 above are equivalent. A semitopology permits richer structure, because we do not insist that intersections of open sets be open, and thus it discriminates more finely between the definitions:

LEMMA 11.1.3. *Suppose* $(\mathsf{P}, \mathsf{Open})$ *is a semitopology and* $\varnothing \neq D \subseteq P \in \mathsf{Open}$. *Then:*

1. *If D is strongly dense in P then D is weakly dense in P.*
2. *In a topology, the reverse implication holds; but*
3. *in a semitopology the reverse implication need not hold: it may be that D is weakly dense but not strongly dense in P.*

Proof. We consider each part in turn:

1. If a nonempty open set is a subset of P then it intersects with P. It follows that if D intersects every nonempty open set that intersects P, then it certainly intersects every nonempty open set that is a subset of P.

2. Suppose $(\mathsf{P}, \mathsf{Open})$ is a topology and suppose D is weakly dense in P and $O \between P$. Then $\varnothing \neq O \cap P \between P$, and because (this being a topology) $O \cap P \in \mathsf{Open}$ we have that $O \cap P \between D$ and so $O \between D$ as required.

3. It suffices to provide a counterexample. Consider the top-right semitopology in Figure 3.1 and take $D = \{0\}$ and $P = \{0, 1\}$. Then D is weakly dense in P (because D intersects $\{0\}$ and $\{0, 1\}$) but D is not strongly dense in P (because D does not intersect $\{1, 2\}$). □

We can rearrange the definitions to obtain more abstract characterisations of weakly and strongly dense:

PROPOSITION 11.1.4. *Suppose* $(\mathsf{P}, \mathsf{Open})$ *is a semitopology and suppose* $\varnothing \neq D \subseteq P \in \mathsf{Open}$ *(D need not be open). Then:*

1. *D is weakly dense in P if and only if $interior(P \setminus D) = \varnothing$.*
2. *D is strongly dense in P if and only if $|D| = |P|$.*

Proof. For each part we prove two implications:

1. $interior(P \setminus D) = \varnothing$ means precisely that there is no nonempty open subset of $P \setminus D$, i.e. that every nonempty subset of P intersects D. But this is just the definition of D being weakly dense in P from Definition 11.1.2(1).

2. Since $D \subseteq P$, also $|D| \subseteq |P|$.

   To prove $|P| \subseteq |D|$ it suffices to prove $\mathsf{P} \setminus |D| \subseteq \mathsf{P} \setminus |P|$. By Corollary 5.1.10 $\mathsf{P} \setminus |D|$ is the union of the open sets that do not intersect D, and $\mathsf{P} \setminus |P|$ is the union of the open sets that do not intersect P. So $\mathsf{P} \setminus |D| \subseteq \mathsf{P} \setminus |P|$ when for every open set $O \in \mathsf{Open}$, if O does not intersect D then O does not intersect P. This is just the contrapositive of the property of D being strongly dense in P from Definition 11.1.2(2). □

COROLLARY 11.1.5. *Suppose* $(\mathsf{P}, \mathsf{Open})$ *is a strongly chain-complete semitopology and* $\varnothing \neq D \subseteq P \in \mathsf{Open}$ *(D need not be open). Then the following are equivalent:*

1. *D is weakly dense in P.*
2. *D intersects every atom $\varnothing \lessdot A \subseteq P$ in P.*

In symbols using Definitions 9.5.7 and 8.4.7 we can write:

$$D \text{ weakly dense in } P \quad \Longleftrightarrow \quad D \between \mathit{Atoms}(P).$$

Proof. We prove two implications:

- Suppose D is weakly dense in P. By Definition 11.1.2(1) this means that D intersects every open $O \subseteq P$. In particular, D intersects every atom $\varnothing \lessdot A \subseteq P$.

- Conversely, suppose D intersects every atom $\varnothing \lessdot A \subseteq P$ and suppose $O \subseteq P$ is open. By Corollary 9.5.9 (since P is strongly chain-complete) there exists an atom $\varnothing \lessdot A \subseteq O$ and by assumption $D \between A$, thus also $D \between O$ as required. □

11.2 Dense subsets of topen sets

LEMMA 11.2.1. *Suppose* $(\mathsf{P}, \mathsf{Open})$ *is a semitopology and suppose* $T \in \mathsf{Topen}$ *and* $O \in \mathsf{Open}_{\neq\varnothing}$ *and* $O \subseteq T$*. Then O is strongly dense in T.*

In words: any nonempty open subset of a topen set is strongly dense.

Proof. Suppose $T \between O' \in \mathsf{Open}$. Thus $O \between T \between O'$ and by transitivity of T (since T is topen; see Definition 3.2.1) we have $O \between O'$ as required. □

COROLLARY 11.2.2. *Suppose* $(\mathsf{P}, \mathsf{Open})$ *is a semitopology and suppose* $\varnothing \neq D \subseteq T \in \mathsf{Topen}$. *Then* precisely one *of the following holds:*

- *D is weakly dense in T.*
- *$T \setminus D$ is a strongly dense neighbourhood in T (Definition 11.1.2(3)).*

As a corollary, precisely one *of the following holds:*

- *D is a strongly dense neighbourhood in T.*
- *$T \setminus D$ is weakly dense in T.*

Proof. For the first part we reason as follows:

- If D is weakly dense in T then by Proposition 11.1.4(1) $\mathit{interior}(T \setminus D) = \varnothing$, so following Definition 11.1.2(3) $T \setminus D$ is not a strongly dense neighbourhood.

- If D is not weakly dense in T then by Proposition 11.1.4(1) $\mathit{interior}(T \setminus D) \neq \varnothing$. By Lemma 11.2.1 (since T is topen) $\mathit{interior}(T \setminus D)$ is strongly dense in T, thus so is $T \setminus D$. It follows from Definition 11.1.2(3) that $T \setminus D$ is a strongly dense neighbourhood in T, as required.

For the corollary, write $D = T \setminus D'$, so that $D' = T \setminus D$. We use the first part of this result, for D'. □

COROLLARY 11.2.3. *Suppose* $(\mathsf{P}, \mathsf{Open})$ *is a strongly chain-complete semitopology and suppose* $\varnothing \neq D \subseteq T \in \mathsf{Topen}$. *Then the following are equivalent:*

- D *is a strongly dense neighbourhood in* T.
- $interior(D) \neq \varnothing$.
- D *contains an atom, or in symbols:* $\varnothing \lessdot A \subseteq D$.

Proof. We prove a cycle of implications:

- If D is a strongly dense neighbourhood in T then $interior(D) \neq \varnothing$ direct from Definition 11.1.2(3).
- If $interior(D) \neq \varnothing$ then there exists an atom $\varnothing \lessdot A \subseteq interior(D) \subseteq D$ by Corollary 9.5.9 (since T is strongly chain-complete).
- If $\varnothing \lessdot A \subseteq D$ then using Lemma 11.2.1 (since T is topen) D is strongly dense in T. ☐

11.3 Explaining kernels

NOTATION 11.3.1. Suppose $(\mathsf{P}, \mathsf{Open})$ is a semitopology and $\varnothing \neq D \subseteq P \subseteq \mathsf{P}$. Then:

1. Call D **minimally weakly dense in** P when:
   - D is weakly dense in P, and
   - if $\varnothing \neq D' \subseteq D$ and D' is weakly dense in P, then $D' = D$.
2. Call D a **minimally strongly dense open subset of** P when:
   - $D \in \mathsf{Open}$,
   - D is a strongly dense subset of P, and
   - if $\varnothing \neq D' \subseteq D$ and D' is a strongly dense open subset of P, then $D' = D$.

Recall from Definition 9.5.7(2) that $Atoms(P) = \{A \in \mathsf{Open} \mid \varnothing \lessdot A \subseteq P\}$.

PROPOSITION 11.3.2. *Suppose* $(\mathsf{P}, \mathsf{Open})$ *is a strongly chain-complete semitopology and suppose* $P \in \mathsf{Open}$. *Then:*

1. $\bigcup Atoms(P)$ *is equal to the sets union of the minimal weakly dense subsets of* P.[2]
2. *If furthermore* P *is transitive (so that* $P \in \mathsf{Topen}$*) then* $\bigcup Atoms(P)$ *is equal to the sets union of the minimal strongly dense neighbourhoods in* P.
3. *If* $p \in \mathsf{P}$ *is regular then* $ker(p)$ *is equal to the union of the minimal weakly dense subsets of* $K(p)$ *and also to the union of the minimal strongly dense neighbourhoods in* $K(p)$.

Proof. We consider each part in turn:

[2]This sentence is potentially confusing, because $\bigcup Atoms(P)$ is itself a sets union of the atoms in P. So what is being asserted is that the sets union of the atoms in P, is equal to the sets union of the minimal weakly dense subsets of P. However, these minimal weakly dense subsets are not necessarily atoms, and the atoms are not necessarily minimal weakly dense subsets.

1. For brevity, write $\mathcal{A} = Atoms(P)$. By Corollary 11.1.5 (since **P** is strongly chain-complete), if D is weakly dense in P then so is $D \cap \bigcup \mathcal{A}$. Thus, the *minimal* weakly dense subsets of P are all contained in $\bigcup \mathcal{A}$.

   It remains to show that every $p \in S$ is contained in *some* minimal weakly dense subset of P. Fix one such p. We need some notation: if $X \subseteq P$ then write $\langle X \rangle_P$ for the union of the set of atoms in P that intersect X. We now argue by transfinite induction (a clear and accessible presentation is in [ELW], or see [Joh87, Section 6]) to generate a minimal weakly dense $D \subseteq P$ that contains p:

   - Set $p_0 = p$ and set A_0 to be some atom in P that contains p; one such exists, since we chose $p \in \bigcup \mathcal{A}$.
   - Suppose $p_{\alpha'}$ and $A_{\alpha'}$ are defined for all $\alpha' < \alpha$. If $\bigcup \mathcal{A} \subseteq \langle \{p_{\alpha'} \mid \alpha' < \alpha\} \rangle_P$ — so that $\bigcup \mathcal{A} = \langle \{p_{\alpha'} \mid \alpha' < \alpha\} \rangle_P$ — then stop. Otherwise, pick an atom A_α of P that is not a subset, and choose some $p_\alpha \in A_\alpha \setminus \langle \{p_{\alpha'} \mid \alpha' < \alpha\} \rangle_P$.
     Note that $A_\alpha \subseteq \langle \{p_{\alpha'} \mid \alpha' \leq \alpha\} \rangle_P$.
   - Continue transfinitely until we stop, and set $D = \{p_{\alpha'} \mid \alpha' < \alpha\}$. By construction, $\langle D \rangle_P = \bigcup \mathcal{A}$.

   Then D is weakly dense in P by Corollary 11.1.5, and it is minimal because if we remove any $p_{\alpha'}$ from D to obtain a smaller $D' = D \setminus \{p_{\alpha'}\}$ then by construction $A_{\alpha'} \not\subseteq \langle D' \rangle_P$.

2. It follows from Corollary 11.2.3 that D is a minimal strongly dense neighbourhood in P precisely when D is equal to an atom in P. The result follows.

3. $K(p)$ is transitive by Definition 4.1.4(3), and $ker(p) = \bigcup Atoms(K(p))$ by Definition 10.1.2(1). We use parts 1 and 2 of this result. □

REMARK 11.3.3. Proposition 11.3.2(3) gives some independent explanation for why $ker(p)$ — the atoms in the community of p, as studied in Section 10 — is interesting. $ker(p)$ identifies where the minimal weakly dense and strongly dense subsets of $K(p)$ are located.

11.4 Unifying is-transitive and is-strongly-dense-in

It turns out that transitivity and denseness are closely related: in this Subsection we explore their relationship.

REMARK 11.4.1. Consider the following three notions:

1. *D is strongly dense in P* from Definition 11.1.2(2).
2. *P is transitive* from Definition 3.2.1(1).
3. *P is strongly transitive* from Definition 3.7.5(1).

Notice that while the definitions are different, they share a 'family resemblance'. Can we identify a common ancestor for them; some definition that naturally subsumes them into a most general principle?

Yes: it is easy to see that items 1 and 3 above are very closely related — see Lemma 11.4.2 — and then we will prove that all three definitions listed above are special instances of a general definition — see Definition 11.4.3 and Proposition 11.4.6.

LEMMA 11.4.2. *Suppose* $(\mathsf{P}, \mathsf{Open})$ *is a semitopology and* $P \subseteq \mathsf{P}$. *Then the following are equivalent:*[3]

- *P is strongly transitive.*
- *$O \cap P$ is strongly dense in P, for every $O \in \mathsf{Open}$ such that $O \between P$ (meaning that $O \cap P \neq \varnothing$).*

In words we can say:

> *P is strongly transitive when every nontrivial open intersection with P is strongly dense.*

Proof. Unpacking Definition 3.7.5(1), P is strongly transitive when $O \between P \between O'$ implies $O \cap P \between O' \cap P$. Unpacking Definition 11.1.2(2), $O \cap P$ is strongly dense in P when $P \between O'$ implies $O \cap P \between O'$ — and this is clearly equivalent to $O \cap P \between O' \cap P$. The result now follows by routine reasoning. □

We can generalise the notion of strongly dense from Definition 11.1.2(2) from $D \subseteq P$ to *any* D.

DEFINITION 11.4.3. Suppose $(\mathsf{P}, \mathsf{Open})$ is a semitopology and $D, P \subseteq \mathsf{P}$.
Call D **strongly dense for** P when

$$\forall O \in \mathsf{Open}. P \between O \Longrightarrow D \between O.$$

We state the obvious:

LEMMA 11.4.4. $\varnothing \neq D \subseteq P$ *is strongly dense* in P *in the sense of Definition 11.1.2(2) if and only if it is strongly dense* for P *in the sense of Definition 11.4.3.*

Proof. The definitions are identical where they overlap. The only difference is that Definition 11.1.2(2) assumes a nonempty subset of P, whereas Definition 11.4.3 assumes a nonempty set that intersects (but is not necessarily a subset of) P. □

LEMMA 11.4.5. *Suppose* $(\mathsf{P}, \mathsf{Open})$ *is a semitopology and* $D, P \subseteq \mathsf{P}$. *Then the following are equivalent:*

- *D is strongly dense for P.*
- $|P| \subseteq |D|$.[4]

Proof. Suppose D is strongly dense for P. Then $O \in \mathsf{Open}$ does not intersect with P if and only if O does not intersect with D, and it follows (just as in the proof of Proposition 11.1.4(2)) that $\mathsf{P} \setminus |D| \subseteq \mathsf{P} \setminus |P|$, and so that $|P| \subseteq |D|$.

Conversely, if $|P| \subseteq |D|$ then $\mathsf{P} \setminus |D| \subseteq \mathsf{P} \setminus |P|$ and it follows that if $O \in \mathsf{Open}$ does not intersect with P then O does not intersect with D, and thus that D is strongly dense for P. □

PROPOSITION 11.4.6. *Suppose* $(\mathsf{P}, \mathsf{Open})$ *is a semitopology and* $T \subseteq \mathsf{P}$. *Then the following are equivalent:*

[3] Cf. also Lemma 3.7.10.

[4] Compare with Proposition 5.3.2.

- *T is transitive.*
- *O is strongly dense for T, for every $T \between O \in \mathsf{Open}$.*[5]

Proof. We unpack Definition 11.4.3 and see that a condition that T is transitive with respect to every $T \between O \in \mathsf{Open}$ is precisely what Definition 3.2.1(1) asserts, namely: for every $O \in \mathsf{Open}$ such that $O \between T$,

$$\forall O' \in \mathsf{Open}. T \between O' \Longrightarrow O \between O'. \qquad \square$$

REMARK 11.4.7. In topology it makes less sense to talk about D being dense in P for $D \not\subseteq P$, since we can just consider $D \cap P$ — and if D and P are open then so is $D \cap P$. In semitopology the following happens:

- The notion of *dense in* splits into two distinct concepts (*weakly dense in* and *strongly dense in*), as we saw in Definition 11.1.2 and the subsequent discussion.
- The notion of *strongly dense in* generalises to a notion that we call *strongly dense for*, which has the same definition but just weakens a precondition that $D \subseteq P$.

Given the above, we then see from Lemma 11.4.2 and Proposition 11.4.6 that the notions of *transitive* and *strongly transitive* from Definitions 3.2.1(1) and 3.7.5(1) lend themselves to being naturally expressed in terms of strongly-dense-for.

11.5 Towards a continuous extension result

REMARK 11.5.1. Topology has a family of results on *continuous extensions* of functions: a nice historical survey is in [Gut22]. Here is an example, adapted from [Erd18, Theorem 24.1.15]:

> Suppose $f : B \to \mathbb{R}$ is uniformly continuous and suppose B is a dense subset of A. Then f has a unique extension to a continuous function $g : A \to \mathbb{R}$.

This is true in the world of topologies: but what might correspond to this in the semitopological world?

A direct translation to semitopologies seems unlikely,[6] because we have seen from Definition 11.1.2 and Lemma 11.1.3 and the subsequent discussion and results how the notion of 'is dense in' behaves differently for semitopologies in general, so that the very premise of the topological result above is now up for interpretation.[7]

But, the impossibility of a direct translation only opens up an even more interesting question: whether we can find definitions and well-behavedness conditions on semitopological spaces that reflect the spirit of the corresponding topological results, but without assuming that intersections of open sets are open.

We shall see that this is possible and we propose a suitable result below in Definition 11.5.2. However, before we come to that, we will sketch a design space of *failing* definitions and counterexamples — and so put our working definition in its proper design context.

---

[5]Compare with Lemma 11.4.2.

[6]Except trivially that we can restrict to those semitopologies that are also topologies (i.e. for which intersections of open sets are open).

[7]There are other differences. For instance we care a lot about value assignments — maps to discrete semitopologies — rather than maps to $\mathbb{R}$. Of course we could try to generalise from value assignments to more general examples, but as we shall see, even this 'simple' case of value assignments is more than rich enough to raise some canonical questions.

We map to semitopologies of values, so (the spirit of) uniform continuity is automatic, and we concentrate (to begin with) on being strongly dense in rather than weakly dense in, since by Lemma 11.1.3(1) the former implies the latter:

1. **Candidate definition 1.**

   Suppose $(\mathsf{P}, \mathsf{Open})$ is a semitopology and suppose $f : \mathsf{P} \to \mathsf{Val}$ is a value assignment that is continuous on $D \subseteq \mathsf{P}$, and suppose D is a strongly dense subset of P. Then f has a unique extension to a continuous function $g : \mathsf{P} \to \mathsf{Val}$.

   This does not work:

   - Take $(\mathsf{P}, \mathsf{Open})$ to be the top-left example in Figure 3.1 and
   - $\mathsf{Val} = \{0, 1\}$ with the discrete semitopology.
   - Define $f : \mathsf{P} \to \mathsf{Val}$ such that $f(0) = 0$ and $f(1) = 1$ and $f(2) = 1$ and
   - set $D = \{0, 2\}$ and $P = \mathsf{P}$.

   Note that D is a strongly dense open subset of P, and f is continuous on D.

   However, f cannot be continuously extended to a g that is continuous at 1. We note that 1 is conflicted and intertwined with two distinct topens: $\{0\}$ and $\{2\}$. Looking at this example we see that Candidate definition 1 is unreasonable: *of course* we cannot extend f continuously to 1, because 1 is intertwined with two distinct topen sets on which f takes distinct values. The natural solution is just to exclude conflicted points since they may be, as the terminology suggests, conflicted:

2. **Candidate definition 2.**

   Suppose $(\mathsf{P}, \mathsf{Open})$ is a semitopology and suppose $f : \mathsf{P} \to \mathsf{Val}$ is a value assignment that is continuous on $D \subseteq \mathsf{P}$, and suppose D is a strongly dense subset of P. Then f has a unique extension to a function $g : \mathsf{P} \to \mathsf{Val}$ that is continuous at all *unconflicted* points.

   This does not work:

   - Take $(\mathsf{P}, \mathsf{Open})$ to be the semitopology in Figure 7.1 and
   - $\mathsf{Val} = \{0, 1, 2\}$ with the discrete semitopology.
   - Define $f : \mathsf{P} \to \mathsf{Val}$ such that $f(0) = 0$ and $f(1) = 1$ and $f(2) = 2$ and $f(*) = 0$ and
   - set $D = \{0, 1, 2\}$ and $P = \mathsf{P}$.

   Note that D is a strongly dense open subset of P and f is continuous on D.

   Note that $*$ is unconflicted (because $*_{\between} = \{*, 1\}$). However, f cannot be continuously extended to a g that is continuous at $*$.

3. Candidate definition 2 is even more telling than it might appear. Note that $*$ is unconflicted and quasiregular (Definition 4.1.4(5)), because $K(*) = \{1\} \neq \varnothing$.

   Thus we could not even rescue Candidate definition 2 above by insisting that points be not only unconflicted but also quasiregular (i.e. unconflicted and having a nonempty community). The next natural step up from this is to be unconflicted and weakly regular, which by Corollary 5.4.13 leads us to regular points.

We can now state a definition and result that work:

DEFINITION 11.5.2. Suppose $f, g : (\mathsf{P}, \mathsf{Open}) \to \mathsf{Val}$ are value assignments (Definition 2.1.3(2)) and suppose $P \subseteq \mathsf{P}$.

1. Say that g **continuously extends** f to regular points in P when:
   - If f is continuous at $p \in P$ then $f(p) = g(p)$.
   - g is continuous on every regular $p \in P$ (Definition 4.1.4(3)).
2. Say that g is a **unique continuous extension** of f to regular points in P when for any other continuous extension g' of f to P, we have $g(p) = g'(p)$ for every regular $p \in P$.

REMARK 11.5.3 (Justification for regular points). Note that 'continuously extends' and 'uniquely' in Definition 11.5.2 both apply to to *regular* points in P only. By the examples in Remark 11.5.1 it would not be reasonable to expect unique continuous extensions on non-regular points. This gives a retrospective justification for the theories of topens and regular points that we develop (see Definitions 3.2.1 and 4.1.4): regularity arises as a natural condition for a semitopological continuous extension result.[8]

PROPOSITION 11.5.4. *Suppose* $(\mathsf{P}, \mathsf{Open})$ *is a semitopology and* $f : (\mathsf{P}, \mathsf{Open}) \to \mathsf{Val}$ *is a value assignment and suppose* $D, P \subseteq \mathsf{P}$. *Then:*

1. *If f is continuous on D then f can be continuously extended to all regular points in* P.
2. *If D is strongly dense for P (Definition 11.4.3) then this extension is unique on P in the sense of Definition 11.5.2(2).*

Proof. Choose some fixed but arbitrary *default value* $v \in \mathsf{Val}$ and for this choice of v define g by cases as follows:

- *Suppose f is continuous at p.*

  We set $g(p) = f(p)$.

- *Suppose f is not continuous at p and $K(p) \between D$.*

  Choose some $d \in K(p) \cap D$ and set $g(p) = f(d)$.

  If p is intertwined with two points d and d' then (because p is regular and so unconflicted by Theorem 6.2.2) $d \between d'$ and their open neighbourhoods of continuity intersect, so that $f(d) = f(d')$.

- *Suppose f is not continuous at p and $K(p) \not\between D$.*

  If D is strongly dense for P then this case cannot happen because $P \between K(p)$ so by the strong dense property $D \between K(p)$ (see Definition 11.1.2(2)).

  Otherwise, we set $g(p) = v$ (so $g(p)$ is the fixed but arbitrary default value).

We now show that if $p \in \mathsf{P}$ is regular, then g is continuous at p. The proof is again by cases:

[8] This is not an exclusive claim. Other reasonable conditions or generalisations might also exist, for example the *hypertwined* points discussed starting from Definition 22.6.6(3). But being hypertwined is a more complex notion, and note by Remark 22.6.20 that intertwined and hypertwined amount to the same thing on regular points. So it seems likely that Definition 11.5.2 is a canonical point in the design space. Investigating this further is future work.

- If f is continuous at p then $g(p) = f(p)$ and so g is continuous at p.

- If f is not continuous at p and $d \in K(p) \cap D$ then $g(p) = f(d)$. Thus (since we assumed that p is regular) $p \in K(p) \subseteq g^{-1}(g(p))$, so that g is continuous at p.

- If f is not continuous at p and $K(p) \cap D = \varnothing$ then $g(p) = v$. Thus (since we assumed that p is regular) $p \in K(p) \subseteq g^{-1}(v)$, so that g is continuous at p.

If D is strongly dense for P then uniqueness follows by routine reasoning from the above, using Theorem 3.2.2. □

In view of Lemma 11.4.5 we can more succinctly rephrase Proposition 11.5.4 as follows:

Corollary 11.5.5. *Suppose* $(\mathsf{P}, \mathsf{Open})$ *is a semitopology and* $f : (\mathsf{P}, \mathsf{Open}) \to \mathsf{Val}$ *is a value assignment. Then if* f *is continuous on* $D \subseteq \mathsf{P}$, *then* f *can be continuously extended to all regular points in* $|D|$.

Proof. Direct from Proposition 11.5.4 taking $P = |D|$ and using Lemma 11.4.5. □

Remark 11.5.6.

1. There are a few subtleties to Corollary 11.5.5. The result actually tells us that there exists an open set $O \in \mathsf{Open}$ such that $D \subseteq |D| \subseteq O$, and f continuously extends to some $g : \mathsf{P} \to \mathsf{Val}$ that is continuous at O. This is because if g is continuous at $p \in |D|$, then it is by definition continuous on some open neighbourhood of p.

2. Similarly, the condition that f be continuous on D is equivalent to insisting that f be continuous on an *open* D.

3. The condition of D being strongly dense in P is required for uniqueness in Proposition 11.5.4(2). Being weakly dense is not enough. For, consider the semitopology illustrated in Figure 11.1, such that:

   - $\mathsf{P} = \{0, 1, 2, 3\}$.
   - Open is generated by $D = \{0\}$, $\{0, 1\}$, $P = \{0, 1, 2\}$, and $\{2, 3\}$.

   Then we have $D \subseteq P \subseteq \mathsf{P}$, and we even have that $D, P \in \mathsf{Open}$ and every point in the space is regular, making this is a particularly well-behaved example. This semitopology is *not* a topology, because $P \cap \{2, 3\} \notin \mathsf{Open}$; we will exploit this fact in a moment. The reader can check that

   - D is weakly dense in P (because D intersects every open subset of P), but
   - D is not strongly dense in P (because D does not intersect $\{2, 3\}$), and

   the value assignment $f : \mathsf{P} \to \mathbb{B}$ mapping 0 to $\bot$ and every other point to $\top$ has two continuous extensions to all of P: g mapping all points to $\bot$, and g' mapping 0 and 1 to $\bot$ and 2 and 3 to $\top$.

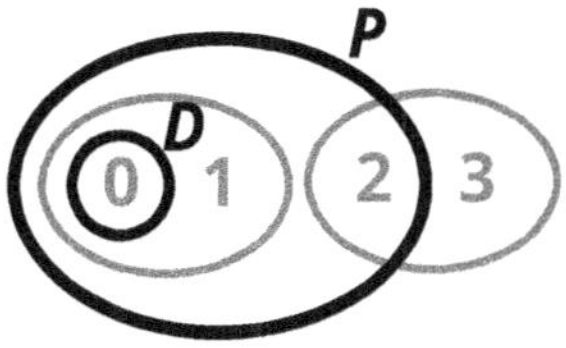

Figure 11.1: A weakly dense subset is not enough for uniqueness (Remark 11.5.6(3))

11.6 Kernels determine values of continuous extensions

REMARK 11.6.1. In Subsection 11.5 we considered continuous extensions in a semitopological context. We concluded with Corollary 11.5.5, which showed how to extend a value assignment $f : \mathsf{P} \to \mathsf{Val}$ that is continuous on some D, to a g that is continuous on D and on the regular points in $|D|$. We also discussed why this result is designed as it is and why it seems likely to be optimal within a certain design space as outlined in Remark 11.5.1.

However, our study of semitopologies is motivated by systems that may be distributed over several participants. This means that we also care about intermediate continuous extensions of f; i.e. about g that continuously extend f but not necessarily on all of $|D|$.

The mathematics in this subsection is in some sense a 'pointwise' dual to the 'setwise' mathematics in Subsection 11.5. Perhaps surprisingly, we shall see that when developed pointwise, the details are different: contrast Theorem 11.6.8 and Corollary 11.6.11 with Proposition 11.5.4 and Corollary 11.5.5; they are similar, but they are not the same.

Recall that:

- A point p is *regular* when its community (which is the interior of its intertwined points) is a topen neighbourhood of p; see Definition 4.1.4(3).
- $\mathit{ker}(p)$ is the union of the kernel atoms of p (minimal nonempty open sets in the community of p); see Definition 10.1.2(1).
- A value assignment $f : \mathsf{P} \to \mathsf{Val}$ is a mapping from P to some set of values Val having the discrete semitopology; see Definition 2.1.3.

We now consider how the value of f on kernel atoms influences the value of f at regular points.

DEFINITION 11.6.2. Suppose $(\mathsf{P}, \mathsf{Open})$ is a semitopology and $f, g : \mathsf{P} \to \mathsf{Val}$ are value assignments (Definition 2.1.3).[9]

1. Call f **confident at** $p \in \mathsf{P}$ when f is continuous on some atom $\varnothing \lessdot A \subseteq \mathit{ker}(p)$.
2. Call f **unanimous at** $p \in \mathsf{P}$ when f is continuous on all of $K(p)$.
3. We generalise Definition 11.5.2 and write $f \leq g$, and call g a **(partial) continuous extension of** f, when for every $p \in \mathsf{P}$, if f is continuous at p then g is continuous at p and $f(p) = g(p)$.[10]

[9]This definition makes sense for f mapping P to any semitopology Q, but (for now) we will only care about the case when Q is a discrete semitopology so that f is a value assignment.

[10]Definition 11.5.2 is interested in a g that continuously extends f all at once from D to $|D|$, which is fine mathematically but less helpful computationally. The definition here refines this concept and is interested in the space of all possible g such that $f \leq g$, which more accurately reflects how g might be computed, in stages, on a network.

It is routine to check that $\leq$ is a preorder (reflexive, transitive).

REMARK 11.6.3. Intuitively, f is 'confident' at p when the value we obtain if we continuously extend f to p, is already determined. It may be that a result has been determined by some part of the system, but not yet fully propagated to the whole system.[11]

We make this formal in Theorem 11.6.8(3). We start with an easy lemma:

LEMMA 11.6.4. *Suppose* $(\mathsf{P}, \mathsf{Open})$ *is a strongly chain-complete semitopology and* $f, g : \mathsf{P} \to \mathsf{Val}$ *are value assignments. Then:*

1. *If* $f \leq g$ *and* f *is confident/unanimous at* p *then* g *is also confident/unanimous at* p.
2. *If* f *is unanimous at* $p \in \mathsf{P}$ *then it is confident at* p.

Proof. By simple arguments from the definitions:

1. Just unpacking definitions: if f is continuous on some $\varnothing \lessdot A \subseteq ker(p)$ then so is g; and if f is continuous on $K(p)$ then so is g.

2. Suppose f is unanimous at p, meaning that f is continuous on $K(p)$. Then f is also continuous on some kernel atom in $K(p)$ — some such kernel atom exists by Corollary 9.5.9, since P is strongly chain-complete. Thus f is confident at p, as required. □

For brevity in what follows, a notation will be useful:

NOTATION 11.6.5. Suppose P is a set and f is some function on P and $\varnothing \neq A \subseteq \mathsf{P}$. Suppose further that it is known that f is constant on A. In symbols:

$$\exists c. \forall p \in A. f(p) = c.$$

Then we may write $f(A)$ for the unique constant value that $f(p)$ takes as p ranges over A.

REMARK 11.6.6. Suppose $(\mathsf{P}, \mathsf{Open})$ is a semitopology and $f : \mathsf{P} \to \mathsf{Val}$ is a value assignment. Then:

1. Suppose f is confident at $p \in \mathsf{P}$.

   By Definition 11.6.2 f is continuous on some kernel atom $\varnothing \lessdot A \subseteq ker(p)$. By Lemma 10.2.5 A is transitive, so by Theorem 3.2.2 (since f is continuous on A) f is constant on A, and thus it makes sense to use Notation 11.6.5 and write $f(A)$ to denote the (unique) value of f on A.

2. Likewise if f is unanimous at p then we can sensibly write $f(K(p))$.

3. Just for this paragraph call f *doubly confident* at p when f is continuous on two distinct kernel atoms $\varnothing \lessdot A \neq A' \subseteq ker(p)$ of p. Suppose f is doubly confident at p and suppose p is regular; so by the previous paragraph $f(A)$ and $f(A')$ are both well-defined. Now $A \between A'$ by Lemma 10.1.4, so $f(A) = f(A')$.

[11] For instance: f may know the results of an election, but not yet have told point p; whereas some $g \geq f$ may represent a state in which this result has been correctly propagated to p. Similarly, technology allows us to determine the weather tomorrow based on weather data that was collected this morning, but that is not the same thing as knowing what the weather will be: a supercomputer needs to run calculations, and the data needs to be broadcast, and put on a webpage and sent down a cable and rendered to a computer screen, and so on.

Thus being doubly confident at p is the same as just being confident, provided that p is regular so that all of its kernel atoms intersect.

Remark 11.6.6 brings us to a notation:

DEFINITION 11.6.7. Suppose $(\mathsf{P}, \mathsf{Open})$ is a semitopology and $f : \mathsf{P} \to \mathsf{Val}$ is a value assignment and $p \in \mathsf{P}$ is a regular point. Then define $\lim_p f$ the **limit of f at p** by

$$\lim_p f = f(A)$$

where $\varnothing \lessdot A \subseteq ker(p)$ is some/any (by Remark 11.6.6(3) writing 'some/any' makes sense) kernel atom of p on which f is continuous. The justification for calling this value the *limit* of f at p is below in Theorem 11.6.8, culminating with part 3 of that result.

Recall that Theorem 3.2.2 asserted that continuous value assignments are constant on transitive sets. We can now prove a more general result along the same lines:

THEOREM 11.6.8. *Suppose that:*

- $(\mathsf{P}, \mathsf{Open})$ *is a semitopology.*
- $p \in \mathsf{P}$ *is regular.*
- $f, g : \mathsf{P} \to \mathsf{Val}$ *are value assignments to some set of values* Val.

Then:

1. *If f is confident at p (Definition 11.6.2(1)) then*

$$f \leq g \quad \textit{implies} \quad \lim_p f = \lim_p g.$$

2. *If f is confident at p and also f is continuous at p (Definition 2.2.1) then*

$$f(p) = \lim_p f.$$

3. *Combining parts 1 and 2 of this result, if f is confident at p and g is continuous at p then*

$$f \leq g \quad \textit{implies} \quad g(p) = \lim_p f.$$

   In words: the limit value of an f confident at p, is the value of any sufficiently continuous extension of f — where 'sufficiently continuous' means 'continuous at p'.

4. *As a corollary using Lemma 11.6.4, if f is unanimous at p (Definition 11.6.2(2)) and g is continuous at p, then*

$$f \leq g \quad \textit{implies} \quad g(p) = \lim_p f = f(K(p)).$$

Proof. We reason as follows:

1. Since f is confident at p, there exists some kernel atom $\varnothing \lessdot A \subseteq ker(p)$ on which f is continuous (by Remark 11.6.6(3) it does not matter which one). Since $f \leq g$, g is also continuous at A. It follows from Definition 11.6.7 that $\lim_p f = \lim_p g$.

2. Since f is confident at p, f is continuous on some kernel atom $\varnothing \lessdot A \subseteq ker(p)$. Since f is continuous at p, f is continuous on some open neighbourhood $p \in O \in \mathsf{Open}$. By Lemma 10.3.1 $O \between A$, and using Corollary 3.2.3 we have that $f(p) = f(O) = f(A) = \lim_p f$ as required.

3. Direct from parts 1 and 2 of this result, using Lemma 11.6.4(1) to note that g is confident at p because f is and $f \leq g$.

4. Suppose f is unanimous at p. Then by Lemma 11.6.4 f is confident at p, and we use part 3 of this result. □

NOTATION 11.6.9. Suppose f is a function on sets, and X is a set. Recall that we write $f|_X$ for the function obtained by **restricting** f to X (so that $domain(f|_X) = domain(f) \cap X$).

REMARK 11.6.10. We can use Theorem 11.6.8 to obtain a result that seems to us similar in spirit to Arrow's theorem [Fey14] from social choice theory, in the sense that $ker(p)$ is identified as a 'dictator set' for $K(p)$ (the technical details seem to be different):

COROLLARY 11.6.11. *Suppose that:*

- $(\mathsf{P}, \mathsf{Open})$ *is a semitopology.*
- $p \in \mathsf{P}$ *is regular.*
- $f, f' : \mathsf{P} \to \mathsf{Val}$ *are value assignments to some set of values* Val.
- f *and* f' *are continuous and confident at* p.

Then

$$f|_{ker(p)} = f'|_{ker(p)} \quad \textit{implies} \quad f(p) = f'(p).$$

In words we can say:

Confident continuous values at regular points are determined by their kernel.

Note that we assume that f and f' are equal on $ker(p)$, but they do not need to be continuous on all of $ker(p)$; they only need to be continuous (and confident) at p.

Proof. By confidence of f and f' there exist

- a kernel atom $\varnothing \lessdot A \subseteq ker(p)$ on which f is continuous and so (as discussed in Remark 11.6.6) on which f is constant with value $f(A) = \lim_p f$, and
- a kernel atom $\varnothing \lessdot A' \subseteq ker(p)$ on which f' is continuous and so constant with value $f'(A') = \lim_p f'$.

It may be that $A \neq A'$, but by Lemma 10.1.4 they intersect — in symbols: $A \between A'$ — so that $\lim_p f = \lim_p f'$.[12]

We use Theorem 11.6.8 and the above to reason as follows:

$$\begin{aligned} f(p) &= \lim_p f && \text{Theorem 11.6.8(2)} \\ &= \lim_p f' && \lim_p f{=}f(A),\ \lim_p f'{=}f'(A'),\ A \between A' \\ &= f'(p) && \text{Theorem 11.6.8(2)} \end{aligned}$$

□

[12] If we only had $f|_O = f'|_O$ for some open set O that intersects the kernel (so $O \between ker(p)$), then the reasoning would break down at this point. We would still know that $A \between A'$ but we would not necessarily know that $O \between (A \cap A')$ so that $f(A) = f(A')$ and $\lim_p f = \lim_p f'$. (Remember that we have not assumed continuity on all of $ker(p)$, so f and f' might not be constant on $ker(p)$.)

Part II

Semiframes: algebra and duality

Semiframes: compatible complete semilattices 12

12.1 Complete join-semilattices, and morphisms between them

REMARK 12.1.1 (Setting the scene). We have studied point-set semitopologies; now the challenge is to give an algebraic account of them.

A straightforward reading of the definition of a semitopology in Definition 1.2.2 is that a semitopology is a complete semilattice (under sets unions) in a powerset, so the algebraic version of this should be just a complete semilattice. This turns out to be wrong: what we need is a *compatible* complete semilattice, which we call a *semiframe*. Precise details are in Definition 12.3.1.

We proceed in three steps:

1. Between here and Section 15 we develop a duality between semitopologies and semiframes, which deliberately echoes the classic duality between topologies and frames.
2. Then in Section 16 we give algebraic versions of the antiseparation properties.
3. Finally, in Section 17 we (briefly) consider alternative representations, using graphs.

Taken together, this gives a fairly comprehensive algebraic treatment of the material we have seen thus far, complementing the point-set approach taken until now.

REMARK 12.1.2. Something amazing will happen below. We have the compatibility relation $*$ in Subsection 12.2. This arises naturally in two ways: is is a natural algebraic abstraction of $\between$ in semitopologies (sets intersection) which we have used to express well-behavedness properties such as intertwinedness and regularity, but it will also be key to our categorical duality result.

These motivations for $*$ are independent: the categorical duality does not require regularity, and the regularity properties do not require a duality — and yet when we study both well-behavedness and duality, the same structures emerge.

We recall some (mostly standard) definitions and facts:

DEFINITION 12.1.3.

1. A **poset** $(\mathsf{X}, \leq)$ is a set X of **elements** and a relation $\leq\ \subseteq \mathsf{X}\times\mathsf{X}$ that is transitive, reflexive, and antisymmetric.
2. A poset $(\mathsf{X}, \leq)$ is a **complete join-semilattice** when every $X \subseteq \mathsf{X}$ (X may be empty or equal to all of X) has a least upper bound — or **join** — $\bigvee X \in \mathsf{X}$.
   All the semilattices we consider will be join (rather than meet) semilattices, so we may omit the word 'join' and just call this a *complete semilattice* henceforth.
3. If $(\mathsf{X}, \leq)$ is a complete semilattice then we may write
   $$\bot_{\mathsf{X}} = \bigvee \varnothing.$$
   By the least upper bound property, $\bot_{\mathsf{X}} \leq x$ for every $x \in \mathsf{X}$.
4. If $(\mathsf{X}, \leq)$ is a complete semilattice then we may write
   $$\top_{\mathsf{X}} = \bigvee \mathsf{X}.$$
   By the least upper bound property, $x \leq \top_{\mathsf{X}}$ for every $x \in \mathsf{X}$.

DEFINITION 12.1.4. Suppose $(\mathsf{X}', \leq')$ and $(\mathsf{X}, \leq)$ are complete join-semilattices. Define a **morphism** $g : (\mathsf{X}', \leq') \to (\mathsf{X}, \leq)$ to be a function $\mathsf{X}' \to \mathsf{X}$ that commutes with joins, and sends $\top_{\mathsf{X}'}$ to $\top_{\mathsf{X}}$. That is:

1. If $X' \subseteq \mathsf{X}'$ then $g(\bigvee X') = \bigvee_{x' \in X'} g(x')$.
2. $g(\top_{\mathsf{X}'}) = \top_{\mathsf{X}}$.

REMARK 12.1.5. In Definition 12.1.3(2) we insist that $g(\top_{\mathsf{X}'}) = \top_{\mathsf{X}}$; i.e. we want our notion of morphism to preserve the top element.

This does not follow from Definition 12.1.3(1), because g need not be surjective onto X, so we need to add this as a separate condition. Contrast with $g(\bot_{\mathsf{X}}) = \bot_{\mathsf{X}'}$, which does follow from Definition 12.1.3(1), because $\bot_{\mathsf{X}}$ is the least upper bound of $\varnothing$.

We want $g(\top_{\mathsf{X}'}) = \top_{\mathsf{X}}$ because our intended model is that $(\mathsf{X}, \leq) = (\mathsf{Open}, \subseteq)$ is the semilattice of open sets of a semitopology $(\mathsf{P}, \mathsf{Open})$, and similarly for $(\mathsf{X}', \leq')$, and g is equal to f^{-1} where $f : (\mathsf{P}, \mathsf{Open}) \to (\mathsf{P}', \mathsf{Open}')$ is a continuous function.

We recall a standard result:

LEMMA 12.1.6. *Suppose $(\mathsf{X}, \leq)$ is a complete join-semilattice. Then:*

1. *If $x_1, x_2 \in \mathsf{X}$ then $x_1 \leq x_2$ if and only if $x_1 \vee x_2 = x_2$.*
2. *If $f : (\mathsf{X}, \leq) \to (\mathsf{X}', \leq')$ is a semilattice morphism (Definition 12.1.4) then f is a **monotone morphism**: if $x_1 \leq x_2$ then $f(x_1) \leq f(x_2)$, for every $x_1, x_2 \in \mathsf{X}$.*

Proof. We consider each part in turn:

1. Suppose $x_1 \leq x_2$. By the definition of a least upper bound, this means precisely that x_2 is a least upper bound for $\{x_1, x_2\}$. It follows that $x_1 \vee x_2 = x_2$. The converse implication follows just by reversing this reasoning.

2. Suppose $x_1 \leq x_2$. By part 1 of this result $x_1 \vee x_2 = x_2$, so $f(x_1 \vee x_2) = f(x_2)$. By Definition 12.1.4 $f(x_1) \vee f(x_2) = f(x_2)$. By part 1 of this result $f(x_1) \leq f(x_2)$. □

REMARK 12.1.7. As the reader may know, *frames* and *locales* are the same thing: the category of locales is just the categorical opposite of the category of frames. So every time we write 'semiframe', the reader can safely read 'semilocale'; these are two names for essentially the same structure up to reversing arrows. The literature on frames and locales is huge: the interested reader can consult two classic texts [Joh86, MM92]; more recent (and very readable) presentations include [PP12, PP21].

12.2 The compatibility relation

Definition 12.2.1 is a simple idea, but so far as we are aware it is novel:

DEFINITION 12.2.1. Suppose $(\mathsf{X}, \leq)$ is a complete semilattice. A **compatibility relation** $* \subseteq \mathsf{X} \times \mathsf{X}$ is a relation on X such that:

1. $*$ is *symmetric*, so if $x, x' \in \mathsf{X}$ then
$$x * x' \quad \text{if and only if} \quad x' * x.$$
2. $*$ is a **properly reflexive relation**,[1] by which we mean
$$\forall x \in \mathsf{X} \setminus \{\bot_{\mathsf{X}}\}. x * x.$$
Note that it will follow from the axioms of a compatibility relation that $x * x \iff x \neq \bot_{\mathsf{X}}$; see Lemma 12.3.7(2).
3. $*$ satisfies a **distributive law**, that if $x \in \mathsf{X}$ and $X' \subseteq \mathsf{X}$ then
$$x * \bigvee X' \iff \exists x' \in X'. x * x'.$$

Thus we can say:

> a compatibility relation $* \subseteq \mathsf{X} \times \mathsf{X}$ is a symmetric properly reflexive completely distributive relation on X.

When $x * x'$ holds, we may call x and x' **compatible elements**.

REMARK 12.2.2. The compatibility relation $*$ is what it is, but we take a moment to discuss some intuitions, and to put it in the context of some natural generalisations:

1. We can think of $*$ as an *abstract intersection*.

   It lets us observe whether x and x' intersect — but without having to explicitly represent this intersection as a meet $x \wedge x'$ in the semilattice itself.

   We call $*$ a *compatibility relation* following an intuition of $x, x' \in \mathsf{X}$ as observations, and $x * x'$ holds when there is some possible world at which it is possible to observe x and x' together. More on this in Example 12.3.3.

[1] 'Properly reflexive' is a loose riff on terminologies like 'proper subset of' or 'proper ideal of a ring'. We might also call this 'non-$\bot$ reflexive', which is descriptive, but perhaps a bit of a mouthful.

2. We can think of $*$ as a *generalised intersection*; so our notion of semiframe in Definition 12.3.1 is an instance of a frame with a *generalised* meet.

   We will concentrate on the case where $x * x'$ measures whether x and x' intersect, but there are other possibilities. Here are some natural ways to proceed:

   a) $(\mathsf{X}, \leq)$ is a complete join-semilattice and $* : (\mathsf{X} \times \mathsf{X}) \to \mathsf{X}$ is any commutative distributive map. For concreteness, we can set $x * x' \in \{\bot_{\mathsf{X}}, \top_{\mathsf{X}}\} \subseteq \mathsf{X}$.
   b) $(\mathsf{X}, \leq)$ is a complete join-semilattice and $* : (\mathsf{X} \times \mathsf{X}) \to \mathbb{N}$ is any commutative distributive map. We think of $x * x'$ as returning the *size* of the intersection of x and x'.
   c) Any complete join-semilattice $(\mathsf{X}, \leq)$ is of course a (generalised) semiframe by taking $x * x' = \bigvee\{x'' \mid x'' \leq x,\ x'' \leq x'\}$.
   d) We can generalise further, in more than one direction. We would take $(\mathsf{X}, \leq)$ and $(\mathsf{X}', \leq')$ to be complete join-semilattices and $* : (\mathsf{X} \times \mathsf{X}) \to \mathsf{X}'$ to be any commutative distributive map (which generalises the above). We could also take X to be a cocomplete symmetric monoidal category [Mac71, Section VII]: a category with all colimits and with a (symmetric) monoid action $*$ that distributes over (commutes with) colimits.

See also Remark 23.3.7.

LEMMA 12.2.3. *Suppose $(\mathsf{X}, \leq)$ is a complete semilattice and suppose $* \subseteq \mathsf{X} \times \mathsf{X}$ is a compatibility relation on X. Then:*

1. *$*$ is monotone on both arguments.*
   That is: if $x_1 * x_2$ and $x_1 \leq x_1'$ and $x_2 \leq x_2'$, then $x_1' * x_2'$.
2. *If $x_1, x_2 \in \mathsf{X}$ have a non-$\bot$ lower bound $\bot_{\mathsf{X}} \lneq x \leq x_1, x_2$, then $x_1 * x_2$.*
   In words we can write: $*$ reflects non-$\bot$ lower bounds.
3. *The converse implication to part 2 need not hold: it may be that $x_1 * x_2$ (x_1 and x_2 are compatible) but the greatest lower bound of $\{x_1, x_2\}$ is $\bot$.*

Proof. We consider each part in turn:

1. We argue much as for Lemma 12.1.6(1). Suppose $x_1 * x_2$ and $x_1 \leq x_1'$ and $x_2 \leq x_2'$. By Lemma 12.1.6 $x_1 \vee x_1' = x_1'$ and $x_2 \vee x_2' = x_2'$. It follows using distributivity and commutativity (Definition 12.2.1(3&1)) that $x_1 * x_2$ implies that $(x_1 \vee x_1') * (x_2 * x_2')$, and thus that $x_1' * x_2'$ as required.

2. Suppose $\bot_{\mathsf{X}} \neq x \leq x_1, x_2$, so x is a non-$\bot_{\mathsf{X}}$ lower bound. By assumption $*$ is properly reflexive (Definition 12.2.1(2)) so (since $x \neq \bot_{\mathsf{X}}$) $x * x$. By part 1 of this result it follows that $x_1 * x_2$ as required.

3. It suffices to provide a counterexample. Define $(\mathsf{X}, \leq, *)$ by:

   - $\mathsf{X} = \{\bot, 0, 1, \top\}$.
   - $\bot \leq 0, 1 \leq \top$ and $\neg(0 \leq 1)$ and $\neg(1 \leq 0)$.
   - $x * x'$ for every $\bot \neq x, x' \in \mathsf{X}$.

   We note that $0 * 1$ but the greatest lower bound of $\{0, 1\}$ is $\bot$. We will revisit a slightly more elaborate version of this counterexample in Figure 13.1. □

12.3 The definition of a semiframe

DEFINITION 12.3.1. A **semiframe** is a tuple $(\mathsf{X}, \leq, *)$ such that

1. $(\mathsf{X}, \leq)$ is a complete semilattice (Definition 12.1.3), and
2. $*$ is a compatibility relation on it (Definition 12.2.1).

Slightly abusing terminology, we can say that

semiframe = *compatible complete semilattice*.

Semiframes are new, so far as we know, but they are a natural idea. We consider some elementary ways to generate examples, starting with arguably the simplest possible instance:

EXAMPLE 12.3.2 (The empty semiframe). Suppose $(\mathsf{X}, \leq, *)$ is a semiframe.

1. If X is a singleton set, so that $\mathsf{X} = \{\bullet\}$ for some element $\bullet$, then we call $(\mathsf{X}, \leq, *)$ the **empty semiframe** or **singleton semiframe**.
   Then necessarily $\bullet = \bot_{\mathsf{X}} = \top_{\mathsf{X}}$ and $\bullet \leq \bullet$ and $\neg(\bullet * \bullet)$.
2. If X has more than one element then we call $(\mathsf{X}, \leq, *)$ a **nonempty semiframe**. Then necessarily $\bot_{\mathsf{X}} \neq \top_{\mathsf{X}}$.

Thus, $(\mathsf{X}, \leq, *)$ is nonempty if and only if $\bot_{\mathsf{X}} \neq \top_{\mathsf{X}}$. We call a singleton semiframe *empty*, because this corresponds to the semiframe of open sets of the empty topology, which has no points and one open set, $\varnothing$.

Example 12.3.3 continues Remark 12.2.2:

EXAMPLE 12.3.3.

1. Suppose $(\mathsf{P}, \mathsf{Open})$ is a semitopology. Then the reader can check that the *semiframe of open sets* $(\mathsf{P}, \subseteq, \between)$ is a semiframe. We will study this example in detail; see Definition 12.3.4 and Lemma 12.3.5.
2. Suppose $(\mathsf{X}, \leq, \bot, \top)$ is a frame (a complete lattice such that meets distribute over arbitrary joins). Then $(\mathsf{X}, \leq, *)$ is a semiframe, where $x * x'$ when $x \wedge x' \neq \bot$.[2]
3. Take $\mathsf{X} = \{\bot, 0, 1, \top\}$ with $\bot \leq 0 \leq \top$ and $\bot \leq 1 \leq \top$ (so 0 and 1 are incomparable). There are two possible semiframe structures on this, characterised by choosing $0 * 1$ or $\neg(0 * 1)$.
4. See also the semiframes used in Lemmas 13.2.7.

Definition 12.3.4 is just an example of semiframes for now, though we will see much more of it later:

DEFINITION 12.3.4 (**Semitopology $\rightarrow$ semiframe**). Suppose $(\mathsf{P}, \mathsf{Open})$ is a semitopology. Define the **semiframe of open sets** $\mathsf{Fr}(\mathsf{P}, \mathsf{Open})$ (cf. Example 12.3.3(1)) by:

1. $\mathsf{Fr}(\mathsf{P}, \mathsf{Open})$ has elements open sets $O \in \mathsf{Open}$.
2. $\leq$ is subset inclusion.
3. $*$ is $\between$ (sets intersection).

[2] Just being a complete lattice is not enough; it has to be distributive as well. Consider $\omega{+}1 = \mathbb{N} \cup \{\omega\}$ with its usual ordering, augmented with an element d such that $0 \leq d \leq \omega$. Then $\omega = \bigvee \mathbb{N}$ and $d * \omega$, but $\neg(d * n)$ for every $n \in \mathbb{N}$.

We may write

$$(\mathsf{Open}, \subseteq, \between) \quad \text{as a synonym for} \quad \mathsf{Fr}(\mathbf{P}, \mathsf{Open}).$$

LEMMA 12.3.5. *Suppose* $(\mathbf{P}, \mathsf{Open})$ *is a semitopology. Then* $(\mathsf{Open}, \subseteq, \between)$ *is indeed a semiframe.*

Proof. As per Definition 12.3.1 we must show that Open is a complete semilattice (Definition 12.1.3) and $\between$ is a compatibility relation (Definition 12.2.1) — symmetric, properly reflexive, and distributive and satisfies a distributive law that if $O \between \bigcup \mathcal{O}'$ then $O \between O'$ for some $O' \in \mathcal{O}'$. These are all facts of sets. □

REMARK 12.3.6. Definition 12.3.4 and Lemma 12.3.5 are the start of our development. Once we have built more machinery, we will have a pair of translations:

- Definition 12.3.4 and Lemma 12.3.5 go from semitopologies to semiframes.
- Definition 13.4.3 and Lemma 13.4.4 go from semiframes to semitopologies.

These translations are part of a dual pair of functors between categories of semitopologies and semiframes, as described in Definitions 15.1.1 and 15.2.1 and Proposition 15.3.7.

Semitopologies are (relatively) concrete: we have concrete points and open sets that are sets of points. Semiframes are more abstract: we have a join-complete semilattice, and a compatibility relation. The duality we are about to build will show how these two worlds interact and reflect each other.

We conclude with a simple technical lemma:

LEMMA 12.3.7. *Suppose* $(\mathbf{X}, \leq, *)$ *is a semiframe (a complete semilattice with a compatibility relation) and* $x \in \mathbf{X}$*. Then:*

1. $\neg(x * \bot_{\mathbf{X}})$ *and in particular* $\neg(\bot_{\mathbf{X}} * \bot_{\mathbf{X}})$.
2. $x * x$ *if and only if* $x \neq \bot_{\mathbf{X}}$.
3. $x * \top_{\mathbf{X}}$ *if and only if* $x \neq \bot_{\mathbf{X}}$.
4. $\top_{\mathbf{X}} * \top_{\mathbf{X}}$ *holds precisely if* $\mathbf{X}$ *is nonempty (Example 12.3.2).*

Proof. We consider each part in turn:

1. Recall from Definition 12.1.3(3) that $\bot_{\mathbf{X}} = \bigvee \varnothing$. By distributivity (Definition 12.2.1(3))
$$x * \bot_{\mathbf{X}} \Longleftrightarrow \exists x' \in \varnothing . x * x' \Longleftrightarrow \bot.$$
2. We just combine part 1 of this result with Definition 12.2.1(2).
3. Suppose $x \neq \bot_{\mathbf{X}}$. Then $\bot_{\mathbf{X}} \lneq x \leq x \leq \top_{\mathbf{X}}$, and by Lemma 12.2.3(2) $x * \top_{\mathbf{X}}$.

   Suppose $x = \bot_{\mathbf{X}}$. Then $\neg(x * \top_{\mathbf{X}})$ by combining commutativity of $*$ (Definition 12.2.1(1)) with part 1 of this result.
4. If $\mathbf{X}$ is nonempty then by Example 12.3.2 $\bot_{\mathbf{X}} \neq \top_{\mathbf{X}}$ and so $\top_{\mathbf{X}} * \top_{\mathbf{X}}$ holds by part 2 of this result. However, in the degenerate case that $\mathbf{X}$ has one element then $\top_{\mathbf{X}} = \bot_{\mathbf{X}}$ and $\top_{\mathbf{X}} * \top_{\mathbf{X}}$ does not hold. □

REMARK 12.3.8. Recall from [DP02, Definition 5.22, page 128] that if X is a lattice, then the **pseudocomplement** to $x \in \mathsf{X}$ is $x^* = \bigvee\{x' \in \mathsf{X} \mid x' \wedge x = \bot\}$. A semiframe $(\mathsf{X}, \leq, *)$ naturally supports a notion of pseudocomplement for $x \in \mathsf{X}$, given by

$$x^c = \bigvee\{x' \in \mathsf{X} \mid \neg(x' * x)\}.$$

It is easy to prove that $\neg(x^c * x)$, arguing by contradiction: if $x^c * x$ then $\bigvee\{x' \mid \neg(x' * x)\} * x$, and by distributivity (Definition 12.2.1(3)) there exists $x' \in \mathsf{X}$ such that $x' * x$ and $\neg(x' * x)$, a contradiction.

Note that it may be that $(x^c)^c \not\leq x$. For example, in the semiframe illustrated in Figure 13.1, $0^c = \bigvee\{1, 2, 3\} = \top$ and $(0^c)^c = \bot \not\leq 0$ (this behaviour will be familiar to the reader who has seen, for example, double negation in intuitionistic logic).

x^c and related constructions will be useful later, in Definition 16.4.2 and Lemma 16.4.4.

Semifilters & abstract points 13

13.1 The basic definition, and discussion

Definition 13.1.1. Suppose $(\mathsf{X}, \leq, *)$ is a semiframe and suppose $F \subseteq \mathsf{X}$. Then:

1. Call F **prime** when for every $x, x' \in \mathsf{X}$,
$$x \curlyvee x' \in F \quad \text{implies} \quad x \in F \vee x' \in F.$$
2. Call F **completely prime** when for every (possibly empty) $X \subseteq \mathsf{X}$,
$$\bigvee X \in F \quad \text{implies} \quad \exists x \in X. x \in F.$$
(This condition is used in Lemma 13.3.2, which is needed for Lemma 13.4.2.)
3. Call F **up-closed** when $x \in F$ and $x \leq x'$ implies $x' \in F$.
4. Call F **compatible** when its elements are **pairwise compatible**, by which we mean that $x * x'$ for every $x, x' \in F$.
5. A **semifilter** is a nonempty, up-closed, compatible subset $F \subseteq \mathsf{X}$.
6. Call $F \subseteq \mathsf{X}$ a **maximal semifilter** when it is a semifilter and is contained in no strictly greater semifilter.
7. An **abstract point** is a completely prime semifilter.
8. Write
$$\mathsf{Points}(\mathsf{X}, \leq, *)$$
for the set of abstract points of $(\mathsf{X}, \leq, *)$.

Notation 13.1.2. We will generally write $F \subseteq \mathsf{X}$ for a subset of X that is intended to be a semifilter, or for which in most cases of interest F is a semifilter. We will generally write $P \subseteq \mathsf{X}$ when the subset is intended to be an abstract point, or when in most cases of interest P is an abstract point.

REMARK 13.1.3. *Note on design:* The notion of semifilter from Definition 13.1.1 is, obviously, based on the standard notion of filter [Joh86, I.2.2, page 12]. We just replace the *closure under binary meets* condition

'if $x, x' \in F$ then $x \wedge x' \in F$'

with a weaker *compatibility condition*

'if $x, x' \in F$ then $x * x'$'.

This is in keeping with our move from frames to semiframes, which weakens from $\wedge$ to the compatibility relation $*$.

Note that a semifilter or abstract point need not be directed:

1. Consider $\mathit{nbhd}(0)$ in the (semiframes of open sets of the) semitopologies in the left-hand and middle examples in Figure 5.1. In both cases, $\{0,1\}, \{0,2\} \in \mathit{nbhd}(0)$ but $\{0\} \notin \mathit{nbhd}(0)$ because $\{0\}$ is not an open set.
2. Consider $\{0,1,2\}$ with the discrete semitopology (so every set is open). Then the set of all two- or three-element subsets $\{\{0,1\},\{1,2\},\{2,0\},\{0,1,2\}\}$ is a semifilter, but it is not closed under sets intersections because it does not contain $\{0\}$, $\{1\}$, or $\{2\}$.

This second example is particularly interesting. As the reader may know, the intuition of a filter in topology is a set of *approximations*. But this example is clearly not approximating anything — after all, we are in the discrete semitopology and there is no need to approximate anything since we can just take a singleton set! This suggests that a better intuition for semiframe is a set of *collaborations*; in this case, of 0 with 1, 1 with 2, and 2 with 0.

Thus in particular, the standard result in frames that a proper finite filter[1] has a non-$\bot$ least element (obtained as the finite meet of all the elements in the filter), does not hold for semifilters in semiframes. See also Remark 13.1.6 and Proposition 13.2.8(1).

EXAMPLE 13.1.4. Suppose $(\mathsf{X}, \leq, *)$ is a semiframe. We recall some (standard) facts about abstract points, which carry over from topologies and frames:

1. Suppose $(\mathsf{P}, \mathsf{Open})$ is a semitopology and $(\mathsf{X}, \leq, *) = (\mathsf{Open}, \subseteq, \between)$. By Lemma 12.3.5, $(\mathsf{X}, \leq, *)$ is a semiframe.
   If $p \in \mathsf{P}$ then the *neighbourhood system*
   $$\mathit{nbhd}(p) = \{O \in \mathsf{Open} \mid p \in O\}$$
   from Definition 5.4.4 is an abstract point: see Proposition 14.2.1. Intuitively, $\mathit{nbhd}(p)$ abstractly represents p as the set of all of its open approximations in Open.
2. Suppose $(\mathsf{P}, \mathsf{Open})$ is a semitopology. Then $(\mathsf{Open}, \subseteq, \between)$ could contain an abstract point that is not the neighbourhood semifilter $\mathit{nbhd}(p)$ of a point $p \in \mathsf{P}$.
   Set $\mathsf{X} = \{\mathbb{Q}\} \cup \{(\pi\text{-}q, \pi\text{+}q) \subseteq \mathbb{Q} \mid q \in \mathbb{Q}_{\geq 0}\}$ (the set of all symmetric open intervals around π in the rational numbers $\mathbb{Q}$), and set $\leq\, = \,\subseteq$ and $* = \between$.
   Set $P = \mathsf{X} \setminus \{\varnothing\}$ to be the set of all *nonempty* symmetric open intervals around π. Note that $\pi \notin \mathbb{Q}$, but P is a set of open sets 'approximating' π.

[1] Recall that a proper filter is a filter that does not contain $\bot$.

3. We mention one more (standard) example. Consider $\mathbb{N}$ with the **final segment semitopology** such that opens are either $\varnothing$ or sets $n_{\geq} = \{n' \in \mathbb{N} \mid n' \geq n\}$. Then $\{n_{\geq} \mid n \in \mathbb{N}\}$ is an abstract point. Intuitively, this approximates a point at infinity, which we can understand as ω.

LEMMA 13.1.5. *Suppose* $(\mathsf{X}, \leq, *)$ *is a semiframe and suppose* $F \subseteq \mathsf{X}$ *is compatible. Then* $\bot_{\mathsf{X}} \notin F$.

Proof. By compatibility, $x * x$ for every $x \in F$. We use Lemma 12.3.7(1). □

REMARK 13.1.6. We continue Remark 13.1.3.

As the reader may know, a semiframe still has greatest lower bounds, because we can build them as $x \wedge x' = \bigvee\{x'' \mid x'' \leq x,\ x'' \leq x'\}$. It is just that this greatest lower bound may be unhelpful. To see why, consider again the examples in Figure 5.1. In the left-hand and middle examples in Figure 5.1, the greatest lower bound of $\{0, 1\}$ and $\{0, 2\}$ exists in the semiframe of open sets: but it is $\varnothing$ the empty set in the left-hand and middle example, not $\{0\}$. In the right-hand example, the greatest lower bound of $\{0, *, 1\}$ and $\{0, *, 2\}$ is $\{0\}$, not $\{0, *\}$.

So the reader could ask whether perhaps we should add the following weakened meet-closure condition to the definition of semifilters (and thus to abstract points):

If $x, x' \in F$ *and* $x \wedge x' \neq \bot$ *then* $x \wedge x' \in F$.

Intuitively, this insists that semifilters are closed under *non-*$\bot$ greatest lower bounds. However, there are two problems with this:

- It would break our categorical duality proof in the construction of g° in Lemma 15.3.3; see the discussion in Remark 15.3.4. This technical difficulty may be superable, but …
- … the condition is probably not what we want anyway. It would mean that the set of open neighbourhoods of $*$ in the right-hand example of Figure 5.1, would not be a semifilter, because it contains $\{0, *, 1\}$ and $\{0, *, 2\}$ but not its (non-$\varnothing$) greatest lower bound $\{0\}$.

13.2 Properties of semifilters

13.2.1 Things that are familiar from filters

LEMMA 13.2.1. *Suppose* $(\mathsf{X}, \leq, *)$ *is a semiframe and* $F \subseteq \mathsf{X}$ *is a semifilter. Then:*

1. $\top_{\mathsf{X}} \in F$.
2. $\bot_{\mathsf{X}} \notin F$.

Proof. We consider each part in turn:

1. By nonemptiness (Definition 13.1.1(7)) F is nonempty, so there exists some $x \in F$. By definition $x \leq \top_{\mathsf{X}}$. By up-closure (Definition 13.1.1(3)) $\top_{\mathsf{X}} \in F$ follows.

2. By assumption in Definition 13.1.1(4) elements in F are pairwise compatible (so $x * x$ for every $x \in F$). We use Lemma 13.1.5. □

LEMMA 13.2.2. *Suppose* $(\mathsf{X}, \leq, *)$ *is a semiframe. It is possible for a semifilter* $F \subseteq \mathsf{X}$ *to be completely prime but not maximal.*

Proof. We give a standard example (which also works for frames and filters). Take $\mathsf{P} = \{0,1\}$ and $\mathsf{Open} = \{\varnothing, \{0\}, \{0,1\}\}$. Then $P' = \{\{0,1\}\}$ is an abstract point, but it is not a maximal semifilter (it is not even a maximal abstract point) since P' is contained in the strictly larger semifilter $\{\{0\}, \{0,1\}\}$ (which is itself also a strictly larger abstract point). □

LEMMA 13.2.3. *If $(\mathsf{X}, \leq, *)$ is a finite semiframe (meaning that X is finite) then the properties of*

- *being a prime semifilter (Definition 13.1.1(1)) and*
- *being a completely prime semifilter (Definition 13.1.1(2)),*

coincide.

Proof. This is almost trivial, except that if $X = \varnothing$ in the condition for being completely prime then we get that $\bot_{\mathsf{X}} \notin P$ — but we know that anyway from Lemma 13.2.1(2), from the compatibility condition on semifilters. □

LEMMA 13.2.4. *Suppose $(\mathsf{X}, \leq, *)$ is a semiframe. Then:*

1. *The union of an ascending chain of semifilters in X, is a semifilter in X.*
2. *As a corollary, every semifilter $F \subseteq \mathsf{X}$ is contained in some maximal semifilter $F' \subseteq \mathsf{X}$ (assuming Zorn's lemma).*

Proof. We consider each part in turn:

1. By a straightforward verification of the conditions of being a semifilter from Definition 13.1.1(5).

2. Direct application of Zorn's lemma. □

REMARK 13.2.5.

1. Lemma 13.2.1(2) has a small twist to it. In the theory of *filters*, it does not follow from the property of being nonempty, up-closed, and closed under finite meets, that $\bot_{\mathsf{X}} \notin F$; this must be added as a distinct condition if required.
   In contrast, we see in the proof of Lemma 13.2.1(2) that for semifilters, $\bot_{\mathsf{X}} \notin F$ follows from the compatibility condition.
2. Lemma 13.2.3 matters in particular to us here, because we are particularly interested in abstracting the behaviours of finite semitopologies, because our original motivation for looking at both of these structures comes from looking at real networks, which are finite.[2]

[2]This is carefully worded. We care about *abstracting* properties of finite semitopologies, but we should not restrict to considering *only* semitopologies and semiframes that are actually finite! See Remark 9.4.7.

13.2.2 Things that are different from filters

REMARK 13.2.6. Obviously, by definition semifilters are necessarily compatible but not necessarily closed under meets. But aside from this fact, we have so far seen semiframes and semifilters behave more-or-less like frames and filters, modulo small details like that mentioned in Remark 13.2.5(1).

But there are also differences, as we will now briefly explore. In the theory of (finite) frames, the following facts hold:

1. *Every proper filter F has a greatest lower bound x, and $F = x^{\leq} = \{x' \mid x \leq x'\}$.* Just take $x = \bigwedge F$ the meet of all of its (finitely many) elements. This is not $\bot$, by the filter's finite intersection property.
2. *Every proper filter can be extended to a maximal filter.*[3] Just extend using Zorn's lemma (as in Lemma 13.2.4).
3. *Every maximal filter is completely prime.* It is a fact of finite frames that a maximal filter is prime,[4] and since we assume the frame is finite, it is also completely prime.
4. *Every non-$\bot$ element $x \neq \bot_{\mathsf{X}}$ in a finite frame is contained in some abstract point.* Just form $\{x' \mid x \leq x'\}$, observe it is a filter, form a maximal filter above it, and we get an abstract point.
5. *As a corollary, if the frame is nonempty (so $\bot \neq \top$; see Example 12.3.2) then it has at least one abstract point.*

In Lemma 13.2.7 and Proposition 13.2.8 we consider some corresponding *non-properties* of (finite) semiframes.

LEMMA 13.2.7. *Suppose $(\mathsf{X}, \leq, *)$ is a semiframe. It is possible for $\mathsf{Points}(\mathsf{X}, \leq, *)$ to be empty, even if $(\mathsf{X}, \leq, *)$ is nonempty (Example 12.3.2(2)). This is possible even if X is finite, and even if X is infinite.*

Proof. It suffices to provide an example. We define a semiframe as below, and as illustrated in Figure 13.1:

- $\mathsf{X} = \{\bot, 0, 1, 2, 3, \top\}$.
- Let $x \leq x'$ when $x = x'$ or $x = \bot$ or $x' = \top$.
- Let $x * x'$ when $x \wedge x' \neq \bot$.[5]

Then $(\mathsf{X}, \leq, *)$ has no abstract points.

For suppose P is one such. By Lemma 13.2.1 $\top \in P$. Note that $\top = 0 \vee 1 = 2 \vee 3$. By assumption P is completely prime, we know that $0 \in P \vee 1 \in P$, and also $2 \in P \vee 3 \in P$. But this is impossible because 0, 1, 2, and 3 are not compatible.

For the infinite case, we just increase the width of the semiframe by taking $\mathsf{X} = \{\bot\} \cup \mathbb{N} \cup \{\top\}$. $\square$

[3] A proper filter is a filter that does not contain $\bot$. A maximal filter is a filter that is maximal amongst proper filters.

[4] A succinct proof is in Wikipedia [Wik24b].

[5] Unpacking what that means, we obtain this: $x \neq \bot \wedge x = x'$ or $x \neq \bot \wedge x' = \top$ or $x' \neq \bot \wedge x = \top$. This definition for $*$ is what we need for our counterexample, but other choices for $*$ also yield valid semiframes. For example, we can set $x * x'$ when $x, x' \neq \bot$.

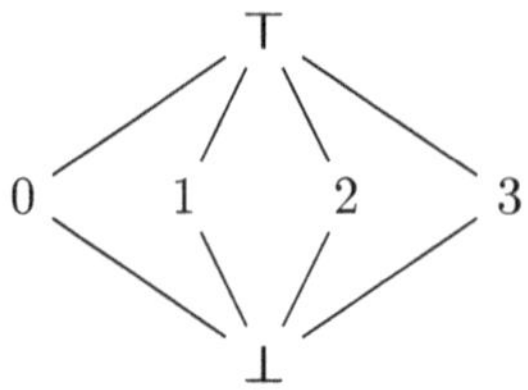

Figure 13.1: A semiframe with no abstract points (Lemma 13.2.7)

PROPOSITION 13.2.8. *Suppose* $(\mathsf{X}, \leq, *)$ *is a semiframe and* $F \subseteq \mathsf{X}$ *is a semifilter. Then:*

1. *It is not necessarily the case that* F *has a non-*$\bot$ *greatest lower bound (even if* X *is finite).*
2. *Every semifilter can be extended to a maximal semifilter, but …*
3. *… this maximal semifilter is not necessarily prime (even if* X *is finite).*
4. *There may exist a non-*$\bot$ *element* $x \neq \bot_{\mathsf{X}}$ *that is contained in no abstract point.*

Proof. We consider each part in turn:

1. Consider $(pow(\{0,1,2\}), \subseteq, \between)$ and take

   $$F = \{\{0,1\},\ \{1,2\},\ \{0,2\},\ \{0,1,2\}\}.$$

   The greatest lower bound of F is $\varnothing$.

2. This is Lemma 13.2.4.

3. F from part 1 of this result is maximal, and it cannot be extended to a point $P \supseteq F$. Figure 13.1 gives another counterexample, and in a rather interesting way: the semitopology has four maximal semifilters $\{i, \top\}$ for $i \in \{0,1,2,3\}$, but by Lemma 13.2.7 it has no prime semifilters at all.[6]

4. We just take $x = 0 \in \mathsf{X}$ from the example in Lemma 13.2.7 (see Figure 13.1). Since this semiframe has no abstract points at all, there is no abstract point that contains x. □

REMARK 13.2.9. For now, we will just read Proposition 13.2.8 as a caution not to assume that semiframes and semifilters behave like frames and filters. Sometimes they do, and sometimes they don't; we have to check.

We now proceed to build our categorical duality, culminating with Theorem 15.4.1. Once that machinery is constructed, we will continue our study of the fine structure of semifilters in Section 16.

[6] See also a discussion of the design of the notion of semifilter in Remarks 13.1.6 and 15.3.4.

13.3 Sets of abstract points

DEFINITION 13.3.1. Suppose $(\mathsf{X}, \leq, *)$ is a semiframe and recall $\mathsf{Points}(\mathsf{X}, \leq, *)$ from Definition 13.1.1(7). Define a map $Op : \mathsf{X} \to pow(\mathsf{Points}(\mathsf{X}, \leq, *))$ by

$$Op(x) = \{P \in \mathsf{Points}(\mathsf{X}, \leq, *) \mid x \in P\}.$$

LEMMA 13.3.2. *Suppose $(\mathsf{X}, \leq, *)$ is a semiframe and $X \subseteq \mathsf{X}$. Then*

$$Op(\bigvee X) = \bigcup_{x \in X} Op(x).$$

In words: we can say that Op commutes with joins, and that Op commutes with taking least upper bounds.

Proof. Suppose $P \in \mathsf{Points}(\mathsf{X}, \leq, *)$. We reason as follows:

$$\begin{aligned} P \in Op(\bigvee X) &\iff \bigvee X \in P && \text{Definition 13.3.1} \\ &\iff \exists x \in X. x \in P && \text{Definition 13.1.1(2)} \\ &\iff P \in \bigcup_{x \in X} Op(x) && \text{Definition 13.3.1} \end{aligned}$$

□

PROPOSITION 13.3.3. *Suppose $(\mathsf{X}, \leq, *)$ is a semiframe and $x, x' \in \mathsf{X}$. Then:*

1. *If $x \leq x'$ then $Op(x) \subseteq Op(x')$.*
2. *If $Op(x) \between Op(x')$ then $x * x'$.*
3. *$Op(\top_{\mathsf{X}}) = \mathsf{Points}(\mathsf{X}, \leq, *)$ and $Op(\bot_{\mathsf{X}}) = \varnothing$.*
4. *$Op(\bigvee X) = \bigcup_{x \in X} Op(x)$ for $X \subseteq \mathsf{X}$.*

Proof. We consider each part in turn:

1. *We prove that $x \leq x'$ implies $Op(x) \subseteq Op(x')$.*

   Suppose $x \leq x'$, and consider some abstract point $P \in Op(x)$. By Definition 13.3.1 $x \in P$, and by up-closure of P (Definition 13.1.1(3)) $x' \in P$, so by Definition 13.3.1 $P \in Op(x')$. P was arbitrary, and it follows that $Op(x) \subseteq Op(x')$.

2. *We prove that $Op(x) \between Op(x')$ implies $x * x'$.*

   Suppose there exists an abstract point $P \in Op(x) \cap Op(x')$. By Definition 13.3.1 $x, x' \in P$, and by compatibility of P (Definition 13.1.1(4)) $x * x'$.

3. Unpacking Definition 13.3.1, it suffices to show that $\top_{\mathsf{X}} \in P$ and $\bot_{\mathsf{X}} \notin P$, for every abstract point $P \in \mathsf{Points}(\mathsf{X}, \leq, *)$. This is from Lemma 13.2.1(1).

4. This is just Lemma 13.3.2. □

REMARK 13.3.4. Proposition 13.3.3 carries a clear suggestion that $(\{Op(x) \mid x \in \mathsf{X}\}, \subseteq, \between)$ is trying, in some sense, to be an isomorphic copy of $(\mathsf{X}, \leq, *)$. Lemma 13.3.5 notes that it may not quite manage this, because there may not be enough points (indeed, there may not be any abstract points at all). This will (just as for topologies and frames) lead us to the notion of a *spatial* semiframe in Definition 14.1.2 and Proposition 14.1.4.

LEMMA 13.3.5. *The converse implications in Proposition 13.3.3(1&2) need not hold. That is:*

1. *There exists a semiframe* $(\mathsf{X}, \leq, *)$ *and* $x, x' \in \mathsf{X}$ *such that* $Op(x) \subseteq Op(x')$ *yet* $x \not\leq x'$.
2. *There exists a semiframe* $(\mathsf{X}, \leq, *)$ *and* $x, x' \in \mathsf{X}$ *such that* $x * x'$ *yet* $Op(x) \not\between Op(x')$.

Proof. The example from Lemma 13.2.7 (as illustrated in Figure 13.1) is a counterexample for both cases:

- $Op(0) \subseteq Op(1)$ because both are equal to the empty set, yet $0 \not\leq 1$; and
- $\top * \top$ yet $Op(\top) \not\between Op(\top)$. □

13.4 The semitopology of abstract points

Recall from Definition 13.1.1(7) that an abstract point in a semiframe $(\mathsf{X}, \leq, *)$ is a nonempty up-closed compatible completely prime subset of X, and recall from Definition 13.3.1 that

$$Op(x) = \{P \in \mathsf{Points}(\mathsf{X}, \leq, *) \mid x \in P\},$$

or in words: $Op(x)$ is the set of abstract points that contain x.

DEFINITION 13.4.1. Suppose $(\mathsf{X}, \leq, *)$ is a semiframe. Then define $\mathsf{Op}(\mathsf{X}, \leq, *)$ by

$$\mathsf{Op}(\mathsf{X}, \leq, *) = \{Op(x) \mid x \in \mathsf{X}\}.$$

LEMMA 13.4.2. *Suppose* $(\mathsf{X}, \leq, *)$ *is a semiframe. Then:*

1. $\mathsf{Op}(\mathsf{X}, \leq, *)$ *from Definition 13.4.1 is closed under arbitrary sets union.*
2. *As a corollary,* $(\mathsf{Op}(\mathsf{X}, \leq, *), \subseteq)$ *(in words:* $\mathsf{Op}(\mathsf{X}, \leq, *)$ *ordered by subset inclusion) is a complete join-semilattice.*

Proof. Part 1 is just Lemma 13.3.2. The corollary part 2 is just a fact, since $\mathsf{Op}(\mathsf{X}, \leq, *) \subseteq pow(\mathsf{Points}(\mathsf{X}, \leq, *))$, and sets union is the join (least upper bound) in the powerset lattice. □

Recall from Definition 12.3.4 and Lemma 12.3.5 that we showed how to go from a semitopology $(\mathsf{P}, \mathsf{Open})$ to a semiframe $(\mathsf{Open}, \subseteq, \between)$. We now show how to go in the other direction:

DEFINITION 13.4.3 (**Semiframe → semitopology**). Suppose $(\mathsf{X}, \leq, *)$ is a semiframe. Define the **semitopology of abstract points** $\mathsf{St}(\mathsf{X}, \leq, *)$ by

$$\mathsf{St}(\mathsf{X}, \leq, *) = \big(\mathsf{Points}(\mathsf{X}, \leq, *), \mathsf{Op}(\mathsf{X}, \leq, *)\big).$$

Unpacking this a little:

1. The set of points of $\mathsf{St}(\mathsf{X}, \leq, *)$ is the set of abstract points $\mathsf{Points}(\mathsf{X}, \leq, *)$ from Definition 13.1.1(7) — namely, the completely prime nonempty up-closed compatible subsets of X.[7]

[7]There are no guarantees in general about *how many* abstract points exist; e.g. Lemma 13.2.7 gives an example of a semiframe that has no abstract points at all and so maps to the empty semitopology. Later on in Definition 14.1.2 we consider conditions to ensure the existence of abstract points.

2. Open sets $\mathsf{Opens}(\mathbf{X},\leq,*)$ are the $Op(x)$ from Definition 13.3.1:

$$Op(x)=\{P\in\mathsf{Points}(\mathbf{X},\leq,*)\mid x\in P\}.$$

LEMMA 13.4.4. *Suppose* $(\mathbf{X},\leq,*)$ *is a semiframe. Then* $\mathsf{St}(\mathbf{X},\leq,*)$ *from Definition 13.4.3 is indeed a semitopology.*

Proof. From conditions 1 and 2 of Definition 1.2.2, we need to check that $\mathsf{Op}(\mathbf{X},\leq,*)$ contains $\varnothing$ and $\mathsf{Points}(\mathbf{X},\leq,*)$ and is closed under arbitrary unions. This is from Proposition 13.3.3(3&4). □

Recall from Definitions 13.4.3 and 12.3.4 that $\mathsf{St}(\mathbf{X},\leq,*)$ is a semitopology, and $\mathsf{Fr}\,\mathsf{St}(\mathbf{X},\leq,*)$ is a semiframe each of whose elements is the set of abstract points of $(\mathbf{X},\leq,*)$ that contain some $x\in\mathbf{X}$:

LEMMA 13.4.5. *Suppose* $(\mathbf{X},\leq,*)$ *is a semiframe. Then* $Op:(\mathbf{X},\leq,*)\rightarrow\mathsf{Fr}\,\mathsf{St}(\mathbf{X},\leq,*)$ *is surjective.*

Proof. Direct from Definition 13.4.3(2). □

We conclude with Definition 13.4.6 and Proposition 13.4.7, which are standard properties of the construction in Definition 13.4.3.

DEFINITION 13.4.6. Suppose $(\mathbf{P},\mathsf{Open})$ is a semitopology and $p,p'\in\mathbf{P}$. Define $p\mathrel{\mathring{=}}p'$ by

$$p\mathrel{\mathring{=}}p' \quad\text{when}\quad \forall O\in\mathsf{Open}.p\in O\Longleftrightarrow p'\in O.$$

We recall some standard terminology from topology:

1. Call p and p' **topologically indistinguishable** when $p\mathrel{\mathring{=}}p'$.
2. Call p and p' **topologically distinguishable** when $\neg(p\mathrel{\mathring{=}}p')$ (so there exists some $O\in\mathsf{Open}$ such that $p\in O\wedge p'\notin O$ or $p\notin O\wedge p'\in O$).
3. Call $(\mathbf{P},\mathsf{Open})$ a T_0 **space** when points are topologically indistinguishable precisely when they are equal, or in symbols: $\mathring{=}\;=\;=$.

PROPOSITION 13.4.7. *Suppose* $(\mathbf{X},\leq,*)$ *is a semiframe. Then* $\mathsf{St}(\mathbf{X},\leq,*)$ *(Definition 13.4.3) is a* T_0 *space.*

Proof. Suppose $P,P'\in\mathsf{Points}(\mathbf{X},\leq,*)$. Unpacking Definition 13.1.1(7), this means that P and P' are completely prime nonempty up-closed compatible subsets of $\mathbf{X}$.

It is immediate that $P=P'$ implies $P\mathrel{\mathring{=}}P'$.

Now suppose $P\mathrel{\mathring{=}}P'$ in $\mathsf{St}(\mathbf{X},\leq,*)$; to prove $P=P'$ it would suffice to show that $x\in P\Longleftrightarrow x\in P'$, for arbitrary $x\in\mathbf{X}$. By Definition 13.4.3(2), every open set in $\mathsf{St}(\mathbf{X},\leq,*)$ has the form $Op(x)$ for some $x\in\mathbf{X}$. We reason as follows:

$$\begin{array}{r@{\;}l@{\qquad}l} x\in P & \Longleftrightarrow P\in Op(x) & \text{Definition 13.3.1}\\ & \Longleftrightarrow P'\in Op(x) & P,P'\ \text{top. indisting.}\\ & \Longleftrightarrow x\in P' & \text{Definition 13.3.1} \end{array}$$

Since x was arbitrary and $P,P'\subseteq\mathsf{Open}$, it follows that $P=P'$ as required. □

Spatial semiframes & sober semitopologies 14

14.1 Definition of spatial semiframes

REMARK 14.1.1. We continue Remark 13.3.4. We saw in Example 13.1.4(2&3) that there may be *more* abstract points than there are concrete points, and in Remark 13.3.4 that there may also be *fewer*.

In the theory of frames, the condition of being *spatial* means that the abstract points and concrete points correspond. We imitate this terminology for a corresponding definition on semiframes:

DEFINITION 14.1.2 (**Spatial semiframe**). Call a semiframe $(\mathsf{X}, \leq, *)$ **spatial** when:

1. $Op(x) \subseteq Op(x')$ implies $x \leq x'$, for every $x, x' \in \mathsf{X}$.
2. $x * x'$ implies $Op(x) \between Op(x')$, for every $x, x' \in \mathsf{X}$.

REMARK 14.1.3. Not every semiframe is spatial, just as not every frame is spatial. Lemma 13.2.7 gives an example of a semiframe that is not spatial because it has no points at all, as illustrated in Figure 13.1.

We check that the conditions in Definition 14.1.2 correctly strengthen the implications in Proposition 13.3.3 to become logical equivalences:

PROPOSITION 14.1.4. *Suppose* $(\mathsf{X}, \leq, *)$ *is a spatial semiframe and* $x, x' \in \mathsf{X}$. *Then:*

1. $x \leq x'$ *if and only if* $Op(x) \subseteq Op(x')$.
2. $x * x'$ *if and only if* $Op(x) \between Op(x')$.
3. $x = x'$ *if and only if* $Op(x) = Op(x')$.
4. $Op(\top_{\mathsf{X}}) = \mathsf{Points}(\mathsf{X}, \leq, *)$ *and* $Op(\bot_{\mathsf{X}}) = \varnothing$.
5. $Op(\bigvee X) = \bigcup_{x \in X} Op(x)$ *for* $X \subseteq \mathsf{X}$.

Proof. We consider each part in turn:

1. *We prove that* $x \leq x'$ *if and only if* $Op(x) \subseteq Op(x')$.

   The right-to-left implication is direct from Definition 14.1.2(1). The left-to-right implication is Proposition 13.3.3(1).

2. *We prove that* $x * x'$ *if and only if* $Op(x) \between Op(x')$.

   The left-to-right implication is direct from Definition 14.1.2(2). The right-to-left implication is Proposition 13.3.3(2).

3. *We prove that* $x = x'$ *if and only if* $Op(x) = Op(x')$.

   If $x = x'$ then $Op(x) = Op(x')$ is immediate. If $Op(x) = Op(x')$ then $Op(x) \subseteq Op(x')$ and $Op(x') \subseteq Op(x)$. By part 1 of this result (or direct from Definition 14.1.2(1)) $x \leq x'$ and $x' \leq x$. By antisymmetry of $\leq$ it follows that $x = x'$.

4. This is just Proposition 13.3.3(3)

5. This is just Lemma 13.3.2. □

Definition 14.1.5 will be useful in a moment:[1]

DEFINITION 14.1.5. Suppose $(\mathsf{X}, \leq, *)$ and $(\mathsf{X}', \leq', *')$ are semiframes. Then an **isomorphism** between them is a function $g : \mathsf{X} \to \mathsf{X}'$ such that:

1. g is a bijection between X and X'.
2. $x_1 \leq x_2$ if and only if $g(x_1) \leq g(x_2)$.
3. $x_1 * x_2$ if and only if $g(x_1) * g(x_2)$.

LEMMA 14.1.6. *Suppose* $(\mathsf{X}, \leq, *)$ *and* $(\mathsf{X}', \leq', *')$ *are semiframes and* $g : \mathsf{X} \to \mathsf{X}'$ *is an isomorphism between them. Then* $g(\bot_{\mathsf{X}}) = g(\bot_{\mathsf{X}'})$ *and* $g(\top_{\mathsf{X}}) = \top_{\mathsf{X}'}$.

Proof. By construction $\bot_{\mathsf{X}} \leq x$ for every $x \in \mathsf{X}$. It follows from Definition 14.1.5(2) that $g(\bot_{\mathsf{X}}) \leq g(x)$ for every $x \in \mathsf{X}$; but g is a bijection, so $g(\bot_{\mathsf{X}}) \leq x'$ for every $x' \in \mathsf{X}'$. It follows that $g(\bot_{\mathsf{X}}) = \bot_{\mathsf{X}'}$.

By similar reasoning we conclude that $g(\top_{\mathsf{X}}) = \top_{\mathsf{X}'}$. □

REMARK 14.1.7. Suppose $(\mathsf{X}, \leq, *)$ is a semiframe and recall from Definition 13.4.1 that $\mathsf{Op}(\mathsf{X}, \leq, *) = \{Op(x) \mid x \in \mathsf{X}\}$. Then the intuitive content of Proposition 14.1.4 is that a semiframe $(\mathsf{X}, \leq, *)$ is spatial when $(\mathsf{X}, \leq, *)$ is isomorphic (in the sense made formal by Definition 14.1.5) to $(\mathsf{Op}(\mathsf{X}, \leq, *), \subseteq, \between)$.

And, because $Op(\top_{\mathsf{X}}) = \mathsf{Points}(\mathsf{X}, \leq, *)$ we can write a slogan:

> *A semiframe is spatial when it is (up to isomorphism) generated by its abstract points.*

We will go on to prove in Proposition 14.2.4 that every semitopology generates a spatial semiframe — and in Theorem 15.4.1 we will tighten and extend the slogan above to a full categorical duality.

[1] More on this topic later on in Definition 15.2.1, when we build the category of semiframes.

14.2 The neighbourhood semifilter $nbhd(p)$

14.2.1 The definition and basic lemma

Recall from Definition 5.4.4 that $nbhd(p) = \{O \in \mathsf{Open} \mid p \in O\}$.

PROPOSITION 14.2.1. *Suppose* $(\mathbf{P}, \mathsf{Open})$ *is a semitopology and* $p \in \mathbf{P}$ *and* $O \in \mathsf{Open}$. *Then:*

1. *$nbhd(p)$ (Definition 5.4.4) is an abstract point (a completely prime semifilter) in the semiframe* $\mathsf{Fr}(\mathbf{P}, \mathsf{Open})$ *(Definition 12.3.4). In symbols:*
$$nbhd : \mathbf{P} \to \mathsf{Points}(\mathsf{Fr}(\mathbf{P}, \mathsf{Open})).$$
2. *The following are equivalent:*
$$nbhd(p) \in Op(O) \quad \Longleftrightarrow \quad O \in nbhd(p) \quad \Longleftrightarrow \quad p \in O.$$
3. *We have an equality:*
$$nbhd^{-1}(Op(O)) = O.$$

Proof. We consider each part in turn:

1. From Definition 13.1.1(7), we must check that $nbhd(p)$ is a nonempty, completely prime, up-closed, and compatible subset of Open when considered as a semiframe as per Definition 12.3.4. All properties are by facts of sets; we give brief details:
   - $nbhd(p)$ is nonempty because $p \in \mathbf{P} \in \mathsf{Open}$.
   - $nbhd(p)$ is completely prime because it is a fact of sets that if $P \subseteq \mathsf{Open}$ and $p \in \bigcup P$ then $p \in O$ for some $O \in P$.
   - $nbhd(p)$ is up-closed because it is a fact of sets that if $p \in O$ and $O \subseteq O'$ then $p \in O'$.
   - $nbhd(p)$ is compatible because if $p \in O$ and $p \in O'$ then $O \between O'$.
2. By Definition 13.3.1, $Op(O)$ is precisely the set of abstract points P that contain O, and by part 1 of this result $nbhd(p)$ is one of those points. By Definition 5.4.4, $nbhd(p)$ is precisely the set of open sets that contain p. The equivalence follows.
3. We reason as follows:
$$\begin{aligned} p \in nbhd^{-1}(Op(O)) &\Longleftrightarrow nbhd(p) \in Op(O) && \text{Fact of function inverse} \\ &\Longleftrightarrow p \in O && \text{Part 2 of this result} \end{aligned}$$
□

COROLLARY 14.2.2. *Suppose* $(\mathbf{P}, \mathsf{Open})$ *is a semitopology and* $O, O' \in \mathsf{Open}$. *Then:*

1. $Op(O) \subseteq Op(O')$ *if and only if* $O \subseteq O'$.
2. $Op(O) \between Op(O')$ *if and only if* $O \between O'$.
3. *As a corollary,* $nbhd^{-1}(\varnothing) = \varnothing$ *and* $nbhd^{-1}(\mathsf{Points}(\mathsf{Open}, \subseteq, \between)) = \mathbf{P}$*; i.e.* $nbhd^{-1}$ *maps the bottom/top element to the bottom/top element.*

Proof. We consider each part in turn:

1. If $Op(O) \subseteq Op(O')$ then $nbhd^{-1}(Op(O)) \subseteq nbhd^{-1}(Op(O'))$ by facts of inverse images, and $O \subseteq O'$ follows by Proposition 14.2.1(3).

   If $O \subseteq O'$ then $Op(O) \subseteq Op(O')$ by Proposition 13.3.3(1).

2. If $O \between O'$ then there exists some point $p \in \mathsf{P}$ with $p \in O \cap O'$. By Proposition 14.2.1(1) $nbhd(p)$ is an abstract point, and by Proposition 14.2.1(2) $nbhd(p) \in Op(O) \cap Op(O')$; thus $Op(O) \between Op(O')$.

   If $Op(O) \between Op(O')$ then $O \between O'$ by Proposition 13.3.3(2).

3. Routine from Proposition 13.3.3(3) (or from Lemma 14.1.6). □

14.2.2 Application to semiframes of open sets

PROPOSITION 14.2.3. *Suppose* $(\mathsf{P}, \mathsf{Open})$ *is a semitopology. Then:*

1. *$nbhd^{-1}$ bijects open sets of* $\mathsf{St}(\mathsf{Open}, \subseteq, \between)$ *(as defined in Definition 13.4.3(2)), with open sets of* $(\mathsf{P}, \mathsf{Open})$*, taking $Op(O)$ to O.*
2. *$nbhd^{-1}$ is an isomorphism between the semiframe of open sets of* $\mathsf{St}(\mathsf{Open}, \subseteq, \between)$*, and the semiframe of open sets of* $(\mathsf{P}, \mathsf{Open})$ *(Definition 14.1.5).*

Proof. We consider each part in turn:

1. Unpacking Definition 13.4.3(2), an open set in $\mathsf{St}\,\mathsf{Fr}(\mathsf{P}, \mathsf{Open})$ has the form $Op(O)$ for some $O \in \mathsf{Open}$. By Proposition 14.2.1(3) $nbhd^{-1}(Op(O)) = O$, and so $nbhd^{-1}$ is surjective and injective.

2. Unpacking Definition 14.1.5 it suffices to check that:

   - $nbhd^{-1}$ is a bijection, and maps $Op(O)$ to O.
   - $Op(O) \subseteq Op(O')$ if and only if $O \subseteq O'$.
   - $Op(O) \between Op(O')$ if and only if $O \between O'$.

   The first condition is part 1 of this result; the second and third are from Corollary 14.2.2. □

PROPOSITION 14.2.4. *Suppose* $(\mathsf{P}, \mathsf{Open})$ *is a semitopology. Then the semiframe* $\mathsf{Fr}(\mathsf{P}, \mathsf{Open}) = (\mathsf{Open}, \subseteq, \between)$ *from Definition 12.3.4 is spatial.*

Proof. The properties required by Definition 14.1.2 are that $Op(O) \subseteq Op(O')$ implies $O \subseteq O'$, and $O \between O'$ implies $Op(O) \between Op(O')$. Both of these are immediate from Proposition 14.2.3(2). □

14.2.3 Application to characterise T_0 spaces

LEMMA 14.2.5. *Suppose* $(\mathsf{P}, \mathsf{Open})$ *is a semitopology and* $p, p' \in \mathsf{P}$*. Then the following are equivalent:*

1. $nbhd(p) = nbhd(p')$ *(cf. also Lemma 16.7.1)*
2. $\forall O \in \mathsf{Open}. p \in O \iff p \in O'$

3. $p \mathrel{\mathring{=}} p'$ *(Definition 13.4.6: p and p' are topologically indistinguishable in* $(\mathsf{P}, \mathsf{Open})$*).*
4. $nbhd(p) \mathrel{\mathring{=}} nbhd(p')$ *($nbhd(p)$ and $nbhd(p')$ are topologically indistinguishable as — by Proposition 14.2.1(1) — abstract points in* $\mathsf{St}\,\mathsf{Fr}(\mathsf{P}, \mathsf{Open})$*).*

Proof. Equivalence of parts 1 and 2 is direct from Definition 5.4.4. Equivalence of parts 2 and 3 is just Definition 13.4.6(1). Equivalence of parts 4 and 1 is from Proposition 13.4.7. □

Corollary 14.2.6. *Suppose* $(\mathsf{P}, \mathsf{Open})$ *is a semitopology. Then the following are equivalent:*

1. $(\mathsf{P}, \mathsf{Open})$ *is* T_0 *(Definition 13.4.6(3)).*
2. $nbhd : \mathsf{P} \to \mathsf{Points}(\mathsf{Open}, \subseteq, \between)$ *is injective.*

Proof. Suppose $(\mathsf{P}, \mathsf{Open})$ is T_0, and suppose $nbhd(p) = nbhd(p')$. By Lemma 14.2.5(1&3) $p \mathrel{\mathring{=}} p'$. By Definition 13.4.6(2) $p = p'$. Since p and p' were arbitrary, $nbhd$ is injective.

Suppose $nbhd$ is injective. Reversing the reasoning of the previous paragraph, we deduce that $(\mathsf{P}, \mathsf{Open})$ is T_0. □

14.3 Sober semitopologies

Recall from Proposition 14.2.4 that if we go from a semitopology $(\mathsf{P}, \mathsf{Open})$ to a semiframe $(\mathsf{Open}, \subseteq, \between)$, then the result is not just any old semiframe — it is a *spatial* one.

We now investigate what happens when we go from a semiframe to a semitopology using Definition 13.4.3.

14.3.1 The definition and a key result

Definition 14.3.1. Call a semitopology $(\mathsf{P}, \mathsf{Open})$ **sober** when every abstract point P of $\mathsf{Fr}(\mathsf{P}, \mathsf{Open})$ — i.e. every completely prime nonempty up-closed compatible set of open sets — is equal to the neighbourhood semifilter $nbhd(p)$ of some unique $p \in \mathsf{P}$. Equivalently, $(\mathsf{P}, \mathsf{Open})$ is sober when $nbhd : \mathsf{P} \to \mathsf{Points}(\mathsf{Fr}(\mathsf{P}, \mathsf{Open}))$ (Definition 13.1.1(7)) is a bijection.

Remark 14.3.2. A bijection is a map that is injective and a surjective. We noted in Corollary 14.2.6 that a space is T_0 when $nbhd$ is injective. So the sobriety condition can be thought of as having two parts:

- $nbhd$ is injective and the space is T_0, so it intuitively contains no 'unnecessary' duplicates of points;
- $nbhd$ is surjective, so the space contains 'enough' points that there is (precisely) one concrete point for every abstract point.[2]

[2] 'Unnecessary' and 'enough' are in scare quotes here because these are subjective terms. For example, if points represent computer servers on a network then we might consider it a *feature* to not be T_0 by having multiple points that are topologically indistinguishable — e.g. for backup, or to reduce latency — and likewise, we might consider it a feature to not have one concrete point for every abstract point, if this avoids redundancies. There is no contradiction here: a computer network based on a small non-sober space with multiple backups of what it has, may be a more efficient and reliable system than one based on a larger non-sober space that does not back up its servers but is full of redundant points. And, this smaller non-sober space may present itself to the user abstractly as the larger, sober space.

Users may even forget about the computation that goes on under the hood of this abstraction, as illustrated by the following *true story:* I had a paper presenting an efficient algorithm rejected because it 'lacked motivation'. Why? Because the algorithm was unnecessary: the reviewer claimed, apparently with a straight face, that guessing the answer until you got it right was computationally equivalent.

We start with a very simple example of sober semitopologies:

LEMMA 14.3.3. *Suppose* P *is any set. Then the discrete semitopology* $(\mathsf{P}, pow(\mathsf{Open}))$ *is sober.*

Proof. Consider an abstract point $P \subseteq \mathsf{Open}$ (completely prime nonempty up-closed and compatible, as per Definition 13.1.1(7)). Then $\mathsf{P} \in P$ and $\mathsf{P} = \bigvee\{\{p\} \mid p \in \mathsf{P}\}$. Since P is completely prime, $\{p\} \in P$ for some $p \in \mathsf{P}$. It follows easily that $P = nbhd(p)$. □

EXAMPLE 14.3.4. We give some more examples of sober and non-sober semitopologies.

1. Take $\mathsf{P} = \{0, 1\}$ and $\mathsf{Open} = \{\varnothing, \{0, 1\}\}$. This has one abstract point $P = \{\{0, 1\}\}$ but two concrete points 0 and 1. It is therefore not sober.

2. Take $\mathsf{P} = \{0, 1\}$ and $\mathsf{Open} = \{\varnothing, \{1\}, \{0, 1\}\}$. This has two abstract points

$$\{\{1\}, \{0, 1\}\} \quad \text{and} \quad \{\{0, 1\}\}$$

   corresponding to two concrete points 0 and 1. It is sober.

3. Take $\mathsf{P} = \mathbb{N}$ with the final topology; so $O \in \mathsf{Open}$ when $O = \varnothing$ or $O = n_{\geq}$ for some $n \in \mathbb{N}$, where $n_{\geq} = \{n' \in \mathbb{N} \mid n' \geq n\}$. Take $P = \{n_{\geq} \mid n \in \mathbb{N}\}$. The reader can check that this is an abstract point (up-closed, completely prime, compatible); however P is not the neighbourhood semifilter of any $n \in \mathbb{N}$. Thus this space is not sober.

4. $\mathbb{R}$ with its usual topology (which is also a semitopology) is sober.

   This is a known result for topologies, but Remark 13.1.3 (and also the later Remark 14.3.11) caution us that we cannot take this for granted, so we sketch the proof.

   Suppose P is an abstract point; we wish to exhibit a unique $p \in \mathbb{R}$ such that $P = nbhd(p)$.

   We cover $\mathbb{R}$ with open intervals of radius 1 by writing $\mathbb{R} = \bigcup\{(r{-}0.5, r{+}0.5) \mid r \in \mathbb{R}\}$, and we use the completely prime property to find (at least one) such open interval that is in P; write it $O_1 \in P$. We then cover O_1 with open intervals of radius at most $1/2$ by writing $O_1 = \bigcup\{O_1 \cap (r{-}0.25, r{+}0.25) \mid r \in O_1\}$, and we iterate to obtain a sequence $(O_i \mid i \in \mathbb{N}) \subseteq P$. This converges to some unique $p \in \mathbb{R}$. We check that $P = nbhd(p)$:

   - Suppose $O \in \mathsf{Open}$ is such that $p \in O$. Because $p \in O$, there exists some ϵ such that $(p{-}\epsilon, p{+}\epsilon) \subseteq O$. It follows that for $i > 1/\epsilon$ we have $O_i \subseteq O$ and thus $O \in P$ by up-closedness.
   - Suppose $O' \in \mathsf{Open}$ is such that $p \notin O'$. For a sufficiently large i we have $O_i \not\between O'$, so by compatibility it follows that $O' \notin P$.

5. $\mathbb{Q}$ is sober. The argument is much as for $\mathbb{R}$ above. We have to work just a little harder because the p we obtain need not be rational, but the arguments on open intervals remain valid. See a brief discussion in point 6 in Subsection 23.4.

PROPOSITION 14.3.5. *Suppose* $(\mathsf{X}, \leq, *)$ *is a semiframe. Then* $\mathsf{St}(\mathsf{X}, \leq, *)$ *from Definition 13.4.3 is a sober semitopology.*

(a) Finite T_0 (and also T_1) semitopology that is not sober (Lemma 14.3.8)

(b) Hausdorff semitopology that is not sober (Lemma 14.3.10)

Figure 14.1: Two counterexamples for sobriety

Proof. We know from Lemma 13.4.4 that $\mathsf{St}(\mathsf{X}, \leq, *)$ is a semitopology. The issue is whether it is sober; thus by Definition 14.3.1 we wish to exhibit every abstract point P of $\mathsf{Fr}\,\mathsf{St}(\mathsf{X}, \leq, *)$ as a neighbourhood semifilter $nbhd(p)$ for some unique abstract point p of $(\mathsf{X}, \leq, *)$. The calculations to do so are routine, but we give details.

Fix some abstract point P of $\mathsf{Fr}\,\mathsf{St}(\mathsf{X}, \leq, *)$. By Definition 13.1.1(7), P is a completely prime nonempty up-closed set of intersecting open sets in the semitopology $\mathsf{St}(\mathsf{X}, \leq, *)$, and by Definition 13.4.3(2) each open set in $\mathsf{St}(\mathsf{X}, \leq, *)$ has the form $Op(x) = \{p \in \mathsf{Points}(\mathsf{X}, \leq, *) \mid x \in p\}$ for some $x \in \mathsf{X}$.

We define $p \subseteq \mathsf{X}$ as follows:

$$p = \{x \in \mathsf{X} \mid Op(x) \in P\} \subseteq \mathsf{X}.$$

By construction we have that $x \in p$ if and only if $Op(x) \in P$, and so

$$\begin{aligned} nbhd(p) &= \{Op(x) \mid p \in Op(x)\} && \text{Definition 5.4.4} \\ &= \{Op(x) \mid x \in p\} && \text{Definition 13.3.1} \\ &= \{Op(x) \mid Op(x) \in P\} && \text{Construction of } p \\ &= P && \text{Fact.} \end{aligned}$$

Now P is completely prime, nonempty, up-closed, and compatible and it follows by elementary calculations using Proposition 14.1.4 that p is also completely prime, nonempty, up-closed, and compatible — so p is an abstract point of $(\mathsf{X}, \leq, *)$.

So we have that

$$p \in \mathsf{Point}(\mathsf{X}, \leq, *) \quad \text{and} \quad P = nbhd(p).$$

To prove uniqueness of p, suppose p' is any other abstract point such that $P = nbhd(p')$. We follow the definitions: $Op(x) \in nbhd(p') \iff Op(x) \in nbhd(p)$, and thus by Definition 5.4.4 $p' \in Op(x) \iff p \in Op(x)$, and thus by Definition 13.3.1 $x \in p' \iff x \in p$, and thus $p' = p$. □

14.3.2 Sober topologies contrasted with sober semitopologies

We will need Notation 14.3.6 for Remark 14.3.7:

NOTATION 14.3.6. Call a closed set **irreducible** when it cannot be written as the union of two proper closed subsets.

(a) Soberification of left-hand example in Figure 14.1

(b) Soberification of right-hand example in Figure 14.1

Figure 14.2: Example soberifications (Remark 14.3.12)

REMARK 14.3.7. Topology has a wealth of separation actions. Three of them are: T_0 (distinct points have distinct neighbourhood (semi)filters); T_1 (distinct points have distinct open neighbourhoods); and T_2, also known as the Hausdorff condition (distinct points have disjoint open neighbourhoods) — see Remark 4.2.1 for formal statements. In the case of topologies, the following is known about sobriety:

1. Every finite T_0 (and thus every finite T_1) topological space is sober.
2. Every T_2/Hausdorff space (including infinite ones) is sober [MM92, page 475, Theorem 3].
3. A topological space is sober if and only if every nonempty irreducible closed set is the closure of a unique point [MM92, page 475].

The situation for semitopologies is different, as we explore in the rest of this Subsection.

LEMMA 14.3.8.

1. *It is not necessarily the case that a finite T_0 semitopology (or even a finite T_1 semitopology) is sober (Definition 14.3.1).*
2. *It is not necessarily the case that if every nonempty irreducible closed set is the closure of a unique point, then a semitopology is sober.*

These non-implications hold even if the semitopology is regular (so $p \in K(p) \in \mathsf{Topen}$ for every p; see Definition 4.1.4(3)).

Proof. We provide a semitopology that is a counterexample for both parts.

Consider the left-hand semitopology illustrated in Figure 14.1, so that:[3]

- $\mathsf{P} = \{0, 1, 2\}$, and
- $\mathsf{Open} = \{\varnothing, \{0,1\}, \{1,2\}, \{0,2\}, \{0,1,2\}\}$.

We note that:

- $(\mathsf{P}, \mathsf{Open})$ is T_0 and T_1.
- $(\mathsf{P}, \mathsf{Open})$ is regular because all points are intertwined, so that $K(p) = \mathsf{P}$ for every $p \in \mathsf{P}$.

[3] Recall that we used this example, for different purposes, in Lemma 3.7.2 and Figure 3.2.

- The nonempty irreducible closed sets are $\{0\}$ (which is the complement of $\{1,2\}$), $\{1\}$, and $\{2\}$. Since these are singleton sets, they are certainly the closures of unique points.

So $(\mathsf{P}, \mathsf{Open})$ is T_0, regular, and irreducible closed sets are the closures of unique points.

We take as our semifilter $P = \mathsf{Open} \setminus \{\varnothing\}$. The reader can check that P is completely prime (Definition 13.1.1(2)), nonempty, up-closed, and compatible (P is also the greatest semifilter); but, P is not the neighbourhood semifilter of 0, 1, or 2 in P. Thus, $(\mathsf{P}, \mathsf{Open})$ is not sober. □

REMARK 14.3.9. The counterexample used in Lemma 14.3.8 generalises, as follows: the reader can check that the *all-but-one* semitopology from Example 2.1.4(5c) on three or more points (so open sets are generated by $\mathsf{P} \setminus \{p\}$ for every $p \in \mathsf{P}$) has similar behaviour.

In topology, every Hausdorff space is sober. In semitopologies, this implication does not hold, and in a rather strong sense:

LEMMA 14.3.10.

1. *It is not necessarily the case that if a semitopology is Hausdorff, then it is sober.*
2. *Every quasiregular Hausdorff semitopology is (discrete and therefore) sober.*

Proof. We consider each part in turn:

1. It suffices to give a counterexample. Consider the right-hand semitopology illustrated in Figure 14.1 (which we also used, for different purposes, in Figure 5.2), so that:

   - $\mathsf{P} = \{0, 1, 2, 3\}$, and
   - Open is generated by $X = \{\{3,0\}, \{0,1\}, \{1,2\}, \{2,3\}\}$.

   This is Hausdorff, but it is not sober: the reader can check that the up-closure $\{3,0\}^{\leq} \subseteq \mathsf{Open}$ is nonempty, up-closed, compatible, and completely prime, but it is not the neighbourhood filter of any $p \in \mathsf{P}$.

2. From Lemmas 4.2.2 (quasiregular Hausdorff is discrete) and 14.3.3 (discrete is sober). □

REMARK 14.3.11. A bit more discussion of Lemma 14.3.10.

1. The space used in the counterexample for part 1 is Hausdorff, T_1, and unconflicted (Definition 6.1.1(2)). It is not quasiregular (Definition 4.1.4(5)) because the community of every point is empty; see Proposition 6.2.1.

2. The implication holds if we add quasiregularity as a condition: every quasiregular Hausdorff space is sober. But, this holds for very bad reasons, because by Lemma 4.2.2 every quasiregular Hausdorff space is discrete.

3. Thus, the non-implication discussed in Lemma 14.3.10 is informative and tells us something interesting about semitopological sobriety. Semitopological sobriety is not just a weak form of topological sobriety, and it has its own distinct personality.

REMARK 14.3.12. We can inject the examples illustrated in Figure 14.1 (used in Lemmas 14.3.8 and 14.3.10) into *soberified* versions of the spaces that are sober and have an isomorphic lattice of open sets, as illustrated in Figure 14.2:

1. The left-hand semitopology has abstract points (completely prime semifilters; see Definition 13.1.1(2)) generated as the $\subseteq$-up-closures of the following sets: $\{A\}$, $\{B\}$, $\{C\}$, $\{A, B\}$, $\{B, C\}$, $\{C, A\}$, and $\{A, B, C\}$. Of these, $\{A, B\}^{\subseteq} = nbhd(0)$, $\{B, C\}^{\subseteq} = nbhd(1)$, and $\{C, A\}^{\subseteq} = nbhd(2)$. The other completely prime semifilters are not generated as the neighbourhood semifilters of any point in the original space, so we add points as illustrated using $\bullet$ in the left-hand diagram in Figure 14.2. This semitopology is sober, and has the same semiframe of open sets.
2. For the right-hand example, we again add a $\bullet$ point for every abstract point in the original space that is not already the neighbourhood semifilter of a point in the original space. These abstract points are generated as the $\subseteq$-up-closures of $\{A\}$, $\{B\}$, $\{C\}$, and $\{D\}$. There is no need to add a $\bullet$ for the abstract point generated as the $\subseteq$-up-closure of $\{A, B\}$, because $\{A, B\}^{\subseteq} = nbhd(0)$. Similarly $\{B, C\}^{\subseteq} = nbhd(1)$, $\{C, D\}^{\subseteq} = nbhd(2)$, and $\{D, A\}^{\subseteq} = nbhd(3)$. Note that $\{A, B, C\}$ does not generate an abstract point because it is not compatible: $A \not\between C$. Similarly for $\{B, C, D\}$, $\{C, D, A\}$, $\{D, A, C\}$, and $\{A, B, C, D\}$.

These soberified spaces are instances of a general construction described in Theorem 15.1.4. And, continuing the observation made in Remark 14.3.11, note that neither of these spaces, with their extra points, are Hausdorff.

Four categories & functors between them 15

15.1 The categories Sober/SemiTop of (sober) semitopologies

DEFINITION 15.1.1.

1. Suppose $(\mathsf{P}, \mathsf{Open})$ and $(\mathsf{P}', \mathsf{Open}')$ are semitopologies and $f : \mathsf{P} \to \mathsf{P}'$ is any function. Then call f a **morphism of semitopologies** when f is continuous, by which we mean (as standard) that
$$O' \in \mathsf{Open}' \quad \text{implies} \quad f^{-1}(O') \in \mathsf{Open}.$$
2. Define SemiTop the **category of semitopologies** such that:
   - objects are semitopologies, and
   - arrows are morphisms of semitopologies (continuous maps on points).[1]
3. Write Sober for the **category of sober semitopologies** and continuous functions between them. By construction, Sober is the full subcategory of SemiTop, on its sober semitopologies.

REMARK 15.1.2. For convenience reading Theorem 15.1.4 we recall some facts:

1. The *semiframe*
$$\mathsf{Fr}(\mathsf{P}, \mathsf{Open}) = (\mathsf{Open}, \subseteq, \between)$$
from Definition 12.3.4 has elements open sets $O \in \mathsf{Open}$, preordered by subset inclusion and with a compatibility relation given by sets intersection.
It is spatial, by Proposition 14.2.4.
2. An abstract point P in $\mathsf{Points}(\mathsf{Fr}(\mathsf{P}, \mathsf{Open}))$ is a completely prime nonempty up-closed compatible subset of Open.
3. $\mathsf{St}\,\mathsf{Fr}(\mathsf{P}, \mathsf{Open})$ is by Definition 13.4.3 a semitopology such that:

[1] A discussion of possible alternatives, for future work, is in Remark 23.3.3. See also Remarks 13.1.3 and 13.1.6.

a) Its set of points is $\mathsf{Points}(\mathsf{Fr}(\mathbf{P},\mathsf{Open}))$; the set of abstract points in $\mathsf{Fr}(\mathbf{P},\mathsf{Open}) = (\mathsf{Open},\subseteq,\between)$, the semilattice of open sets of $(\mathbf{P},\mathsf{Open})$, and
b) Its open sets are given by the $Op(O)$, for $O\in\mathsf{Open}$.

It is sober, by Proposition 14.3.5.

Continuing Remark 15.1.2, a notation will be useful:

NOTATION 15.1.3. Suppose $(\mathbf{P},\mathsf{Open})$ is a semitopology. Then define

$$\mathsf{Soberify}(\mathbf{P},\mathsf{Open}) = \mathsf{St}\,\mathsf{Fr}(\mathbf{P},\mathsf{Open}).$$

We may use $\mathsf{Soberify}(\mathbf{P},\mathsf{Open})$ and $\mathsf{St}\,\mathsf{Fr}(\mathbf{P},\mathsf{Open})$ interchangeably, depending on whether we want to emphasise "this is a sober semitopology obtained from $(\mathbf{P},\mathsf{Open})$" or "this is St acting on $\mathsf{Fr}(\mathbf{P},\mathsf{Open}) = (\mathsf{Open},\subseteq,\between)$".

THEOREM 15.1.4. *Suppose* $(\mathbf{P},\mathsf{Open})$ *is a semitopology. Then*

1. *nbhd* : $\mathbf{P}\to\mathsf{Points}(\mathsf{Fr}(\mathbf{P},\mathsf{Open}))$ *is a morphism of semitopologies from* $(\mathbf{P},\mathsf{Open})$ *to* $\mathsf{St}\,\mathsf{Fr}(\mathbf{P},\mathsf{Open}) = \mathsf{Soberify}(\mathbf{P},\mathsf{Open})$
2. *taking* $(\mathbf{P},\mathsf{Open})$ *to a sober semitopology* $\mathsf{Soberify}(\mathbf{P},\mathsf{Open})$, *such that*
3. $nbhd^{-1}$ *induces a bijection on open sets by mapping* $Op(O)$ *to* O, *and furthermore this is an isomorphism of the semiframes of open sets, in the sense of Definition 14.1.5.*

Proof. We consider each part in turn:

1. Following Definition 15.1.1 we must show that *nbhd* is continuous (inverse images of open sets are open) from $(\mathbf{P},\mathsf{Open})$ to $\mathsf{Soberify}(\mathbf{P},\mathsf{Open})$. So following Definition 13.4.3(2), consider $Op(O)\in\mathsf{Open}(\mathsf{Soberify}(\mathbf{P},\mathsf{Open}))$. By Proposition 14.2.1(3)

$$nbhd^{-1}(Op(O)) = O\in\mathsf{Open}.$$

   Continuity follows.

2. $\mathsf{Soberify}(\mathbf{P},\mathsf{Open})$ is sober by Proposition 14.3.5.

3. This is Proposition 14.2.3. □

REMARK 15.1.5. We can summarise Theorem 15.1.4 as follows:

1. By construction, the kernel of the *nbhd* function (the relation determined by which points it maps to equal elements) is topological indistinguishability $\mathrel{\mathring{=}}$.
2. We can think of $\mathsf{St}\,\mathsf{Fr}(\mathbf{P},\mathsf{Open})$ as being obtained from $(\mathbf{P},\mathsf{Open})$ by
   a) quotienting topologically equivalent points to obtain a T_0 space, and then
   b) adding extra points to make it sober.

   See also the discussion in Remark 14.3.2 about what it means to have 'enough' points.
3. This is done without affecting the semiframe of open sets (up to isomorphism), with the semiframe bijection given by $nbhd^{-1}$.

In this sense, we can view $\mathsf{St}\,\mathsf{Fr}(\mathbf{P},\mathsf{Open})$ as a **soberification** of $(\mathbf{P},\mathsf{Open})$.

15.2 The categories Spatial/SemiFrame of (spatial) semiframes

Definition 15.2.1.

1. Suppose $(\mathsf{X}, \leq, *)$ and $(\mathsf{X}', \leq', *')$ are semiframes (Definition 12.3.1) and $g : \mathsf{X} \to \mathsf{X}'$ is any function. Then call g a **morphism of semiframes** when:
   a) g is a morphism of complete semilattices (Definition 12.1.4).
   b) g is **compatible**, by which we mean that $g(x') * g(x'')$ implies $x' * x''$ for every $x', x'' \in \mathsf{X}'$.
2. We define SemiFrame the **category of semiframes** such that:
   - objects are semiframes, and
   - arrows are morphisms of semiframes.
3. Write Spatial for the **category of spatial semiframes** and semiframe morphisms between them. By construction, Spatial is the full subcategory of SemiFrame, on its spatial semiframes (Definition 14.1.2).

Lemma 15.2.2. *Suppose $(\mathsf{X}, \leq, *)$ is a semiframe. Then $Op : (\mathsf{X}, \leq, *) \to \mathsf{Fr\,St}(\mathsf{X}, \leq, *)$ is a morphism of semiframes and is surjective on underlying sets.*

Proof. Following Definition 15.2.1(1) we must show that

- Op is a semilattice morphism (Definition 12.1.4) — commutes with joins and maps $\top_{\mathsf{X}}$ to $\mathsf{Points}(\mathsf{X}, \leq, *)$) — and
- is compatible with the compatibility relation $*$, and
- we must show that Op is surjective.

We consider each property in turn:

- *Op is a semilattice morphism.*
  $Op(\bigvee X) = \bigvee_{x \in X} Op(x)$ by Lemma 13.3.2, and $Op(\top_{\mathsf{X}}) = \mathsf{Points}(\mathsf{X}, \leq, *)$ by Proposition 13.3.3(3).
- *Op is compatible with $*$.*
  Unpacking Definition 15.2.1(1b), we must show that $Op(x) \between Op(x')$ implies $x * x'$. We use Proposition 13.3.3(2).
- *Op is surjective* … by Lemma 13.4.5. □

15.3 Functoriality of the maps

Definition 15.3.1. Suppose $g : (\mathsf{X}', \leq', *') \to (\mathsf{X}, \leq, *) \in \mathsf{SemiFrame}$. Define a mapping $g^{\circ} : \mathsf{St}(\mathsf{X}, \leq, *) \to \mathsf{St}(\mathsf{X}', \leq', *')$ by

$$\begin{array}{rcl} g^{\circ} : \mathsf{Points}(\mathsf{X}, \leq, *) & \longrightarrow & \mathsf{Points}(\mathsf{X}', \leq', *') \\ P & \longmapsto & P' = \{x' \in \mathsf{X}' \mid g(x') \in P\}. \end{array}$$

Remark 15.3.2. We will show that g° from Definition 15.3.1 is an arrow in SemiTop. We will need to prove the following:

- If $P\in\mathsf{Points}(\mathbf{X},\leq,*)$ then $g^\circ(P)\in\mathsf{Points}(\mathbf{X}',\leq',*')$.
- g° is a morphism of semitopologies.

We do this in Lemmas 15.3.3 and 15.3.6 respectively.

LEMMA 15.3.3 (g° well-defined). *Suppose* $g:(\mathbf{X}',\leq',*')\to(\mathbf{X},\leq,*)\in\mathsf{SemiFrame}$ *and suppose* $P\in\mathsf{Points}(\mathbf{X},\leq,*)$. *Then* $g^\circ(P)$ *from Definition 15.3.1 is indeed in* $\mathsf{Points}(\mathbf{X}',\leq',*')$ *— and thus* g° *is well-defined function from* $\mathsf{Points}(\mathbf{X},\leq,*)$ *to* $\mathsf{Points}(\mathbf{X}',\leq',*')$.

Proof. For brevity write

$$P'=\{x'\in\mathbf{X}'\mid g(x')\in P\}.$$

We must check that P' is a completely prime nonempty up-closed compatible subset of $\mathbf{X}'$. We consider each property in turn:

1. *P' is completely prime.*

   Consider some $X'\subseteq P'$ and suppose $g(\bigvee X')\in P$. By Definition 15.2.1(1a) g is a semilattice homomorphism, so by Definition 12.1.3(2) $g(\bigvee X')=\bigvee_{x'\in X'}g(x')$. Thus $\bigvee_{x'\in X'}g(x')\in P$. By assumption P is completely prime, so $g(x')\in P$ for some $x'\in X'$. Thus $x'\in P'$ for that x'. Since X' was arbitrary, it follows that P' is completely prime.

2. *P' is nonempty.*

   By assumption g is an arrow in $\mathsf{SemiFrame}$ (i.e. a semiframe morphism) and unpacking Definition 15.2.1(1a) it follows that it is a semilattice homomorphism. In particular by Definition 12.1.3(2) $g(\top_{\mathbf{X}'})=\top_{\mathbf{X}}$, and by Lemma 13.2.1(1) $\top_{\mathbf{X}}\in P$. Thus $\top_{\mathbf{X}'}\in P'$, so P' is nonempty.

3. *P' is up-closed.*

   Suppose $x'\in P'$ and $x'\leq x''$. By construction $g(x')\in P$. By Lemma 12.1.6 (because g is a semilattice morphism by Definition 15.2.1(1a)) g is monotone, so $g(x')\leq g(x'')$. By assumption in Definition 13.1.1(3) P is up-closed, so that $g(x'')\in P$ and thus $x''\in P'$ as required.

4. *P' is compatible.*

   Suppose $x',x''\in P'$. Thus $g(x'),g(x'')\in P$. By assumption in Definition 13.1.1(4) P is compatible, so $g(x')*g(x'')$. By compatibility of g (Definition 15.2.1(1b)) it follows that $x'*x''$. Thus P' is compatible. □

REMARK 15.3.4. *Note on design:* If we want to impose further conditions on being an abstract point (such as those discussed in Remark 13.1.6) then Lemma 15.3.3 would need to be extended to show that these further conditions are preserved by the g° operation, so that for $P\in\mathsf{Points}(\mathbf{X},\leq,*)$ an abstract point in $(\mathbf{X},\leq,*)$, $g^\circ(P)=\{x'\in\mathbf{X}'\mid g(x')\in P\}$ is an abstract point in $(\mathbf{X}',\leq',*')$.

For example: consider what would happen if we add the extra condition on semifilters from Remark 13.1.6. Then the P' defined in the proof of Lemma 15.3.3 above might not be closed under this additional condition (it will be if g is surjective). This could be mended by closing P' under greatest lower bounds that are not $\bot$, but that in turn might compromise the property of being completely prime. These comments are not a proof that the problems would be insuperable; but they suggest that complexity would be added. For now, we prefer to keep things simple!

LEMMA 15.3.5. *Suppose* $g : (\mathsf{X}', \leq', *') \to (\mathsf{X}, \leq, *) \in \mathsf{SemiFrame}$, *and suppose* $x' \in \mathsf{X}'$. *Then*

$$(g^\circ)^{-1}(Op(x')) = Op(g(x')).$$

Proof. Consider an abstract point $P \in \mathsf{Point}(\mathsf{Gr}(\mathsf{X}', \leq', *'))$. We just chase definitions:

$$\begin{aligned} P \in (g^\circ)^{-1}(Op(x')) &\iff g^\circ(P) \in Op(x') && \text{Fact of inverse image} \\ &\iff x' \in g^\circ(P) && \text{Definition 13.3.1} \\ &\iff g(x') \in P && \text{Definition 15.3.1} \\ &\iff P \in Op(g(x')). && \text{Definition 13.3.1} \end{aligned}$$

The choice of P was arbitrary, so $(g^\circ)^{-1}(Op(x')) = Op(g(x'))$ as required. □

LEMMA 15.3.6 (g° continuous). *Suppose* $g : (\mathsf{X}', \leq', *') \to (\mathsf{X}, \leq, *) \in \mathsf{SemiFrame}$. *Then* $g^\circ : \mathsf{St}(\mathsf{X}, \leq, *) \to \mathsf{St}(\mathsf{X}', \leq', *')$ *is continuous:*

$$(g^\circ)^{-1}(\mathcal{O}') \in \mathsf{Open}(\mathsf{St}(\mathsf{X}, \leq, *))$$

for every $\mathcal{O}' \in \mathsf{Open}(\mathsf{St}(\mathsf{X}', \leq', *'))$.

Proof. By Definition 13.4.3, $\mathcal{O}' = Op(x')$ for some $x' \in \mathsf{X}'$. By Lemma 15.3.5 $(g^\circ)^{-1}(Op(x')) = Op(g(x'))$. By Definition 13.4.3(2) $Op(g(x')) \in \mathsf{Open}(\mathsf{St}(\mathsf{X}, \leq, *))$. □

PROPOSITION 15.3.7 (Functoriality).

1. *Suppose* $f : (\mathsf{P}, \mathsf{Open}) \to (\mathsf{P}', \mathsf{Open}') \in \mathsf{SemiTop}$ *(so* f *is a continuous map on underlying points).*
   Then f^{-1} *is an arrow* $\mathsf{Fr}(\mathsf{P}', \mathsf{Open}') \to \mathsf{Fr}(\mathsf{P}, \mathsf{Open})$ *in* SemiFrame.
2. *Suppose* $g : (\mathsf{X}', \leq', *') \to (\mathsf{X}, \leq, *) \in \mathsf{SemiFrame}$.
   Then g° *from Definition 15.3.1 is an arrow* $\mathsf{St}(\mathsf{X}, \leq, *) \to \mathsf{St}(\mathsf{X}', \leq', *')$ *in* SemiTop.
3. *The assignments* $f \mapsto f^{-1}$ *and* $g \mapsto g^\circ$ *are* ***functorial*** *— they map identity maps to identity maps, and commute with function composition.*

Proof. We consider each part in turn:

1. Following Definition 15.2.1, we must check that f^{-1} is a morphism of semiframes. We just unpack what this means and see that the required properties are just facts of taking inverse images:
   - f^{-1} *commutes with joins, i.e. with* $\bigcup$.
     This is a fact of inverse images.
   - f^{-1} *maps* $\top_{\mathsf{Fr}(\mathsf{P}', \mathsf{Open}')} = \mathsf{P}'$ *to* $\top_{\mathsf{Fr}(\mathsf{P}, \mathsf{Open})} = \mathsf{P}$.
     This is a fact of inverse images.
   - f^{-1} *is compatible, meaning that* $f^{-1}(O') \between f^{-1}(O'')$ *implies* $O' \between O''$.
     This is a fact of inverse images.

2. We must check that g° is continuous. This is Lemma 15.3.6.

3. Checking functoriality is routine, but we sketch the reasoning anyway:

   - Consider the identity function id on some semitopology $(\mathsf{P}, \mathsf{Open})$. Then id^{-1} should be the identity function on $(\mathsf{Open}, \subseteq, \between)$. It is.
   - Consider $f : (\mathsf{P}, \mathsf{Open}) \to (\mathsf{P}', \mathsf{Open}')$ and $f' : (\mathsf{P}', \mathsf{Open}') \to (\mathsf{P}'', \mathsf{Open}'')$. Then $(f' \circ f)^{-1}$ should be equal to $f^{-1} \circ (f')^{-1}$. It is.
   - Consider the identity function id on $(\mathsf{X}, \leq, *)$. Then id° should be the identity function on $\mathsf{Points}(\mathsf{X}, \leq, *)$. We look at Definition 15.3.1 and see that this amounts to checking that $P = \{x \in \mathsf{X} \mid id(x) \in P\}$. It is.
   - Consider $g : (\mathsf{X}, \leq, *) \to (\mathsf{X}', \leq', *')$ and $g' : (\mathsf{X}', \leq', *') \to (\mathsf{X}'', \leq'', *'')$ and consider some $P'' \in \mathsf{Points}(\mathsf{X}'', \leq'', *'')$. Then $(g' \circ g)^{\circ}(P'')$ should be equal to $(g^{\circ} \circ (g')^{\circ})(P'')$. We look at Definition 15.3.1 and see that this amounts to checking that $\{x \in \mathsf{X} \mid g'(g(x)) \in P''\} = \{x \in \mathsf{X} \mid g(x) \in P'\}$ where $P' = \{x' \in \mathsf{X}' \mid g'(x') \in P''\}$. Unpacking these definitions, we see that the equality does indeed hold. □

15.4 Sober semitopologies are dual to spatial semiframes

We can now state the duality result between Sober and Spatial:

THEOREM 15.4.1. *The maps* St *(Definition 13.4.3) and* Fr *(Definition 12.3.4), with actions on arrows as described in Proposition 15.3.7, form a categorical duality between the categories*

- Sober *of sober semitopologies (Definition 14.3.1) and continuous compatible morphisms between them; and*
- Spatial *of spatial semiframes and morphisms between them (Definition 15.2.1(3)).*

Proof. There are various things to check:

- Proposition 14.3.5 shows that St maps spatial semiframes to sober semitopologies.
- Proposition 14.2.4 shows that Fr maps sober semitopologies to spatial semiframes.
- By Proposition 15.3.7 the maps $f \mapsto f^{-1}$ (inverse image) and $g \mapsto g^{\circ}$ (Definition 15.3.1) are functorial.
- The equivalence morphisms are given by the bijections Op and $nbhd$:
  - Op is from Definition 13.3.1. By Lemma 15.2.2 Op is a morphism $(\mathsf{X}, \leq, *) \to \mathsf{Fr}\,\mathsf{St}(\mathsf{X}, \leq, *)$ in Spatial that is surjective on underlying sets. Injectivity is from Proposition 14.1.4(3).
  - $nbhd$ is from Definition 5.4.4. By Theorem 15.1.4 $nbhd$ is a morphism $(\mathsf{P}, \mathsf{Open}) \to \mathsf{St}\,\mathsf{Fr}(\mathsf{P}, \mathsf{Open})$ in Sober. It is a bijection on underlying sets by the sobriety condition in Definition 14.3.1.

Finally, we must check naturality of Op and $nbhd$, which means (as standard) checking commutativity of the following diagrams:

$$\begin{array}{ccc} (\mathsf{P},\mathsf{Open}) & \xrightarrow{nbhd} & \mathsf{St}\,\mathsf{Fr}(\mathsf{P},\mathsf{Open}) \\ \downarrow f & & \downarrow (f^{-1})^{\circ} \\ (\mathsf{P}',\mathsf{Open}') & \xrightarrow{nbhd} & \mathsf{St}\,\mathsf{Fr}(\mathsf{P}',\mathsf{Open}') \end{array} \qquad \begin{array}{ccc} (\mathsf{X},\leq,*) & \xrightarrow{Op} & \mathsf{Fr}\,\mathsf{St}(\mathsf{X},\leq,*) \\ \downarrow g & & \downarrow (g^{\circ})^{-1} \\ (\mathsf{X}',\leq',*') & \xrightarrow{Op} & \mathsf{Fr}\,\mathsf{St}(\mathsf{X}',\leq',*') \end{array}$$

We proceed as follows:

- Suppose $g : (\mathsf{X}',\leq',*') \to (\mathsf{X},\leq,*)$ in Spatial, so that $g^{\circ} : \mathsf{St}(\mathsf{X},\leq,*) \to \mathsf{St}(\mathsf{X}',\leq',*')$ in Sober and $(g^{\circ})^{-1} : \mathsf{Fr}\,\mathsf{St}(\mathsf{X}',\leq',*') \to \mathsf{Fr}\,\mathsf{St}(\mathsf{X},\leq,*)$ in Spatial. To prove naturality we must check that
$$(g^{\circ})^{-1}(Op(x)) = Op(g(x))$$
for every $x \in \mathsf{X}$. This is just Lemma 15.3.5.

- Suppose $f : (\mathsf{P},\mathsf{Open}) \to (\mathsf{P}',\mathsf{Open}')$ in SemiTop, so that $f^{-1} : \mathsf{Fr}(\mathsf{P}',\mathsf{Open}') \to \mathsf{Fr}(\mathsf{P},\mathsf{Open})$ in Spatial and $(f^{-1})^{\circ} : \mathsf{St}\,\mathsf{Fr}(\mathsf{P},\mathsf{Open}) \to \mathsf{St}\,\mathsf{Fr}(\mathsf{P}',\mathsf{Open}')$ in SemiTop. To prove naturality we must check that
$$(f^{-1})^{\circ}(nbhd(p)) = nbhd(f(p)).$$
We just chase definitions, for an open set $O' \in \mathsf{Open}'$:
$$\begin{array}{rlr} O' \in (f^{-1})^{\circ}(nbhd(p)) & \Longleftrightarrow f^{-1}(O') \in nbhd(p) & \text{Definition 15.3.1} \\ & \Longleftrightarrow p \in f^{-1}(O') & \text{Definition 5.4.4} \\ & \Longleftrightarrow f(p) \in O' & \text{Inverse image} \\ & \Longleftrightarrow O' \in nbhd(f(p)) & \text{Definition 5.4.4.} \end{array}$$

□

REMARK 15.4.2. We review the background to Theorem 15.4.1:

1. A semitopology $(\mathsf{P},\mathsf{Open})$ is a set of points P and a set of open sets $\mathsf{Open} \subseteq pow(\mathsf{P})$ that contains P and is closed under arbitrary (possibly empty) unions (Definition 1.2.2).
2. A morphism between semitopologies is a continuous function, just as for topologies (Definition 15.1.1(1)).
3. A semiframe $(\mathsf{X},\leq,*)$ is a complete join-semilattice $(\mathsf{X},\leq)$ with a properly reflexive distributive *compatibility relation* $*$ (Definition 12.3.1).
4. A morphism between semiframes is a morphism of complete join-semilattices with $\top$ that is compatible with the compatibility relation (Definition 15.2.1(1)).
5. An *abstract point* of a semitopology $(\mathsf{P},\mathsf{Open})$ is a completely prime nonempty up-closed compatible subset $P \subseteq \mathsf{Open}$ (Definition 13.1.1(7)).
6. A semitopology is *sober* when the neighbourhood semifilter map $p \in \mathsf{P} \mapsto nbhd(p) = \{O \in \mathsf{Open} \mid p \in O\}$ is injective and surjective between the points of P and the abstract points of P (Definition 14.3.1).

7. By Theorem 15.1.4, and as discussed in Remark 15.1.5, every (possibly non-sober) semitopology $(\mathsf{P}, \mathsf{Open})$ maps into its *soberification* $\mathsf{St}\,\mathsf{Fr}(\mathsf{P}, \mathsf{Open})$, which has an isomorphic semiframe of open sets. So even if our semitopology $(\mathsf{P}, \mathsf{Open})$ is not sober, there is a standard recipe to make it so.
8. A semiframe is *spatial* when $x \in \mathsf{X} \mapsto Op(x) = \{P \in \mathsf{Point} \mid x \in P\}$ respects $\leq$ and $*$ in senses make formal in Definition 14.1.2 and Proposition 14.1.4.
9. Sober semitopologies and continuous functions between them, and spatial semiframes and semiframe morphisms between them, are categorically dual (Theorem 15.4.1).

REMARK 15.4.3. A *categorical duality* between two categories $\mathbb{C}$ and $\mathbb{D}$ is an equivalence between $\mathbb{C}$ and $\mathbb{D}^{op}$; this is an adjoint pair of functors whose unit and counit are natural isomorphisms. See [Mac71, IV.4].[2]

There are many duality results in the literature. The duality between topologies and frames is described (for example) in [MM92, page 479, Corollary 4]. A duality between distributive lattices and coherent spaces is in [Joh86, page 66]. There is the classic duality by Stone between Boolean algebras and compact Hausdorff spaces with a basis of clopen sets [Sto36, Joh86]. An encyclopedic treatment is in [Car11], with a rather good overview in Example 2.9 on page 17.

Theorem 15.4.1 appends another item to this extensive canon. It also constructively moves us forward in studying semitopologies, because it gives us an algebraic treatment of semitopologies, and a formal framework for studying morphisms between semitopologies. For instance: taking morphisms to be continuous functions is sensible not just because this is also how things work for topologies, but also because this is what is categorically dual to the $\leq/*$-homomorphisms between semiframes (Definition 15.2.1). And of course, if we become interested in different notions of semitopology morphism (a flavour of these is in Remark 23.3.3) then the algebraic framework gives us a distinct mathematical light with which to inspect and evaluate them.

Note what Theorem 15.4.1 does *not* do: it does not give a duality between all semitopologies and all semiframes; it gives a duality between sober semitopologies and spatial semiframes. This in itself is nothing new — the topological duality is just the same — but what is interesting is that our motivation for studying semitopologies comes from practical network systems. These tend to be (finite) non-sober semitopologies — non-sober, because a guarantee of sobriety cannot be enforced, and anyway it is precisely the point of the exercise to achieve coordination, *without* explicitly requiring every possible constellation of cooperating agents to be explicitly represented by a point.

It is true that by Theorem 15.1.4 every non-sober T_0 semitopology can be embedded into a sober one without affecting the semiframe of open sets, but this makes the system to which it corresponds larger, by adding points. Thus the duality in Theorem 15.4.1 is a mathematical statement, but not necessarily a practical one — and this is as expected, because we knew that this is an abstract result. $nbhd$ maps a point to a set of (open) sets; and Op maps an (open) set of points to a set of sets of (open) sets. Of course these might not be computationally optimal.

REMARK 15.4.4. We have constructed an algebraic representation of semitopologies — but this is not the last word on representing semitopologies. Other methodologies are also illuminating, and because our motivation comes from computer systems that are networks of machines, we are particularly interested in representations based on ideas from graphs. We will investigate these in Section 17.

[2]The Wikipedia page is also exceptionally clear [Wik24a].

Well-behavedness conditions, dually 16

We want to understand semifilters better, and in particular we want to understand how properties of semifilters and abstract points correspond to the well-behavedness properties which we found useful in studying semitopologies — for example *topens*, *regularity*, and being *unconflicted* (Definitions 3.2.1, 4.1.4 and 6.1.1).

16.1 (Maximal) semifilters and transitive elements

REMARK 16.1.1 (Semifilters are not filters). We know that semifilters do not necessarily behave like filters. For instance:

1. It is possible for a finite semifilter to have more than one minimal element, because the closure under binary meets condition of filters is replaced by a weaker compatibility condition (see also Remarks 13.1.3 and 13.1.6).
2. There are more semifilters than proper filters — even if the underlying space is a topology. For example, the discrete semitopology on $\{0,1,2\}$ (whose open sets are all subsets of the space) is a topology. Every proper filter in this space is a semifilter, but it also has a semifilter $\big\{\{0,1\},\{1,2\},\{2,0\},\{0,1,2\}\big\}$ (see the top-left diagram in Figure 14.1) and this is not a filter.

More on this in Subsection 13.2.2.

In summary: semifilters are different and we cannot necessarily take their behaviour for granted without checking it. We now examine them in more detail.

We start with some easy definitions and results:

NOTATION 16.1.2. Suppose $(\mathsf{X}, \leq, *)$ is a semiframe and $X, Y \subseteq \mathsf{X}$ and $x \in \mathsf{X}$. Then we generalise $x * y$ to $x * Y$, $X * y$, and $X * Y$ as follows:

1. Write $x * Y$ for $\forall y \in Y. x * y$.
2. Write $X * y$ for $\forall x \in X. x * y$.

3. Write $X * Y$ for $\forall x \in X. \forall y \in Y. x * y$.

We read $x * Y$ as 'x is **compatible** with Y', and similarly for $X * y$ and $X * Y$.

REMARK 16.1.3. We will see later on in Lemma 16.7.1 that $X * X'$ generalises $p \between p'$, in the sense that if $X = nbhd(p)$ and $X' = nbhd(p')$, then $p \between p'$ if and only if $nbhd(p) * nbhd(p')$.

LEMMA 16.1.4 (Characterisation of maximal semifilters). *Suppose* $(\mathsf{X}, \leq, *)$ *is a semiframe and* $F \subseteq \mathsf{X}$ *is a semifilter. Then the following conditions are equivalent:*

1. *F is maximal.*
2. *For every $x \in \mathsf{X}$, $x * F$ if and only if $x \in F$.*

Proof. We prove two implications:

- *Suppose F is a maximal semifilter.*

  Suppose $x \in F$. Then $x * F$ is immediate from Notation 16.1.2(1) and semifilter compatibility (Definition 13.1.1(4)).

  Suppose $x * F$; thus by Notation 16.1.2(1) x is compatible with (every element of) F. We note that the $\leq$-up-closure of $\{x\} \cup F$ is a semifilter (nonempty, up-closed, compatible). By maximality, $x \in F$.

- *Suppose $x * F$ if and only if $x \in F$, for every $x \in \mathsf{X}$.*

  Suppose F' is a semifilter and $F \subseteq F'$. Consider $x' \in F'$. Then $x * F$ by compatibility of F', and so $x \in F$. Thus, $F' \subseteq F$. □

DEFINITION 16.1.5. Suppose $(\mathsf{X}, \leq, *)$ is a semiframe and $x \in \mathsf{X}$. Call x **transitive** when:

1. $x \neq \bot_{\mathsf{X}}$.
2. $x' * x * x''$ implies $x' * x''$, for every $x', x'' \in \mathsf{X}$.

'Being topen' in semitopologies (Definition 3.2.1(2)) corresponds to 'being transitive' in semiframes (Definition 16.1.5):

LEMMA 16.1.6 (Characterisation of topen sets). *Suppose* $(\mathsf{P}, \mathsf{Open})$ *is a semitopology and* $O \in \mathsf{Open}$. *Then the following are equivalent:*

1. *O is topen in $(\mathsf{P}, \mathsf{Open})$ in the sense of Definition 3.2.1(2).*
2. *O is transitive in $(\mathsf{Open}, \subseteq, \between)$ in the sense of Definition 16.1.5.*[1]

Proof. We unpack the definitions and note that the condition for being topen — being a nonempty open set that is transitive for $\between$ — is identical to the condition for being transitive in $(\mathsf{Open}, \subseteq, \between)$ — being a non-$\bot_{\mathsf{Open}}$ element that is transitive for $* = \between$. □

[1] *Confusing terminology alert:* Definition 3.2.1(1) also has a notion of *transitive set*. The notion of transitive set is well-defined for a set that may not be open. In the world of semiframes, we just have elements of the semiframe (which correspond, intuitively, to open sets). Thus *transitive* semiframe elements correspond to (nonempty) transitive open sets of a semitopology, which are called *topens*.

16.2 The compatibility system x^*

DEFINITION 16.2.1. Suppose $(\mathsf{X}, \leq, *)$ is a semiframe and $x \in \mathsf{X}$. Then define x^* the **compatibility system** of x by

$$x^* = \{x' \in \mathsf{X} \mid x' * x\}.$$

LEMMA 16.2.2. *Suppose $(\mathsf{X}, \leq, *)$ is a semiframe and $X \subseteq \mathsf{X}$. Then $(\bigvee X)^* = \bigcup_{x \in X} x^*$.*

Proof. We just follow the definitions:

$$\begin{aligned} y \in (\bigvee X)^* &\iff y * \bigvee X && \text{Definition 16.2.1} \\ &\iff \exists x \in X. y * x && \text{Definition 12.2.1(3)} \\ &\iff \exists x \in X. y \in x^* && \text{Definition 16.2.1} \\ &\iff y \in \bigcup_{x \in X} x^* && \text{Fact of sets} \end{aligned}$$

□

LEMMA 16.2.3. *Suppose $(\mathsf{X}, \leq, *)$ is a semiframe and $x \in \mathsf{X}$ is transitive. Then the following are equivalent for every $y \in \mathsf{X}$:*

$$y * x \quad \iff \quad y \in x^* \quad \iff \quad y * x^*.$$

Proof. We prove a cycle of implications:

- Suppose $y * x$. Then $y \in x^*$ is direct from Definition 16.2.1.
- Suppose $y \in x^*$. Then $y * x^*$ — meaning by Notation 16.1.2(1) that $y * x'$ for every $x' \in x^*$ — follows by transitivity of x.
- Suppose $y * x^*$. By proper reflexivity of $*$ (Definition 12.2.1(2); since $x \neq \bot_{\mathsf{X}}$) $x \in x^*$, and $y * x$ follows. □

PROPOSITION 16.2.4. *Suppose $(\mathsf{X}, \leq, *)$ is a semiframe and suppose $\bot_{\mathsf{X}} \neq x \in \mathsf{X}$. Then the following are equivalent:*

1. *x is transitive.*
2. *x^* is a completely prime semifilter (i.e. an abstract point).*
3. *x^* is a semifilter.*
4. *x^* is compatible.*
5. *x^* is a maximal semifilter.*

Proof. We first prove a cycle of implications between parts 1, 2, 3, and 4:

1. Suppose x is transitive. We need to check that x^* is nonempty, up-closed, compatible, and completely prime. We consider each property in turn:

   - $x * x$ by proper reflexivity of $*$ (Definition 12.2.1(2); since $x \neq \bot_{\mathsf{X}}$), so $x \in x^*$.
   - It follows from monotonicity of $*$ (Lemma 12.2.3(1)) that if $x' \leq x''$ and $x * x'$ then $x * x''$.
   - Suppose $x' * x * x''$. By transitivity of x (Definition 16.1.5), $x' * x''$.
   - Suppose $x * \bigvee X'$; then by distributivity of $*$ (Definition 12.2.1(3)) $x * x'$ for some $x' \in X'$.

2. If x^* is a completely prime semifilter, then it is certainly a semifilter.

3. If x^* is a semifilter, then it is compatible (Definition 13.1.1(5)&4).

4. Suppose x^* is compatible (Definition 13.1.1(4)) and suppose $x' * x * x''$. By Lemma 16.2.3 $x', x'' \in x^*$, and by compatibility of x^* we have $x' * x''$. Thus, x is transitive.

To conclude, we prove two implications between parts 4 and 5:

- Suppose x^* is a semifilter. By equivalence of parts 3 and 1 of this result, x is transitive, and so using Lemma 16.2.3 $x' * x^*$ if and only if $x' \in x^*$. By Lemma 16.1.4, x^* is maximal.
- Clearly, if x^* is a maximal semifilter then it is a semifilter. □

16.3 The compatibility system F^*

16.3.1 Basic definitions and results

DEFINITION 16.3.1. Suppose $(\mathsf{X}, \leq, *)$ is a semiframe and suppose $F \subseteq \mathsf{X}$ (F may be a semifilter, but the definition does not depend on this). Define F^* the **compatibility system** of F by

$$F^* = \{x' \in \mathsf{X} \mid x' * F\}$$

Unpacking Notation 16.1.2(1), and combining with Definition 16.2.1, we can write:

$$F^* = \{x' \in \mathsf{X} \mid x' * F\} = \{x' \in \mathsf{X} \mid \forall x \in F. x' * x\} = \bigcap\{x^* \mid x \in F\}.$$

Lemma 16.3.2 presents one easy and useful example of Definition 16.3.1:

LEMMA 16.3.2. *Suppose* $(\mathsf{P}, \mathsf{Open})$ *is a semitopology and suppose* $p \in \mathsf{P}$ *and* $O' \in \mathsf{Open}$. *Then:*

$$\begin{aligned} O' \in nbhd(p)^* &\iff \forall O \in \mathsf{Open}. p \in O \Longrightarrow O' \between O \\ O' \notin nbhd(p)^* &\iff \exists O \in \mathsf{Open}. p \in O \land O' \not\between O. \end{aligned}$$

Proof. We just unpack Definitions 5.4.4 and 16.3.1. □

LEMMA 16.3.3. *Suppose* $(\mathsf{X}, \leq, *)$ *is a semiframe and* $F \subseteq \mathsf{X}$. *Then* F^* *is up-closed.*

Proof. This is just from Definition 16.3.1 and monotonicity of $*$ (Lemma 12.2.3(1)). □

LEMMA 16.3.4. *Suppose* $(\mathsf{X}, \leq, *)$ *is a semiframe and* $F \subseteq \mathsf{X}$ *is a semifilter. Then:*

1. *If* $x \in F$ *then* $F \subseteq x^*$.
2. *As a corollary,* $F \subseteq F^*$.

Proof. Suppose $x \in F$. By compatibility of F (Definition 13.1.1(4)), $x' * x$ for every $x' \in F$. It follows from Definition 16.2.1 that $F \subseteq x^*$. The corollary is immediate from Definition 16.3.1. □

We can use Lemma 16.3.4 and Definition 16.3.1 to give a more succinct rendering of Lemma 16.1.4:

COROLLARY 16.3.5. *Suppose* $(\mathsf{X}, \leq, *)$ *is a semiframe and* $F \subseteq \mathsf{X}$ *is a semifilter. Then the following are equivalent:*

1. *F is maximal.*
2. *$F^* = F$.*
3. *$F^* \subseteq F$.*

Proof. Equivalence of parts 1 and 2 just repeats Lemma 16.1.4 using Definition 16.3.1. To prove equivalence of parts 2 and 3 we use use Lemma 16.3.4(2). □

16.3.2 Strong compatibility: when F^* is a semifilter

Proposition 16.2.4 relates good properties of x (transitivity) to good properties of its compatibility system x^* (e.g. being compatible). It will be helpful to ask similar questions of F^*. What good properties are of interest for F^*, and what conditions can we impose on F to guarantee them?

DEFINITION 16.3.6. Suppose $(\mathsf{X}, \leq, *)$ is a semiframe. Then:

1. Call $F \subseteq \mathsf{X}$ **strongly compatible** when F^* is nonempty and compatible.
2. Call $(\mathsf{X}, \leq, *)$ **strongly compatible** when every abstract point (completely prime semifilter) $P \subseteq \mathsf{X}$ is strongly compatible.

REMARK 16.3.7. For the reader's convenience we unpack Definition 16.3.6.

1. By Definition 13.1.1(4), F^* is compatible when $x * x'$ for every $x, x' \in F^*$. Combining this with Definition 16.3.1 and Notation 16.1.2, F^* is compatible when $x * F * x'$ implies $x * x'$, for every $x, x' \in \mathsf{X}$. Thus, F is strongly compatible when
$$\forall x, x' \in \mathsf{X}.\ x * F * x' \Longrightarrow x * x'.$$
2. $(\mathsf{X}, \leq, *)$ is strongly compatible when every abstract point $P \in \mathsf{Point}(\mathsf{X}, \leq, *)$ is strongly compatible in the sense just given above.

LEMMA 16.3.8. *Suppose $(\mathsf{P}, \mathsf{Open})$ is a semitopology and $p \in \mathsf{P}$. Recall from Definition 12.3.4(3) and Lemma 12.3.5 that $(\mathsf{Open}, \subseteq, \between)$ is a semiframe. Then the following are equivalent:*

1. *The point $p \in \mathsf{P}$ is hypertransitive in the sense of Definition 6.5.4.*
2. *The semifilter $nbhd(p) \subseteq \mathsf{Open}$ is strongly compatible in the sense of Definition 16.3.6.*

Proof. Remark 16.3.7 notes that the condition in Definition 6.5.4 is precisely the condition for $nbhd(p)$ to be strongly compatible. □

REMARK 16.3.9. Given Lemma 16.3.8, the reader might ask why we do not just call a strongly compatible semifilter 'hypertransitive'.

There is a case for doing so, but caution is required: strong compatibility of semiframes is not *quite* the same thing as hypertransitivity of points. Every point p generates a semifilter $nbhd(p)$, but there may be more semifilters than there are points, and this makes the strong compatibility condition subtly different from the hypertransitivity condition. We shall see the effects of this in Lemma 16.5.3(2), and in Theorem 16.5.4 (see Remark 16.5.5 for a brief discussion), and then again in Definition 16.8.5 where we define a notion of *strongly compatible semitopology* (essentially: all of its semifilters are strongly compatible), which is not the same thing as the space being hypertransitive (essentially: all of its points are hypertransitive).

Therefore, we maintain a terminological distinction: *points* are hypertransitive, *semiframes* are strongly compatible. The notions are related, but not quite the same thing.

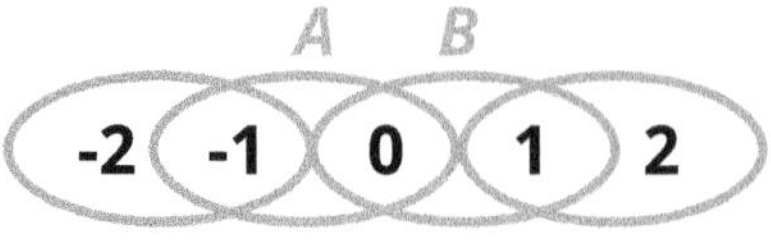

Figure 16.1: Strongly compatible filter that contains no transitive element

LEMMA 16.3.10. *Suppose* $(\mathsf{X}, \leq, *)$ *is a semiframe and suppose* $F \subseteq \mathsf{X}$ *is nonempty. Then the following are equivalent:*

1. F^* *is a semifilter.*
2. F^* *is compatible.*
3. F *is strongly compatible.*

Proof. Equivalence of parts 2 and 3 is just Definition 16.3.6. For equivalence of parts 1 and 2 we prove two implications:

- Suppose F^* is a semifilter. Then F^* is compatible by assumption in Definition 13.1.1(5).
- Suppose F^* is compatible. Then F^* is up-closed by Lemma 16.3.3, and nonempty by Lemma 16.3.4(2) (since F is nonempty). Thus, by Definition 13.1.1(5) F^* is a semifilter. □

LEMMA 16.3.11. *Suppose* $(\mathsf{X}, \leq, *)$ *is a semiframe and suppose* $F \subseteq \mathsf{X}$*. Then it is not necessarily the case that* F^* *is a semifilter.*

This non-implication holds even in strong well-behavedness conditions: that $(\mathsf{X}, \leq, *)$ *is spatial and* F *is an abstract point (a completely prime semifilter).*

Proof. It suffices to provide a counterexample. Let $(\mathsf{P}, \mathsf{Open}) = (\{0, 1, 2\}, \{\varnothing, \{0\}, \{2\}, \mathsf{P}\})$, as illustrated in the top-left semitopology in Figure 3.1. Take $(\mathsf{X}, \leq, *) = (\mathsf{Open}, \subseteq, \between)$ (which is spatial by Proposition 14.2.4) and set $F = \mathit{nbhd}(1) = \{0, 1, 2\}$. Then $\mathit{nbhd}(1)^* = \{\{0\}, \{2\}, \{0, 1, 2\}\}$, and this is not compatible because $\{0\} \not\between \{2\}$.[2] □

REMARK 16.3.12. Lemma 16.3.11 gives an example of a semifilter F that is not strongly compatible (i.e. such that F^* is not a semifilter). Note that in this example both the space and F are well-behaved. This raises the question of finding sufficient (though perhaps not necessary) criteria for strong compatibility. We conclude with Proposition 16.3.13 which provides one such criterion; it will be useful later in Lemma 16.5.3 and Theorem 16.8.7.

Proposition 16.3.13 bears a family resemblance to Theorem 4.2.6 (if a point has a topen neighbourhood then it is regular):

PROPOSITION 16.3.13. *Suppose* $(\mathsf{X}, \leq, *)$ *is a semiframe and* $F \subseteq \mathsf{X}$ *is a semifilter. Then:*

1. *If* F *contains a transitive element then* F *is strongly compatible.*
2. *The converse implication need not hold: it may be that* F *is strongly compatible yet* F *contains no transitive element.*

[2] 1 is also a *conflicted* point; see Example 6.1.2(1). This is no accident: by Lemma 6.5.6(2) if p is conflicted then it is not hypertransitive, and by Lemma 16.3.8 it follows that $\mathit{nbhd}(p)^*$ is not compatible.

Proof. We consider each part in turn:

1. Suppose $x \in F$ is transitive. By Lemma 16.3.10 it would suffice to show that F^* is compatible (Definition 13.1.1(4)). So consider $y * F * y'$. Then $y * x * y'$ and by transitivity $y * y'$. Thus F^* is compatible.

2. It suffices to provide a counterexample. We take, as illustrated in Figure 16.1,

   - $\mathsf{P} = \{\text{-}2, \text{-}1, 0, 1, 2\}$ and
   - we let Open be generated by $\{i, i{+}1\}$ for $\text{-}2 \leq i \leq 1$ (unordered pairs of adjacent numbers).

   Write $A = \{\text{-}1, 0\}$ and $B = \{0, 1\}$ and let F be the up-closure of $\{A, B\}$. Note that A and B are not transitive (i.e. not topen). The reader can check that $F^* = F$ (e.g. $\{1,2\} \notin F^*$ because $\{1,2\} \not\between \{\text{-}1,0\} \in F$), but F contains no transitive element. □

16.4 Semiframe characterisation of community

Remark 16.4.1. We saw the notion of $K(p)$ the *community* of a point in Definition 4.1.4(1). In this Subsection we construct an analogue to it in semiframes. We will give two characterisations: one in Definition 16.4.5, and another in Proposition 16.4.7.

We will mostly be interested in Definition 16.4.2 when F is a semifilter, but the definition does not require this:

Definition 16.4.2. Suppose $(\mathsf{X}, \leq, *)$ is a semiframe and $F \subseteq \mathsf{X}$ and $x \in \mathsf{X}$. Then define $F^c \in \mathsf{X}$, $F^{*c} \in \mathsf{X}$, and $x^{*c} \in \mathsf{X}$ by

$$\begin{aligned} F^c &= \bigvee \{y \in \mathsf{X} \mid y \notin F\} \\ F^{*c} &= (F^*)^c \\ x^{*c} &= (x^*)^c. \end{aligned}$$

Remark 16.4.3. We unpack the definitions of F^{*c} and x^{*c}:

$$\begin{aligned} F^{*c} &= (F^*)^c && \text{Definition 16.4.2} \\ &= \bigvee \{y \in \mathsf{X} \mid y \notin F^*\} && \text{Definition 16.4.2} \\ &= \bigvee \{y \in \mathsf{X} \mid \neg(y * F)\} && \text{Definition 16.3.1} \end{aligned}$$

$$\begin{aligned} x^{*c} &= (x^*)^c && \text{Definition 16.4.2} \\ &= \bigvee \{y \in \mathsf{X} \mid y \notin x^*\} && \text{Definition 16.4.2} \\ &= \bigvee \{y \in \mathsf{X} \mid \neg(y * x)\}. && \text{Definition 16.2.1} \end{aligned}$$

Lemma 16.4.4 will be useful, and gives some intuition for $(\text{-})^c$ and $(\text{-})^{*c}$ by unpacking their concrete meaning in the special case of a semiframe of open sets of a semitopology:

Lemma 16.4.4. *Suppose* $(\mathsf{P}, \mathsf{Open})$ *is a semitopology and* $p \in \mathsf{P}$ *and* $O \in \mathsf{Open}$. *Then:*

1. $nbhd(p)^c = \mathsf{P} \setminus |p|$.
2. $nbhd(p)^{*c} = \mathsf{P} \setminus p_{\between}$.
3. $O^{*c} = \mathsf{P} \setminus |O| = interior(\mathsf{P} \setminus O)$.

Proof. We consider each part in turn:

1. It is a fact of Definition 5.1.2 that $\mathbf{P}\setminus|p| = \bigcup\{O'\in\mathsf{Open} \mid p\notin O'\}$. By Proposition 14.2.1(2) $p\notin O'$ if and only if $O'\notin nbhd(p)$.

2. It is a fact of Definition 3.6.1, which is spelled out in Lemma 5.8.2(2), that $\mathbf{P}\setminus p_{\between} = \bigcup\{O'\in\mathsf{Open} \mid \exists O\in\mathsf{Open}.p\in O\wedge O'\not\between O\}$. By Lemma 16.3.2 $\exists O\in\mathsf{Open}.p\in O\wedge O'\not\between O$ precisely when $O'\notin nbhd(p)^*$.

3. By Definitions 16.4.2 and 16.3.1 we have

$$O^c = (O^*)^{*c} = \bigcup\{O'\in\mathsf{Open} \mid O'\notin O^*\} = \bigcup\{O'\in\mathsf{Open} \mid O'\not\between O\}.$$

The result then follows by routine reasoning on closures (Definition 5.1.2). □

Definition 16.4.5. Suppose $(\mathbf{X},\leq,*)$ is a semiframe and $F\subseteq\mathbf{X}$. Then define $k(F)\in\mathbf{X}$ the **abstract community** of F by

$$k(F) = (F^{*c})^{*c}\in\mathbf{X}.$$

(For a more direct characterisation, see Proposition 16.4.7.)

Proposition 16.4.6. *Suppose* $(\mathbf{P},\mathsf{Open})$ *is a semitopology and* $p\in\mathbf{P}$. *Then*

$$k(nbhd(p)) = K(p).$$

In words: the abstract community of the abstract point $nbhd(p)$ *in* $(\mathsf{Open},\subseteq,\between)$*, is identical to the community of* p.

Proof. We reason as follows:

$$\begin{aligned}
k(nbhd(p)) &= (nbhd(p)^{*c})^{*c} && \text{Definition 16.4.5}\\
&= (\mathbf{P}\setminus p_{\between})^{*c} && \text{Lemma 16.4.4(2)}\\
&= interior(\mathbf{P}\setminus(\mathbf{P}\setminus p_{\between})) && \text{Lemma 16.4.4(3)}\\
&= interior(p_{\between}) && \text{Fact of sets}\\
&= K(p) && \text{Definition 4.1.4(1)}
\end{aligned}$$

□

We can also give a more direct characterisation of the abstract community from Definition 16.4.5:

Proposition 16.4.7. *Suppose* $(\mathbf{X},\leq,*)$ *is a semiframe and* $F\subseteq\mathbf{X}$. *Then*

$$k(F) = \bigvee\{x\in\mathbf{X} \mid x^*\subseteq F^*\},$$

and $k(F)$ *is the greatest element in* $\mathbf{X}$ *such that* $k(F)^*\subseteq F^*$.

Proof. We follow the definitions:

$$\begin{aligned}
(F^{*c})^{*c} &= \bigvee\{x\in\mathbf{X} \mid \neg(x * F^{*c})\} && \text{Remark 16.4.3}\\
&= \bigvee\{x\in\mathbf{X} \mid \neg(x * \bigvee\{y \mid \neg(y * F)\})\} && \text{Remark 16.4.3}\\
&= \bigvee\{x\in\mathbf{X} \mid \neg\exists y\in\mathbf{X}.(x * y \wedge \neg(y * F))\} && \text{Definition 12.2.1(3)}\\
&= \bigvee\{x\in\mathbf{X} \mid \forall y\in\mathbf{X}.y * x \Longrightarrow y * F\} && \text{Fact of logic}\\
&= \bigvee\{x\in\mathbf{X} \mid x^*\subseteq F^*\} && \text{Definitions 16.2.1 \& 16.3.1}
\end{aligned}$$

To see that $k(F)$ is the greatest element such that $k(F)^*\subseteq F^*$, we note from Lemma 16.2.2 that

$$k(F)^* = \bigcup\{x^* \mid x\in\mathbf{X},\ x^*\subseteq F^*\}.$$

□

16.5 Semiframe characterisation of regularity

We now have enough to generalise the notions of quasiregularity, weak regularity, and regularity from semitopologies (Definition 4.1.4 parts 5, 4, and 3) to semiframes:

DEFINITION 16.5.1. Suppose $(\mathsf{X}, \leq, *)$ is a semiframe and $F \subseteq \mathsf{X}$ is a semifilter.

1. Call F **quasiregular** when $k(F) \neq \bot_{\mathsf{X}}$.
   Thus, there exists some $x \in \mathsf{X}$ such that $x^* \subseteq F^*$.
2. Call F **weakly regular** when $k(F) \in F$.
3. Call F **regular** when $k(F) \in F$ and $k(F)$ is transitive.

Lemma 16.5.2 does for semiframes what Lemma 4.1.6 does for semitopologies:

LEMMA 16.5.2. *Suppose* $(\mathsf{X}, \leq, *)$ *is a semiframe and* $F \subseteq \mathsf{X}$ *is a semifilter. Then:*

1. *If F is regular then it is weakly regular.*
2. *If F is weakly regular then it is quasiregular.*

(The converse implications need not hold, and it is possible for F to not be quasiregular: it is convenient to defer the proofs to Corollary 16.6.3.)

Proof. The proofs are easy: If $k(F) \in F$ and $k(F)$ is transitive, then certainly $k(F) \in F$. If $k(F) \in F$ then by Lemma 13.2.1(2) $k(F) \neq \bot_{\mathsf{X}}$. □

LEMMA 16.5.3. *Suppose* $(\mathsf{X}, \leq, *)$ *is a semiframe and* $F \subseteq \mathsf{X}$ *is a semifilter. Then:*

1. *If F is quasiregular and strongly compatible then $k(F)$ is transitive.*
2. *The converse implication need not hold: it is possible for F to be quasiregular and $k(F)$ to be transitive, yet F is not strongly compatible.*
3. *If F is weakly regular and $k(F)$ is transitive then F is strongly compatible.*
4. *If F is weakly regular, then $k(F)$ is transitive if and only if F is strongly compatible.*

Proof. We consider each part in turn:

1. Suppose F is quasiregular and strongly compatible.
   By quasiregularity $\bot_{\mathsf{X}} \neq k(F)$. By Proposition 16.4.7 $k(F)^* \subseteq F^*$. By strong compatibility F^* is a semifilter and so in particular F^* is compatible. It follows from Proposition 16.2.4(1&4) that $k(F)$ is transitive, as required.

2. It suffices to provide a counterexample. Let $(\mathbb{R}, \mathsf{Open})$ be the real numbers with their usual topology, and let $(\mathbb{R}, \mathsf{Open}')$ be the topology generated by $\mathsf{Open} \cup \{\{0\}\}$ — in words: we add $\{0\}$ as an open set.

   Let F be the semifilter of all Open-open neighbourhoods of 0. F^* is the set of Open'-open sets that intersect every Open-open neighbourhood of 0. This is not compatible, because it contains $(0,)$ (the set of numbers strictly less than 0) and $(, 0)$ (the set of numbers strictly greater than 0), and these do not intersect. Using Proposition 16.4.7, we calculate that $k(F) = \{0\}$; this is transitive because it is a singleton set.

   So F is quasiregular, $k(F)$ is transitive, yet F is not strongly compatible.

3. Suppose $k(F)$ is transitive and suppose F is weakly regular, so $k(F)\in F$. By Proposition 16.3.13 F is strongly compatible.

4. From parts 1 and 3 of this result, noting from Lemma 16.5.2 that if F is weakly regular then it is quasiregular. □

THEOREM 16.5.4. *Suppose* $(\mathsf{X}, \leq, *)$ *is a semiframe and* $F \subseteq \mathsf{X}$ *is a semifilter. Then* F *is regular if and only if* F *is weakly regular and strongly compatible. We can write this succinctly as follows:*

Regular = weakly regular + strongly compatible.[3]

Proof. Suppose F is weakly regular and strongly compatible. By Lemma 16.5.3(4) $k(F)$ is transitive, and by Definition 16.5.1(3) F is regular.

For the converse implication we just reverse the reasoning above. □

REMARK 16.5.5. In Theorem 6.5.8 we characterised regularity of points in terms of quasiregularity and being hypertransitive. In view of Lemma 16.3.8 we might expect Theorem 16.5.4 to read 'regular = quasiregular + strongly compatible'. But this is false, as per the discussion in Remark 16.3.9 and the counterexample in Lemma 16.5.3(2). Thus, the semiframes results are subtly different from those governing point-set semitopologies.

16.6 Semiframe characterisation of (quasi/weak)regularity

The direct translation in Definition 16.5.1 of parts 5, 4, and 3 of Definition 4.1.4, along with the machinery we have now built, makes Lemma 16.6.1 easy to prove:

LEMMA 16.6.1. *Suppose* $(\mathsf{P}, \mathsf{Open})$ *is a semitopology and* $p \in \mathsf{P}$. *Recall from Definition 5.4.4 and Proposition 14.2.1(1) that* $nbhd(p) = \{O \in \mathsf{Open} \mid p \in O\}$ *is a (completely prime) semifilter. Then:*

1. *p is quasiregular in the sense of Definition 4.1.4(5) if and only if $nbhd(p)$ is quasiregular in the sense of Definition 16.5.1(1).*
2. *p is weakly regular in the sense of Definition 4.1.4(4) if and only if $nbhd(p)$ is weakly regular in the sense of Definition 16.5.1(2).*
3. *p is regular in the sense of Definition 4.1.4(3) if and only if $nbhd(p)$ is regular in the sense of Definition 16.5.1(3).*

Proof. We consider each part in turn:

1. Suppose p is quasiregular. By Definition 4.1.4(5) $K(p) \neq \varnothing$. By Proposition 16.4.6 $k(nbhd(p)) \neq \varnothing = \bot_{\mathsf{Open}}$. By Definition 16.5.1(1) $nbhd(p)$ is quasiregular.

   The reverse implication follows just reversing the reasoning above.

2. Suppose p is weakly regular. By Definition 4.1.4(4) $p \in K(p)$. By Definition 5.4.4 $K(p) \in nbhd(p)$. By Proposition 16.4.6 $k(nbhd(p)) \in nbhd(p)$ as required.

   The reverse implication follows just reversing the reasoning above.

[3]Compare this slogan with the version for semitopologies in Theorem 6.2.2.

3. Suppose p is regular. By Definition 4.1.4(3) $p\in K(p)\in\mathsf{Topen}$. By Definition 5.4.4 and Proposition 16.4.6 $k(nbhd(p))\in nbhd(p)$. By Proposition 16.4.6 and Lemma 16.1.6 $k(nbhd(p))$ is transitive.

   The reverse implication follows just reversing the reasoning above. □

PROPOSITION 16.6.2. *Suppose* $(\mathbf{P},\mathsf{Open})$ *is a semitopology and* $p\in\mathbf{P}$. *Then*

- *p is quasiregular / weakly regular / regular in* $(\mathbf{P},\mathsf{Open})$ *in the sense of Definition 4.1.4 if and only if*
- *$nbhd(p)$ is quasiregular / weakly regular / regular in* $\mathsf{Soberify}(\mathbf{P},\mathsf{Open})$ *in the sense of Definition 16.5.1.*

Proof. We consider just the case of regularity; quasiregularity and weak regularity are no different.

Suppose p is regular. By Definition 4.1.4(3) $p\in K(p)\in\mathsf{Topen}$. It follows from Lemma 16.1.6 that $K(p)$ is transitive in $(\mathsf{Open},\subseteq,\between)$, and from Proposition 14.2.1(2) that $K(p)\in nbhd(p)$. It follows from Proposition 16.4.6 that $nbhd(p)$ is regular in the sense of Definition 16.5.1(3). □

COROLLARY 16.6.3. *Suppose* $(\mathbf{X},\leq,*)$ *is a semiframe and* $F\subseteq\mathbf{X}$ *is a semifilter. Then the converse implications in Lemma 16.5.2 need not hold: F may be quasiregular but not regular, and it may be weakly regular but not regular, and it may not even be quasiregular.*

Proof. It suffices to provide counterexamples. We easily obtain these by using Proposition 16.6.2 to consider $nbhd(p)$ for $p\in\mathbf{P}$ as used in Lemma 4.1.8. □

16.7 Characterisation of being intertwined

This Subsection continues Remark 16.1.3.

The notion of points being intertwined from Definition 3.6.1(1) generalises in semiframes to the notion of semifilters being compatible:

LEMMA 16.7.1. *Suppose* $(\mathbf{P},\mathsf{Open})$ *is a semitopology and* $p,p'\in\mathbf{P}$. *Then*

$$p\between p' \quad\Longleftrightarrow\quad nbhd(p)*nbhd(p') \quad\Longleftrightarrow\quad nbhd(p)\between nbhd(p').$$

For clarity and precision we unpack this. The following are equivalent:

1. *$p\between p'$ in the semitopology* $(\mathbf{P},\mathsf{Open})$ *(Definition 3.6.1(1)).*
   In words: the point p is intertwined with the point p'.
2. *$nbhd(p)*nbhd(p')$ in the semiframe* $(\mathsf{Open},\subseteq,\between)$ *(Notation 16.1.2(3)).*
   In words: the abstract point $nbhd(p)$ is compatible with the abstract point $nbhd(p')$.
3. *$nbhd(p)\between nbhd(p')$ in the semitopology* $\mathsf{St}(\mathsf{Open},\subseteq,\between)$ *(Definition 3.6.1(1)).*
   In words: the point $nbhd(p)$ is intertwined with the point $nbhd(p')$.

Proof. We unpack definitions:

- By Definition 3.6.1(1) $p\between p'$ when for every pair of open neighbourhoods $p\in O$ and $p'\in O'$ we have $O\between O'$.

- By Notation 16.1.2(3) $nbhd(p) * nbhd(p')$ when for every $O \in nbhd(p)$ and $O' \in nbhd(p')$ we have $O * O'$.
  By Proposition 14.2.1(2) we can simplify this to: $p \in O$ and $p' \in O'$ implies $O * O'$.
- By Definition 3.6.1(1) and Theorem 15.1.4, $nbhd(p) \between nbhd(p')$ when: for every pair of open neighbourhoods $nbhd(p) \in Op(O)$ and $nbhd(p') \in Op(O')$ we have $Op(O) \between Op(O')$.
  By Proposition 14.2.1(2) we can simplify this to: $p \in O$ and $p' \in O'$ implies $Op(O) \between Op(O')$.
  By Proposition 13.3.3(2) we can simplify this further to: $p \in O$ and $p' \in O'$ implies $O * O'$.

But by definition, the compatibility relation $*$ of $(\mathsf{Open}, \subseteq, \between)$ is $\between$, so $O * O'$ and $O \between O'$ are the same assertion. The equivalences follow. □

The property of being intertwined is preserved and reflected when we use $nbhd$ to map to the soberified space:

COROLLARY 16.7.2. *Suppose* $(\mathsf{P}, \mathsf{Open})$ *is a semitopology and* $p, p' \in \mathsf{P}$. *Then* $p \between p'$ *in* $(\mathsf{P}, \mathsf{Open})$ *if and only if* $nbhd(p) \between nbhd(p')$ *in* $\mathsf{Soberify}(\mathsf{P}, \mathsf{Open})$.

Proof. This just reiterates the equivalence of parts 1 and 3 in Lemma 16.7.1. □

PROPOSITION 16.7.3. *Suppose* $(\mathsf{P}, \mathsf{Open})$ *is a semitopology. Then:*

1. *It may be that* $(\mathsf{P}, \mathsf{Open})$ *is unconflicted (meaning that it contains no conflicted points), but the semitopology* $\mathsf{Soberify}(\mathsf{P}, \mathsf{Open})$ *contains a conflicted point.*
2. *It may further be that* $(\mathsf{P}, \mathsf{Open})$ *is unconflicted and* $p \in \mathsf{P}$ *is such that* $nbhd(p)$ *is conflicted in the semitopology* $\mathsf{Soberify}(\mathsf{P}, \mathsf{Open})$.

We can summarise the two assertions above as follows:

1. *Soberifying a space might introduce a conflicted point, even if none was originally present.*
2. *Soberifying a space can make a point that was unconflicted, into a point that is conflicted.*[4]

Proof. It suffices to provide counterexamples.

1. Consider the right-hand semitopology in Figure 14.1; this is unconflicted because every point is intertwined only with itself. The soberification of this space is illustrated in the right-hand semitopology in Figure 14.2. Each of the extra points is intertwined with the two numbered points next to it; e.g. the extra point in the open set A — write it $\bullet_A$ (in-between 3 and 0) — is intertwined with 0 and 3; so $3 \between \bullet_A \between 0$. However, the reader can check that $3 \not\between 0$. Thus, $\bullet_A$ is conflicted.

2. We define $(\mathsf{P}, \mathsf{Open})$ by:

   - $\mathsf{P} = (\text{-}1, 1)$ (real numbers between -1 and 1 exclusive).
   - Open is generated by:
     - All open intervals that do not contain 0; so this is open intervals (r_1, r_2) where $\text{-}1 \leq r_1 < r_2 \leq 0$ or $0 \leq r_1 < r_2 \leq 1$.

[4] If we stretch the English language, we might say that soberifying a space can conflictify one of its points.

– All of the open intervals $(\text{-}1/n, 1/n)$, for $n \geq 2$.

The reader can check that:

- Points in this semitopology are intertwined only with themselves.
- The soberification includes four additional points, corresponding to completely prime semifilters $\text{-}1/0$ generated by $\{(\text{-}1/n, 0) \mid n \geq 2\}$ and $+1/0$ generated by $\{(0, 1/n) \mid n \geq 2\}$, and to the endpoints $\text{-}1$ and $i{+}1$.
- $\text{-}1/0$ and $+1/0$ are intertwined with 0, but are not intertwined with one another.

Thus, 0 is conflicted in $\mathsf{Soberify}(\mathbf{P}, \mathsf{Open})$ but not in $(\mathbf{P}, \mathsf{Open})$. □

REMARK 16.7.4. Proposition 16.7.3 may seem surprising in view of Corollary 16.7.2, but the key observation is that the soberified space may add points to the original space. These points can add conflicting behaviour that is 'hidden' in the completely prime semifilters of the original space.

Thus, Proposition 16.7.3 shows that the property of 'being unconflicted' *cannot* be characterised purely in terms of the semiframe of open sets — if it could be, then soberification would make no difference, by Theorem 15.1.4(3).

There is nothing wrong with that, except that we are interested in well-behavedness conditions on semiframes. We can now look for some other condition — but one having to do purely with open sets — that might play a similar role in the theory of (weak/quasi)regularity of semiframes, as being unconflicted does in theory of (weak/quasi)regularity of semitopologies.

We already saw a candidate for this in Theorem 16.5.4: *strong compatibility*. We examine this next.

16.8 Strong compatibility in semitopologies

REMARK 16.8.1. Note that:

1. Theorem 16.5.4 characterises 'regular' for semiframes as 'weakly regular + strongly compatible'.
2. Theorem 6.2.2 characterises 'regular' for semitopologies as 'weakly regular + unconflicted'.

We know from results like Lemma 16.6.1 and Corollary 16.7.2 that there are accurate correspondences between notions of regularity in semiframes and semitopologies. This is by design, e.g. in Definition 16.5.1; we designed the semiframe definitions so that semiframe regularity and semitopological regularity would match up closely.

Yet there are differences too, since Theorem 16.5.4 uses strong compatibility, and Theorem 6.2.2 uses being unconflicted. What is the difference here, and why does it arise?

One answer is given by Proposition 16.7.3, which illustrates that the condition of 'unconflicted' (which comes from semitopologies) does not sit comfortably with the 'pointless' semiframe definitions. This raises the question of how strong compatibility (which comes from semiframes) translates into the context of semitopologies; and how this relates to being (un)conflicted?

We look into this now; see Remark 16.8.8 for a summary.

LEMMA 16.8.2. *Suppose* $(\mathsf{P}, \mathsf{Open})$ *is a semitopology and* $p \in \mathsf{P}$. *Then the following are equivalent:*

1. *p is hypertransitive (Definition 6.5.4).*
2. *$nbhd(p)$ is strongly compatible.*
3. *$nbhd(p)^*$ is compatible.*
4. *For every $O', O'' \in \mathsf{Open}$, if $O' * nbhd(p) * O''$ then $O' \between O''$.*

(Above, $O' * nbhd(p)$ follows Notation 16.1.2(1) and means that $O' \between O$ for every $p \in O \in \mathsf{Open}$, and similarly for $nbhd(p) * O''$.)

Proof. Equivalence of parts 1 and 2 is just Lemma 16.3.8. Equivalence of parts 2 and 3 is Definition 16.3.6(1). For the equivalence of parts 3 and 4, we just unpack what it means for $nbhd(p)^*$ to be compatible (see Remark 16.3.7). □

Lemma 16.8.3 shows that the situation outlined in Proposition 16.7.3(2) cannot arise if we work with a strongly compatible point instead of an unconflicted one …

LEMMA 16.8.3. *Suppose* $(\mathsf{P}, \mathsf{Open})$ *is a semitopology and* $p, p' \in \mathsf{P}$. *Then the following are equivalent:*

1. *p is hypertransitive in $(\mathsf{P}, \mathsf{Open})$.*
2. *$nbhd(p)$ is hypertransitive in $\mathsf{Soberify}(\mathsf{P}, \mathsf{Open})$ (Notation 15.1.3).*

Proof. Note that from Lemma 16.8.2, p is hypertransitive in $(\mathsf{P}, \mathsf{Open})$ when

$$(\forall O \in \mathsf{Open}. p \in O \Longrightarrow O' \between O \between O'') \quad \text{implies} \quad O' \between O''$$

for every $O', O'' \in \mathsf{Open}$. Also, from Definition 13.4.3(2) and Lemma 16.8.2, $nbhd(p)$ is hypertransitive in $\mathsf{Soberify}(\mathsf{P}, \mathsf{Open})$ when

$$(\forall O \in \mathsf{Open}. nbhd(p) \in Op(O) \Longrightarrow Op(O') \between Op(O) \between Op(O'')) \quad \text{implies} \quad Op(O') \between Op(O'')$$

for every $Op(O')$, $Op(O'') \in \mathsf{Opens}(\mathsf{Soberify}(\mathsf{P}, \mathsf{Open}))$.

Now by Proposition 14.2.1(2), $nbhd(p) \in Op(O)$ if and only if $p \in O$, and by Corollary 14.2.2 $Op(O') \between Op(O)$ if and only if $O' \between O$, and $Op(O) \between Op(O'')$ if and only if $O \between O''$. The result follows. □

…but, the situation outlined in Proposition 16.7.3(1) *can* arise, indeed we use the same counterexample:

LEMMA 16.8.4. *It may be that every point in* $(\mathsf{P}, \mathsf{Open})$ *is hypertransitive, yet* $\mathsf{Soberify}(\mathsf{P}, \mathsf{Open})$ *contains a point that is not hypertransitive.*

Proof. The same counterexample as used in Proposition 16.7.3(1) illustrates a space $(\mathsf{P}, \mathsf{Open})$ such that every point in $(\mathsf{P}, \mathsf{Open})$ is hypertransitive, but $\mathsf{Soberify}(\mathsf{P}, \mathsf{Open})$ contains a point that is not hypertransitive. We note that $\bullet_A$ (the extra point in-between 3 and 0) is not hypertransitive, because both B and D intersect with every open neighbourhood of $\bullet_A$, but B does not intersect with D. □

The development above suggests that we define:

DEFINITION 16.8.5. Call a semitopology $(\mathsf{P}, \mathsf{Open})$ **strongly compatible** when $(\mathsf{Open}, \subseteq, \between)$ is strongly compatible in the sense of Definition 16.3.6(2).

The proof of Proposition 16.8.6 is then very easy:

PROPOSITION 16.8.6. *Suppose* $(\mathsf{P}, \mathsf{Open})$ *is a semitopology. Then the following are equivalent:*

1. $(\mathsf{P}, \mathsf{Open})$ *is strongly compatible in the sense of Definition 16.8.5.*
2. $\mathsf{Soberify}(\mathsf{P}, \mathsf{Open})$ *is strongly compatible in the sense of Definition 16.8.5.*
3. $\mathsf{Soberify}(\mathsf{P}, \mathsf{Open})$ *is strongly compatible in the sense of Definition 16.3.6(2).*
4. $\mathsf{Soberify}(\mathsf{P}, \mathsf{Open})$ *is hypertransitive in the sense of Definition 6.5.4.*

Proof. We unpack Definition 16.8.5 and note that strong compatibility of $(\mathsf{P}, \mathsf{Open})$ is expressed purely as a property of its semiframe of open sets $(\mathsf{Open}, \subseteq, \between)$. By Theorem 15.1.4(3), the semiframe of open sets of $\mathsf{Soberify}(\mathsf{P}, \mathsf{Open})$ is isomorphic to $(\mathsf{Open}, \subseteq, \between)$, via $nbhd^{-1}$. Equivalence of parts 1 and 2 follows.

By Notation 15.1.3 and Remark 15.1.2(3), the points of $\mathsf{Soberify}(\mathsf{P}, \mathsf{Open})$ are just abstract points of $(\mathsf{Open}, \subseteq, \between)$. Equivalence of parts 2 and 3 follows.

Equivalence of part 4 with the other parts follows using Lemmas 16.8.2 and 16.8.3. □

Recall from Definition 4.1.4(7) that $(\mathsf{P}, \mathsf{Open})$ being (weakly) regular means that every point in $(\mathsf{P}, \mathsf{Open})$ is (weakly) regular. Recall from Definition 16.8.5 that $(\mathsf{P}, \mathsf{Open})$ being strongly compatible means that $(\mathsf{Open}, \subseteq, \between) = \mathsf{Fr}(\mathsf{P}, \mathsf{Open})$ is strongly compatible in the sense of Definition 16.3.6(2). We can now prove an analogue of Theorems 6.2.2 and 16.5.4:

THEOREM 16.8.7. *Suppose* $(\mathsf{P}, \mathsf{Open})$ *is a semitopology and* $p \in \mathsf{P}$*. Then the following are equivalent:*

1. $(\mathsf{P}, \mathsf{Open})$ *is regular.*
2. $(\mathsf{P}, \mathsf{Open})$ *is weakly regular and strongly compatible.*

Proof. Suppose $(\mathsf{P}, \mathsf{Open})$ is regular, meaning that every $p \in \mathsf{P}$ is regular.

By Theorems 6.2.2 and 6.5.8 every $p \in \mathsf{P}$ is weakly regular and hypertransitive. (So by Lemma 16.8.2 every $nbhd(p)$ is strongly compatible, and by Lemma 16.8.3 also hypertransitive.) The definition of weak regularity for a space in Definition 4.1.4(7) is pointwise, so it follows immediately that $(\mathsf{P}, \mathsf{Open})$ is weakly regular.

But, the definition of strong compatibility for a space in Definition 16.8.5 is on its semiframe of open sets, which may include abstract points not only of the form $nbhd(p)$. It therefore does not follow immediately that $(\mathsf{P}, \mathsf{Open})$ is strongly compatible; Lemma 16.8.4 contains a counterexample.

We can still prove that $(\mathsf{P}, \mathsf{Open})$ is strongly compatible — but we need to do a bit more work.

Unpacking Definition 16.8.5, we must show that $(\mathsf{Open}, \subseteq, \between)$ is strongly compatible. Unpacking Definition 16.3.6(2), we must show that every abstract point in $(\mathsf{Open}, \subseteq, \between)$ is strongly compatible.

So consider an abstract point $P \subseteq \mathsf{Open}$. By Corollary 4.3.3 P has a topen partition $\mathcal{T}$, which means that: every $T \in \mathcal{T}$ is topen; the elements of $\mathcal{T}$ are disjoint; and $\bigcup \mathcal{T} = \mathsf{P}$.

Now $\bigcup \mathcal{T} = \mathsf{P} \in T$ by Definition 13.1.1(7) and Lemma 13.2.1(1), so by Definition 13.1.1(2) there exists at least one (and in fact precisely one) $T \in \mathcal{T}$ such that $T \in P$. Now T is a transitive element in Open, so by Proposition 16.3.13 $P \subseteq \mathsf{Open}$ is strongly compatible as required. □

Remark 16.8.8. We summarise what we have seen:

1. The notions of (quasi/weak)regularity match up particularly nicely between a semitopology and its soberification as a semiframe (Proposition 16.6.2).
2. We saw in Proposition 16.7.3 that the notions of (un)conflicted point and unconflicted space from Definition 6.1.1(2) are not robust under forming soberification (Notation 15.1.3). From the point of view of a pointless methodology in semitopologies — in which we seek to understand a semitopology $(\mathsf{P}, \mathsf{Open})$ starting from its semiframe structure $(\mathsf{Open}, \subseteq, \between)$ — this is a defect.
3. A pointwise notion of strong compatibility exists; by Lemma 16.8.2 it is actually hypertransitivity from Definition 6.5.4. This is preserved pointwise by soberification (Lemma 16.8.3), but soberification can still introduce *extra* points, and it turns out that the property of a space being pointwise hypertransitive is still not robust under soberification because the extra points need not necessarily be hypertransitive; see Lemma 16.8.4.
4. This motivates the notion of a *strongly compatible* semitopology from Definition 16.8.5; and then Proposition 16.8.6 becomes easy.
   Our larger point (no pun intended) is that the Definition and its corresponding Proposition are natural, *and also* that the other design decisions are *less* natural, as noted above.
   Perhaps somewhat unexpectedly, 'regular = weakly regular + strongly compatible' then works pointwise *and* for the entire space; see Theorem 16.8.7. Thus Definition 16.8.5 has good properties and is natural from a pointless/semiframe/open sets perspective.

Graph representation of semitopologies 17

A substantial body of literature exists studying social networks as connected graphs. A semitopology has the flavour of a social network, in the sense that it models voting and consensus on a distributed system. It is therefore interesting to consider representations of semitopologies as graphs. We will consider two ways to do this:

1. We can map a semitopology to the intersection graph of its open sets. We discuss it in Subsection 17.1. This works well but loses information (Remark 17.1.16).
2. We can use a *subintersection* relation between sets. We discuss this in Subsection 17.2.

17.1 From a semitopology to its intersection graph

We start with a very simple representation of $(\mathsf{P}, \mathsf{Open})$ obtained just as the *intersection graph* of Open (Definition 17.1.1). This is not necessarily the most detailed representation (we spell out why, with examples, in Remark 17.1.16), but it is still simple, direct, and nontrivial:

17.1.1 The basic definition

DEFINITION 17.1.1. Suppose $(\mathsf{P}, \mathsf{Open})$ is a semitopology. Define its **intersection graph** $\mathsf{IntGr}(\mathsf{P}, \mathsf{Open})$ by:

- The nodes of $\mathsf{IntGr}(\mathsf{P}, \mathsf{Open})$ are nonempty open sets $O \in \mathsf{Open}_{\neq\varnothing}$.
- There is an edge $O \leftrightarrow O'$ between O and O' when $O \between O'$.

REMARK 17.1.2.

1. The notion of the *intersection graph* of a set of sets is standard [Wik24c]. The notion used in Definition 17.1.1 is slightly different, in that we exclude the empty set. This technical tweak is mildly useful, to give us Lemma 17.1.4.

2. If $(\mathbf{P}, \mathsf{Open})$ is a semitopology and $\mathsf{IntrGr}(\mathbf{P}, \mathsf{Open})$ is its intersection graph in the sense of Definition 17.1.1, then $O \leftrightarrow O'$ is a synonym for $O \between O'$. But, writing $O \leftrightarrow O'$ suggests that we view O and O' as nodes.

NOTATION 17.1.3. For the rest of this Section we assume a fixed but arbitrary

$$G = \mathsf{IntGr}(\mathbf{P}, \mathsf{Open})$$

that is the open intersection graph of a semitopology $(\mathbf{P}, \mathsf{Open})$.

We start with an easy lemma:

LEMMA 17.1.4. *$O \leftrightarrow O$ always (the graph G is reflexive).*

Proof. From Definition 17.1.1, noting that nodes are *nonempty* open sets $O \in \mathsf{Open}$, and it is a fact of sets intersection that $O \between O$ when O is nonempty. □

17.1.2 The node preorder $\leq$

DEFINITION 17.1.5. Write $O \leq O'$ when for every O'', if $O \leftrightarrow O''$ then $O' \leftrightarrow O''$. In symbols:

$$O \leq O' \quad \text{when} \quad \forall O''.O \leftrightarrow O'' \Longrightarrow O' \leftrightarrow O''.$$

We note that $\leq$ from Definition 17.1.5 is a preorder (reflexive and transitive relation):

LEMMA 17.1.6.

1. *$\leq$ is reflexive: $O \leq O$.*
2. *$\leq$ is transitive: if $O \leq O' \leq O''$ then $O \leq O''$.*

Proof. By routine calculations. □

LEMMA 17.1.7.

1. *If $O \leq O'$ then $O \leftrightarrow O'$.*
2. *It is not in general the case that $O \leftrightarrow O'$ implies $O \leq O'$ (cf. Proposition 17.1.12(1&2)).*

In symbols we can write: $\leq \subseteq \leftrightarrow$ and $\leftrightarrow \not\subseteq \leq$ in general.[1]

Proof. We consider each part in turn:

1. Suppose $O \leq O'$. By Lemma 17.1.4 $O \leftrightarrow O$, and it follows (since $O \leq O'$) that $O' \leftrightarrow O$ as required.

2. It suffices to give a counterexample. Consider the semitopology $(\mathbf{P}, \mathsf{Open})$ where $\mathbf{P} = \{0, 1, 2\}$ and Open is generated by $O = \{0, 1\}$, $O' = \{1, 2\}$, and $O'' = \{0\}$, as illustrated in Figure 17.1. Then $O \leftrightarrow O'$ but $O \not\leq O'$ since $O \leftrightarrow \{0\}$ but $O' \not\leftrightarrow \{0\}$. □

[1] It gets better: see Lemma 17.1.9.

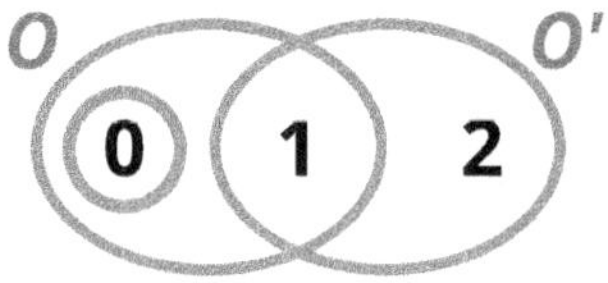

Figure 17.1: $O \between O'$ but $O \not\leq O'$ (Lemma 17.1.7(2))

REMARK 17.1.8. Suppose $O \leq O'$, so that by Lemma 17.1.7 also $O \between O'$. We can illustrate Definition 17.1.5 in the style of a categorical diagram —

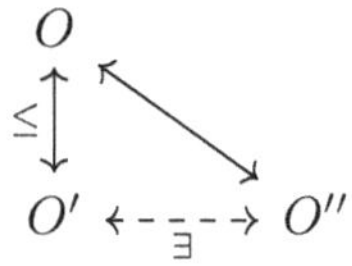

— expressing that $O \leq O'$ holds when every arrow out of O factorises through O'.

LEMMA 17.1.9 ($\leq$ generalises $\subseteq$). *We have:*

1. *If $O \subseteq O'$ then $O \leq O'$.*
2. *The converse implication need not hold: $O \leq O'$ does not necessarily imply $O \subseteq O'$.*[2]

Proof.

1. A fact of sets.

2. It suffices to give a counterexample. Set $\mathsf{P} = \{0, 1\}$ and $\mathsf{Open} = \{\varnothing, \{0\}, \{0, 1\}\}$. This generates a very simple graph G as follows:

$$\{0\} \longleftrightarrow \{0, 1\}$$

The reader can check that $\{0, 1\} \leq \{0\}$, but $\{0, 1\} \not\subseteq \{0\}$. □

17.1.3 Transitive elements

DEFINITION 17.1.10. Call $T \in G$ **transitive** when for every $O, O' \in G$ we have that

$$O \between T \between O' \quad \text{implies} \quad O \between O'.$$

In pictures:

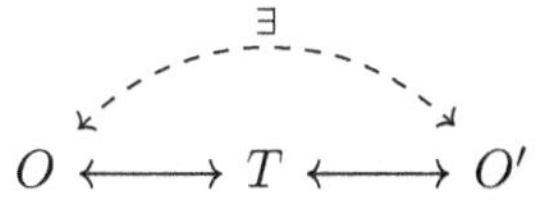

LEMMA 17.1.11. *Suppose $T \in G$ is a node. Then the following are equivalent:*

[2] So to sum up this and Lemma 17.1.7: $\subseteq \subseteq \leq \subseteq \between$, and the inclusion may be strict: $\subseteq \subsetneq \leq$ in general.

- *T is transitive in the sense of Definition 3.2.1(1).*[3]
- *T is transitive in the sense of Definition 17.1.10.*

Proof. Just unfolding Definitions 17.1.1 and 3.2.1. □

Lemmas 17.1.7 and 17.1.9 suggest that $\leq$ generalises subset inclusion $\leq$. In this light, Proposition 17.1.12 makes transitive elements look like singleton sets:

PROPOSITION 17.1.12. *The following conditions are equivalent:*

1. *T is transitive.*
2. $\forall O.(T \between O \Longrightarrow T \leq O)$.
3. $\forall O.(T \between O \Longleftrightarrow T \leq O)$.

Proof. For the top equivalence, we prove two implications:

- *The top-down implication.*

  Suppose T is transitive and suppose $T \between O$. To prove $T \leq O'$ it suffices to consider any O' and show that $T \between O'$ implies $O \between O'$.

  But this is just from transitivity and the fact that $\between$ is symmetric: $O \between T \between O'$ implies $O \between O'$.

- *The bottom-up implication.*

  Suppose for every O, if $T \between O$ then $T \leq O$, and suppose $O \between T \between O'$. Because $T \leq O$ and $T \between O'$, we have $O \between O'$ as required.

The lower equivalence then follows from Lemma 17.1.7. □

COROLLARY 17.1.13.

1. *If T is transitive then T is $\leq$-minimal. That is:*

$$O \leq T \quad \textit{implies} \quad T \leq O.$$

2. *It is possible for T to be $\leq$-least (and thus $\leq$-minimal) and not transitive.*

Proof.

1. Suppose T is transitive and $O \leq T$. By Lemma 17.1.7(1) $O \between T$ and by Proposition 17.1.12(2) $T \leq O$.

2. It suffices to provide a counterexample. Consider the semitopology illustrated in Figure 5.2. It is a fact that A is not transitive, yet A is $\leq$-least: $A \not\leq B$ (because $B \between D$ yet $A \not\between D$) and similarly $A \not\leq C$, and $A \not\leq D$ (because $A \not\between D$). □

DEFINITION 17.1.14. Suppose $O, O' \in G$. Define $O \approx O'$ when $O \leq O' \wedge O' \leq O$, and in this case call O and O' **extensionally equivalent**. It is easy to see from Definition 17.1.5 that

$$O \approx O' \Longleftrightarrow \forall O''.(O \between O'' \Longleftrightarrow O' \between O'').$$

[3]Equivalently, T is topen (Definition 3.2.1(2)), since every node in G is a nonempty open.

COROLLARY 17.1.15.

1. *If T and T' are transitive (Definition 17.1.10) then the following are equivalent:*

$$T \leq T' \iff T' \leq T \iff T \between T' \iff T \approx T'.$$

2. *As a corollary, if T and T' are transitive then $T \between T'$ if and only if $T \approx T'$.*

Proof. The left-hand equivalence is from Corollary 17.1.13 (since T and T' are transitive). The middle equivalence is from Proposition 17.1.12. The right-hand equivalent follows from the left-hand and middle equivalences using Definitions 17.1.5 and 17.1.14.

The corollary just repeats the right-hand equivalence. □

REMARK 17.1.16 (Intersection graph loses information). The proof of Corollary 17.1.15(2) is not hard but it tells us something useful: the intersection graph identifies intersecting topens, and thus identifies a topen with the (by Corollary 3.5.3) unique maximal topen that contains it.

Consider a semitopology $(\mathsf{P}, \mathsf{Open})$ and its intersection graph $\mathsf{IntGr}(\mathsf{P}, \mathsf{Open})$, and consider some regular point $p \in K(p)$. Recall from Theorem 4.2.6 that $K(p)$ is the greatest topen (transitive open) neighbourhood of p. Putting Corollary 17.1.15(2), Theorem 4.2.6, and Lemma 3.4.3 together, we have that $K(p)$ — when considered as a node in the intersection graph of open sets — is extensionally equivalent to each of its topen subsets, and also (equivalently) to any topen set that it intersects with.

So, if we were to build a functor from intersection graphs back to semitopologies, by forming a notion of abstract point and mapping a node to the set of abstract points that contain it, then Corollary 17.1.15 tells us that this will map all connected transitive nodes down to a single point. Thus, our intersection graph representation from Definition 17.1.1 *loses information.*

It is easy to generate examples of this kind of information loss. The following clearly different semitopologies give rise to isomorphic intersection graphs, namely: the full graph on three points, representing three pairwise intersecting nonempty open sets.

1. $\mathsf{P} = \{0, 1, 2\}$ and $\mathsf{Open} = \{\varnothing, \{0\}, \{0, 1\}, \{0, 1, 2\}\}$.
2. $\mathsf{P}' = \{0, 1, 2\}$ and $\mathsf{Open}' = \{\varnothing, \{0, 1\}, \{1, 2\}, \{0, 1, 2\}\}$.

See Figure 17.2; the intersection graph isomorphism is illustrated on the right (where we equate $\{0\}$ with $\{1, 2\}$). The left-hand and middle examples in the figure are of intersecting topens, consistent with Corollary 17.1.15(2).

Whether this behaviour is a feature or a bug depends on what we want to accomplish — but for our purposes of modelling networks, we may care to also consider a representation that retains more information. In the next Subsection we consider a slightly more elaborate graph representation, which is more discriminating.

17.2 From a semiframe to its subintersection graph

REMARK 17.2.1. In Remark 17.1.16 we gave a natural representation of a semitopology as its intersection graph. We noted in Corollary 17.1.15 that this identifies open sets up to a notion of extensional equivalence $\approx$ given in Definition 17.1.14, and because topen sets are extensionally equivalent if and only if they intersect by Corollary 17.1.15, the intersection graph representation of semitopologies identifies two topens precisely when they intersect.

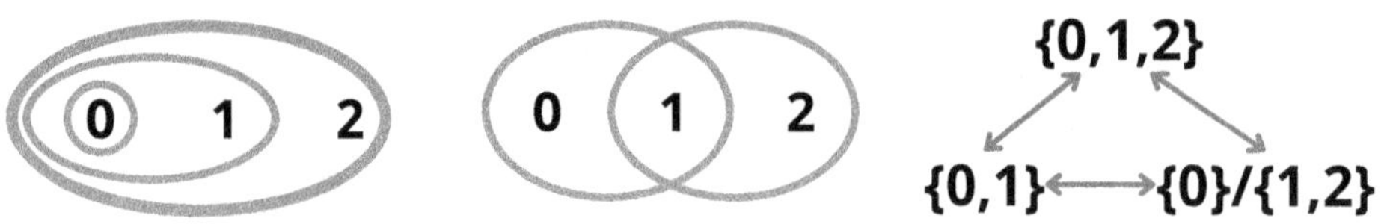

Figure 17.2: Semitopologies with isomorphic intersection graphs (Remark 17.1.16)

This is not wrong — intersection topen sets *are* extensionally equivalent, after all — but suppose we want to retain a bit more detail. How can we proceed?

17.2.1 The subintersection relation $\ltimes$

REMARK 17.2.2. Notice that the notion of semiframe $(\mathsf{X}, \leq, *)$ from Definition 12.3.1 is based on *two* structures on X:

- a semilattice relation $\leq$, and
- a compatibility relation $*$.

Correspondingly, our notion of semitopology observes *two* properties of open sets: whether O is a subset of O', and whether O intersects O'.

We can ask whether these two structures can be obtained from a single relation. The answer is yes (if we are also allowed to observe equality): we can combine $\leq$ and $*$ into a single relation and so obtain a graph structure, without the loss of information we noted of intersection graphs. The definition is as follows:

DEFINITION 17.2.3.

1. Suppose P is a set and $X \subseteq \mathsf{P}$.[4] Define X^c the **complement of** X by $X^c = \mathsf{P} \setminus X$.

2. Suppose P is a set and $X, Y \subseteq \mathsf{P}$. Define a relation $X \ltimes Y$, read 'X **properly subintersects** Y', as follows:
$$X \ltimes Y \quad \text{when} \quad X \between Y \land X^c \between Y.$$
When $X \ltimes Y \lor X = Y$, we say that X **subintersects** Y.

3. Suppose $(\mathsf{X}, \leq, *)$ is a semiframe and $x, y \in \mathsf{X}$. Define a relation $x \ltimes y$, and say that x **properly subintersects** y, when
$$x \ltimes y \quad \text{when} \quad x * y \land y \not\leq x.$$

EXAMPLE 17.2.4. Set $P = \{0, 1, 2\}$. Then:

1. $\{0\}$ properly subintersects $\{0, 1\}$, because $\{0\} \between \{0, 1\}$ and $\{0, 1, 2\} \setminus \{0\} = \{1, 2\} \between \{0, 1\}$. Similarly, $\{1\}$ properly subintersects $\{0, 1\}$.
2. $\{0, 2\}$ properly subintersects $\{0, 1\}$, because $\{0, 2\} \between \{0, 1\}$ and $\{0, 1, 2\} \setminus \{0, 2\} = \{1\} \between \{0, 1\}$.

[4] We will be most interested in the case that P is the set of points of a semitopology, but the definition does not depend on this.

3. $\{2\}$ does not subintersect $\{0,1\}$, because $\{2\} \not\between \{0,1\}$.
4. $\{0,1\}$ does not subintersect $\{0\}$ or $\{1\}$, because $\{2\} \not\between \{0\}$ and $\{2\} \not\between \{1\}$ and $\{2\} \not\between \{0,1\}$.
5. $\{0,1\}$ subintersects itself, but not properly because $\{0,1,2\} \setminus \{0,1\} \not\between \{0,1\}$. More generally X subintersects itself by definition, but it does not properly subintersect itself (think: $X \subsetneq Y$ vs. $X \subseteq Y$).

REMARK 17.2.5 (One property, and three non-properties). It is easy to show that $X \ltimes Y$ is positive (covariant) in its second argument: if $X \ltimes Y$ and $Y \subseteq Y'$ then $X \ltimes Y'$. However, $X \ltimes Y$ is neither positive nor negative in its first argument, and it does not commute with intersection in its right argument.

Take $X, Y \subseteq \{0,1,2,3\}$. Then:

1. It is not the case that $X \ltimes Y \wedge X \subseteq X'$ implies $X' \ltimes Y$.
   Take $X = \{0\}$ and $Y = \{0,1\}$ and $X' = \{0,1\}$.
2. It is not the case that $X \ltimes Y \wedge X' \subseteq X$ implies $X' \ltimes Y$.
   Take $X = \{0,1\}$ and $Y = \{1,2\}$ and $X' = \{0\}$.
3. It is not the case that $X \ltimes Y \wedge X \ltimes Y'$ implies $X \ltimes (Y \cap Y')$.
   Take $X = \{0,1\}$ and $Y = \{1,2\}$ and $Y' = \{1,3\}$.

LEMMA 17.2.6. *Suppose* P *is a set and* $X, Y \subseteq \mathsf{P}$. *Then the following are equivalent:*

1. $X \ltimes Y$ *in the sense of Definition 17.2.3(2).*
2. $X \between Y \wedge Y \not\subseteq X$.

In other words: $X \ltimes Y$ *for* X *and* Y *considered as sets as per Definition 17.2.3(2), precisely when* $X \ltimes Y$ *for* X *and* Y *considered as elements in the semiframe* $(pow(\mathsf{P}), \subseteq, \between)$ *as per Definition 17.2.3(3).*

Proof. Routine, using the fact of sets that $Y \between X^c$ if and only if $Y \not\subseteq X$. □

COROLLARY 17.2.7. *Suppose* $(\mathsf{X}, \leq, *)$ *is a spatial semiframe and* $x, y \in \mathsf{X}$. *Then*

$$x \ltimes y \quad \textit{if and only if} \quad Op(x) \ltimes Op(y).$$

Proof. Suppose $(\mathsf{X}, \leq, *)$ is a spatial semiframe. By Proposition 14.1.4(2&1) $x * y$ if and only if $Op(x) \between Op(y)$, and $x \leq y$ if and only if $Op(x) \subseteq Op(y)$. We use Lemma 17.2.6. □

17.2.2 Recovering $\leq$ and $*$ from $\ltimes$

REMARK 17.2.8. We can recover $\subseteq$ and $\between$ from $\ltimes$. We can also recover $\leq$ and $*$. We consider the construction for semiframes and $\leq$ and $*$, because it is the more general setting; the proofs for the concrete instance of $\subseteq$ and $\between$ are identical:

PROPOSITION 17.2.9. *Suppose* $(\mathsf{X}, \leq, *)$ *is a semiframe and suppose* $x, y \neq \bot_{\mathsf{X}}$. *Then:*

1. $x * y$ *if and only if* $x = y \vee x \ltimes y \vee y \ltimes x$.
2. $x \leq y$ *if and only if* $x = y \vee (x \ltimes y \wedge \neg(y \ltimes x))$.

Proof. We consider each implication in turn:

- *We show that $x * y$ implies $x = y \lor x \ltimes y \lor y \ltimes x$.*

  Suppose $x * y$. By antisymmetry of $\leq$, either $x = y$ or $y \not\leq x$ or $x \not\leq x$. The result follows.

- *We show that $x = y \lor x \ltimes y \lor y \ltimes x$ implies $x * y$.*

  By reversing the reasoning of the previous case.

- *We show that $x = y \lor (x \ltimes y \land \neg(y \ltimes x))$ implies $x \leq y$.*

  Suppose $x = y \lor (x \ltimes y \land \neg(y \ltimes x))$.

  If $x = y$ then $x \leq y$ and we are done. If $x \neq y$ then we unpack Definition 17.2.3(3) and simplify as follows:

$$x \ltimes y \land \neg(y \ltimes x) \iff x * y \land y \not\leq x \land (\neg(x * y) \lor x \leq y) \iff$$
$$x * y \land y \not\leq x \land x \leq y \Longrightarrow x \leq y$$

- *We show that $x \leq y$ implies $x = y \lor (x \ltimes y \land \neg(y \ltimes x))$.*

  Suppose $x \leq y$. By assumption $x \neq \bot$, so by Lemma 12.2.3(2) $x * y$. If $x = y$ then we are done; otherwise by antisymmetry $y \not\leq x$ and again we are done. □

Remark 17.2.10. It follows from the above that a semiframe $(\mathsf{X}, \leq, *)$ can be presented as a graph $\mathsf{Gr}(\mathsf{X}, \leq, *)$ such that:

- Nodes of the graph are elements $x \in \mathsf{X}$ such that $x \neq \bot_{\mathsf{X}}$.
- There is an edge $x \to x'$ when $x \ltimes x'$ in the sense of Definition 17.2.3(3) — that is, when $x * x' \land x' \not\leq x$.

Similarly we can present a semitopology $(\mathsf{P}, \mathsf{Open})$ as a graph $\mathsf{Gr}(\mathsf{P}, \mathsf{Open})$ such that:

- Nodes of the graph are nonempty open sets $\varnothing \neq O \in \mathsf{Open}$.
- There is an edge $O \to O'$ when $O \ltimes O'$ in the sense of Definition 17.2.3(2) — that is, when $O \between O' \land O^c \between O'$.

These presentations are equivalent in the following sense: if we start from $(\mathsf{P}, \mathsf{Open})$ and consider it as a semiframe $(\mathsf{Open}, \subseteq, \between)$ (which is spatial by Proposition 14.2.4) and then map to a graph, then we get the same graph as if we just map direct from $(\mathsf{P}, \mathsf{Open})$ to the graph. In symbols we can write:

$$\mathsf{Gr}(\mathsf{P}, \mathsf{Open}) = \mathsf{Gr}(\mathsf{Open}, \subseteq, \between).$$

By Proposition 17.2.9, the mapping from $(\mathsf{X}, \leq, *)$ to $\mathsf{Gr}(\mathsf{X}, \leq, *)$ loses no information; we can view the graph as just a different way of presenting the same structure.

Remark 17.2.11. Although $(\mathsf{X}, \leq, *)$ and $\mathsf{Gr}(\mathsf{X}, \leq, *)$ are in one-to-one correspondence as discussed in Remark 17.2.10, the representations suggest different notions of morphism.

- For a semiframe $(\mathsf{X}, \leq, *)$, the natural notion of morphism is a $\leq/*$-preserving map, in a suitable sense as defined in Definition 15.2.1(1): $x \leq x'$ implies $g(x) \leq g(x')$, and $g(x) * g(x')$ implies $x * x'$.

- For a graph $(\mathsf{G}, \ltimes)$, the natural notion of morphism is some notion of $\ltimes$-preserving map, and this is *not* necessarily the same as a $\leq/*$-preserving map, because $\ltimes$ uses $\not\leq$.
  If we still want to preserve notions of lattice structure and semifilter, then we look at how $\leq$ is defined from $\ltimes$ in Proposition 17.2.9, and see that it uses both $\ltimes$ and $\neg\ltimes$, and so in this case we may want a notion of morphism such that $x \ltimes x'$ *if and only if* $g(x) \ltimes g(x')$. Looked at from the point of view of $\leq$ and $*$, this suggests that $x \leq x'$ if and only if $g(x) \leq g(x')$, and $x * x'$ if and only if $g(x) * g(x')$.
  Investigating the design space here is future work. See also Remark 23.3.3.

Part III

Logic and computation

Three-valued logic 18

18.1 Three-valued logic, valuations, and continuity

REMARK 18.1.1 (Setting the scene). Suppose $(\mathsf{P}, \mathsf{Open})$ is a semitopology and suppose $p, p' \in \mathsf{P}$. Then two foundational questions of semitopologies, as we apply them, are:

1. Determine whether $p \between p'$. That is, determine whether every open neighbourhood of p intersects with every open neighbourhood of p'.[1]
2. Determine whether a function f on P, whose values are known at p and p', can be continuously extended to a continuous function on the whole space.[2]

Both questions are hard to answer if we just look at continuous functions from P to the discrete space $\{\mathbf{f}, \mathbf{t}\}$. For example: the only continuous maps to $\{\mathbf{f}, \mathbf{t}\}$ from the top-left semitopology in Figure 3.1 ($\mathsf{P} = \{0, 1, 2\}$ and Open is generated by $\{0\}$ and $\{2\}$) are $\lambda p.\mathbf{t}$ and $\lambda p.\mathbf{f}$. This is so uninformative that it does not even distinguish between that semitopology and a space with just one point.

An answer from topology is to work with the Sierpiński space that we mentioned in Example 5.8.4. As is known, continuous functions in $\mathsf{P} \to \mathsf{Sk}$ biject with Open.[3] However, and perhaps surprisingly, this is still not enough! As we noted in Remarks 2.1.7 and 12.3.6 a semitopology is not *just* its collection of open sets. We also need to know how they intersect.

It turns out that what we need is the semitopology $\mathbf{3}$, as illustrated in Figure 18.1. We can think of this as $\{\mathbf{f}, \mathbf{t}\}$ augmented with a third truth-value b (for 'both'), or as two Sierpiński spaces glued end-to-end. Intuitively, a map $f : \mathsf{P} \to \mathbf{3}$ can encode whether a point is in some O, in some O', or in both — and this gives us the expressivity to express semitopological properties of interest.

[1]This matters because being intertwined corresponds to the *quorum intersection property* familiar from the traditional theory of consensus.

[2]Continuous extensions to topological closures matter because (within our semitopological framework) taking closures corresponds to propagating consensus. This has been a running theme in our maths; see (for example) Remarks 5.5.1 and 5.5.5 and results including those in Subsections 11.5 and 11.6.

[3]The Wikipedia page [Wik24f] has a brief but clear description.

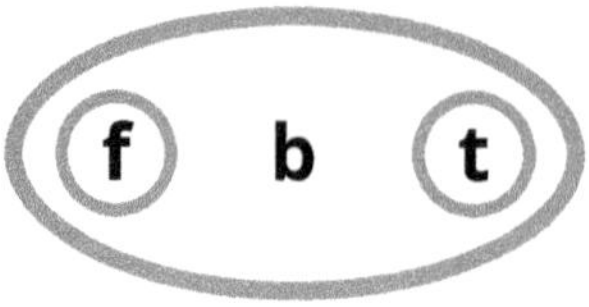

Figure 18.1: The semitopology **3** / the set of truth values of three-valued logic

As a domain of truth-values, **3** will let us use logic to describe properties of semitopologies, including e.g. being intertwined (see Subsection 20.1). This matters in particular because logic is a portal to computation: if we can state something in logic, then we can compute it — using e.g. proof-search or a SAT solver.

Thus the logical material that follows below, while a prototype, is still a substantive step in the direction of turning semitopologies into a practical and applicable specification and computational framework.

18.1.1 The semitopology 3

DEFINITION 18.1.2.

1. Define a semitopology $(\mathbf{3}, \mathsf{Open}_\mathbf{3})$, as illustrated in Figure 18.1, by

$$\mathbf{3} = \{\mathrm{t}, \mathrm{b}, \mathrm{f}\} \quad \text{and} \quad \mathsf{Open}_\mathbf{3} = \{\varnothing, \{\mathrm{t}\}, \{\mathrm{f}\}, \{\mathrm{t}, \mathrm{f}\}, \mathbf{3}\}.$$

   Closed sets of $(\mathbf{3}, \mathsf{Open}_\mathbf{3})$ are

$$\mathsf{Closed}_\mathbf{3} = \{\mathbf{3}, \{\mathrm{t}, \mathrm{b}\}, \{\mathrm{f}, \mathrm{b}\}, \{\mathrm{b}\}, \varnothing\}.$$

   We may write $(\mathbf{3}, \mathsf{Open}_\mathbf{3})$ just as **3**.
2. Suppose P is a set. Call a function $f : \mathsf{P} \to \mathbf{3}$ a **valuation**.
3. Call a valuation $f : \mathsf{P} \to \mathbf{3}$ **continuous** when f is continuous as a map of semitopologies from $(\mathsf{P}, \mathsf{Open})$ to $(\mathbf{3}, \mathsf{Open}_\mathbf{3})$, as per Definition 2.2.1(2) (inverse image of an open set is an open set).

REMARK 18.1.3. **3** is essentially just the top-left example in Figure 3.1, where in that Figure we identify 0 with f, 1 with b, and 2 with t.

LEMMA 18.1.4. *Suppose* $(\mathsf{P}, \mathsf{Open})$ *is a semitopology and* $f : \mathsf{P} \to \mathbf{3}$ *is a valuation. Then the following are equivalent:*

1. *f is continuous.*
2. *$f^{-1}\{\mathrm{t}\}$ and $f^{-1}\{\mathrm{f}\}$ are open sets.*
3. *$f^{-1}\{\mathrm{t}, \mathrm{b}\}$ and $f^{-1}\{\mathrm{f}, \mathrm{b}\}$ are closed sets (see also Definition 18.4.5).*

Proof. By Corollary 5.1.12, f is continuous when the inverse image of every open set is open, and also when the inverse image of every closed set is closed.

It follows from Definition 18.1.2(1) that $\{\{\mathtt{t}\},\{\mathtt{f}\}\}$ is a subbasis for the open sets,[4] and that $\{\{\mathtt{t},\mathtt{b}\},\{\mathtt{f},\mathtt{b}\}\}$ is a subbasis for the closed sets.[5] The result follows by routine calculations. □

18.1.2 Indicator functions and characteristic sets

Definition 18.1.5. Suppose that:

- $(\mathsf{P}, \mathsf{Open})$ is a semitopology.
- $O_{\mathtt{t}}, O_{\mathtt{f}} \in \mathsf{Open}$ and $C_{\mathtt{tb}}, C_{\mathtt{fb}} \in \mathsf{Closed}$.
- $O_{\mathtt{t}} \not\between O_{\mathtt{f}}$ and $C_{\mathtt{tb}} \cup C_{\mathtt{fb}} = \mathsf{P}$ and $O_{\mathtt{t}} \subseteq C_{\mathtt{tb}}$.

Then define **indicator functions**

$$\delta_{O_{\mathtt{t}},O_{\mathtt{f}}},\ \delta_{O_{\mathtt{t}},C_{\mathtt{tb}}},\ \delta_{C_{\mathtt{tb}},C_{\mathtt{fb}}} : \mathsf{P} \to \mathbf{3}$$

as follows:

$$\delta_{O_{\mathtt{t}},O_{\mathtt{f}}}(p) = \begin{cases} \mathtt{t} & p \in O_{\mathtt{t}} \\ \mathtt{b} & p \in \mathsf{P}\setminus(O_{\mathtt{t}} \cup O_{\mathtt{f}}) \\ \mathtt{f} & p \in O_{\mathtt{f}} \end{cases} \qquad \delta_{O_{\mathtt{t}},C_{\mathtt{tb}}}(p) = \begin{cases} \mathtt{t} & p \in O_{\mathtt{t}} \\ \mathtt{b} & p \in C_{\mathtt{tb}}\setminus O_{\mathtt{t}} \\ \mathtt{f} & p \in \mathsf{P}\setminus C_{\mathtt{tb}} \end{cases}$$

$$\delta_{C_{\mathtt{tb}},C_{\mathtt{fb}}}(p) = \begin{cases} \mathtt{t} & p \in \mathsf{P}\setminus C_{\mathtt{fb}} \\ \mathtt{b} & p \in C_{\mathtt{tb}} \cap C_{\mathtt{fb}} \\ \mathtt{f} & p \in \mathsf{P}\setminus C_{\mathtt{fb}} \end{cases}$$

Lemma 18.1.6. *Suppose* $(\mathsf{P}, \mathsf{Open})$ *is a semitopology and suppose*

$$O_{\mathtt{t}}, O_{\mathtt{f}} \in \mathsf{Open}, \quad C_{\mathtt{tb}}, C_{\mathtt{fb}} \in \mathsf{Closed}, \quad O_{\mathtt{t}} \not\between O_{\mathtt{f}}, \quad C_{\mathtt{tb}} \cup C_{\mathtt{fb}} = \mathsf{P}, \quad \textit{and} \quad O_{\mathtt{t}} \subseteq C_{\mathtt{tb}}.$$

Then the indicator functions

$$\delta_{O_{\mathtt{t}},O_{\mathtt{f}}},\ \delta_{O_{\mathtt{t}},C_{\mathtt{tb}}},\ \delta_{C_{\mathtt{tb}},C_{\mathtt{fb}}} \in \mathsf{P} \to \mathbf{3}$$

from Definition 18.1.5 are all continuous from $(\mathsf{P}, \mathsf{Open})$ *to* $\mathbf{3}$.

Proof. By Lemma 18.1.4 it suffices to check that the inverse images of $\{\mathtt{t}\}$ and $\{\mathtt{f}\}$ are open, or that the inverse images of $\{\mathtt{t},\mathtt{b}\}$ and $\{\mathtt{f},\mathtt{b}\}$ are closed. This follows by construction in Definition 18.1.5, noting from Lemma 5.1.9 that a set is open/closed if and only if its complement is closed/open. □

Definition 18.1.7. Suppose $(\mathsf{P}, \mathsf{Open})$ is a semitopology and suppose $f : \mathsf{P} \to \mathbf{3}$. Then define its **characteristic sets**

$$char_{OO}(f),\ char_{OC}(f),\ char_{CC}(f) \in pow(\mathsf{P}) \times pow(\mathsf{P}).$$

as follows:

$$\begin{aligned} char_{OO}(f) &= (f^{-1}\{\mathtt{t}\}, f^{-1}\{\mathtt{f}\}) \\ char_{OC}(f) &= (f^{-1}\{\mathtt{t}\}, f^{-1}\{\mathtt{fb}\}) \\ char_{CC}(f) &= (f^{-1}\{\mathtt{t},\mathtt{b}\}, f^{-1}\{\mathtt{f},\mathtt{b}\}) \end{aligned}$$

[4]By which we mean a set that, together with P, generates all open sets by forming arbitrary, possibly empty, unions.

[5]By which we mean a set that, together with $\varnothing$, generates all closed sets by forming arbitrary, possibly empty, intersections.

REMARK 18.1.8. The pair of sets $f^{-1}\{\mathtt{t},\mathtt{b}\}$ and $f^{-1}\{\mathtt{f},\mathtt{b}\}$ will be particularly useful. Later on in Definition 18.4.5 we will give them names — $f^{\models}$ and $f^{\models\neg}$ — and study their properties.

LEMMA 18.1.9. *Suppose* $(\mathsf{P}, \mathsf{Open})$ *is a semitopology and suppose* $f : \mathsf{P} \to \mathbf{3}$ *is continuous. Then we have:*

1. $char_{OO}(f) = (O_{\mathtt{t}}, O_{\mathtt{f}}) \in \mathsf{Open} \times \mathsf{Open}$ *and* $O_{\mathtt{t}} \between O_{\mathtt{f}}$.
2. $char_{OC}(f) = (O_{\mathtt{t}}, C_{\mathtt{fb}}) \in \mathsf{Open} \times \mathsf{Closed}$ *and* $O_{\mathtt{t}} \subseteq C_{\mathtt{fb}}$.
3. $char_{CC}(f) = (C_{\mathtt{tb}}, C_{\mathtt{fb}}) \in \mathsf{Closed} \times \mathsf{Closed}$ *and* $C_{\mathtt{tb}} \cup C_{\mathtt{fb}} = \mathsf{P}$.

Proof. By Corollary 5.1.12 the inverse image of an open set under a continuous function is open and that of a closed set is closed. The conditions $O_{\mathtt{t}} \between O_{\mathtt{f}}$, $O_{\mathtt{t}} \subseteq C_{\mathtt{fb}}$, and $C_{\mathtt{tb}} \cup C_{\mathtt{fb}} = \mathsf{P}$ come from the fact that $\{\mathtt{t}\} \between \{\mathtt{f}\}$, $\{\mathtt{t}\} \subseteq \{\mathtt{t},\mathtt{b}\}$, and $\{\mathtt{t},\mathtt{b}\} \cup \{\mathtt{b},\mathtt{f}\} = \mathbf{3}$. □

PROPOSITION 18.1.10. *Suppose* $(\mathsf{P}, \mathsf{Open})$ *is a semitopology. Then* $\delta_{\text{-}}$ *and* $char(\text{-})$ *from Definitions 18.1.5 and 18.1.7 determine bijections between:*

1. *Continuous valuations* $f : (\mathsf{P}, \mathsf{Open}) \to \mathbf{3}$.
2. *Ordered pairs of disjoint (possibly empty) open sets:* $(O_{\mathtt{t}}, O_{\mathtt{f}}) \in \mathsf{Open} \times \mathsf{Open}$ *such that* $O_{\mathtt{t}} \between O_{\mathtt{f}}$.
3. *Ordered pairs of an open set contained in a closed set:* $(O_{\mathtt{t}}, C_{\mathtt{tb}}) \in \mathsf{Open} \times \mathsf{Closed}$ *such that* $O_{\mathtt{t}} \subseteq C_{\mathtt{tb}}$. *Note that* $|O_{\mathtt{t}}| \subseteq C_{\mathtt{tb}}$ *by construction, but we do not require an equality.*
4. *Ordered pairs of closed sets whose union is* P*:* $(C_{\mathtt{tb}}, C_{\mathtt{fb}}) \in \mathsf{Closed} \times \mathsf{Closed}$ *such that* $C_{\mathtt{tb}} \cup C_{\mathtt{fb}} = \mathsf{P}$.

Proof. From Lemmas 18.1.6 and 18.1.9 and by routine computations on sets. □

COROLLARY 18.1.11. *Suppose* $(\mathsf{P}, \mathsf{Open})$ *is a semitopology. Then:*

1. *The following are equivalent:*
   a) $O \in \mathsf{Open}$.
   b) *There exists a continuous valuation* $f : \mathsf{P} \to \mathbf{3}$ *such that* $f^{-1}\{\mathtt{t}\} = O$.
   c) *There exists a continuous valuation* $f : \mathsf{P} \to \mathbf{3}$ *such that* $f^{-1}\{\mathtt{f}\} = O$.
2. *The following are equivalent (and see also Definition 18.4.5):*
   a) $C \in \mathsf{Closed}$.
   b) *There exists a continuous* $f : \mathsf{P} \to \mathbf{3}$ *such that* $f^{-1}\{\mathtt{t},\mathtt{b}\} = C$.
   c) *There exists a continuous* $f : \mathsf{P} \to \mathbf{3}$ *such that* $f^{-1}\{\mathtt{f},\mathtt{b}\} = C$.

Proof. Routine using Proposition 18.1.10. For example, we can map O to $\delta_{O_{\mathtt{t}}O_{\mathtt{f}}}(O, \varnothing)$. □

18.2 Three-valued truth-tables

Having three truth-values gives us a great deal of extra structure over the two-valued case. In this Subsection we survey the connectives that will be useful to us later.

$p{\supset}q$	t	b	f
t	t	b	f
b	t	b	$\underline{\mathrm{b}}$
f	t	t	t

$\neg p$	
t	f
b	b
f	t

$p{\equiv}q$	t	b	f
t	t	b	f
b	b	b	b
f	f	b	t

$p{\wedge}q$	t	b	f
t	t	b	f
b	b	b	f
f	f	f	f

$p{\vee}q$	t	b	f
t	t	t	t
b	t	b	b
f	t	b	f

$p{\Rightarrow}q$	t	b	f
t	t	b	f
b	t	b	$\underline{\mathrm{f}}$
f	t	t	t

$p{\Rightarrow}\mathbf{f}$	
t	f
b	f
f	t

$p{\Leftrightarrow}q$	t	b	f
t	t	b	f
b	b	b	f
f	f	f	t

$\mathsf{T}p$	
t	t
b	f
f	f

$\mathsf{TB}p$	
t	t
b	t
f	f

$\mathsf{B}p$	
t	f
b	t
f	f

Above, the vertical axis of a table indicates values for p; the horizontal axis (if nontrivial) denotes values for q.

Figure 18.2: Truth-tables for three-valued logic (Definition 18.2.1)

$$
\begin{aligned}
\neg\neg p &= p \\
p{\vee}q &= \neg(\neg p{\wedge}\neg q) \\
p{\wedge}q &= \neg(\neg p{\vee}\neg q) \\
p{\supset}q &= (\neg p){\vee}q \\
p{\supset}\mathbf{f} &= \neg p \\
p{\equiv}q &= (p{\supset}q){\wedge}(q{\supset}p) = (\neg p{\vee}q){\wedge}(p{\vee}\neg q) \\
p{\Rightarrow}q &= (\mathsf{TB}p){\supset}q = (\neg\mathsf{TB}p){\vee}q = (\mathsf{T}\neg p){\vee}q \\
p{\Leftrightarrow}q &= (p{\Rightarrow}q){\wedge}(q{\Rightarrow}p) \\
\mathsf{T}p &= \neg\mathsf{TB}\neg p = (\neg p){\Rightarrow}\mathbf{f} \\
\mathsf{TB}p &= \neg\mathsf{T}\neg p = \neg(p{\Rightarrow}\mathbf{f}) = (p{\Rightarrow}\mathbf{f}){\Rightarrow}\mathbf{f} \\
\mathsf{B}p &= \mathsf{TB}(p{\wedge}\neg p) = \mathsf{TB}p{\wedge}\mathsf{TB}\neg p = \mathsf{TB}p{\wedge}\neg\mathsf{T}p \\
\mathsf{T}\mathsf{TB}p &= \mathsf{TB}p \\
\mathsf{TB}(p{\Rightarrow}q) &= (\mathsf{TB}p){\Rightarrow}\mathsf{TB}q \\
\mathsf{M}(p\circ q) &= \mathsf{M}p\circ\mathsf{M}q \qquad (\mathsf{M}{\in}\{\mathsf{T},\mathsf{TB}\},\ \circ{\in}\{\wedge,\vee\})
\end{aligned}
$$

Figure 18.3: Some truth-table equivalences (Lemma 18.2.2)

18.2.1 Truth-tables of connectives

DEFINITION 18.2.1. We define

1. unary functions $Q \in \{\neg, \mathsf{T}, \mathsf{TB}, \mathsf{B}\}$ where $Q : \mathbf{3} \to \mathbf{3}$, and
2. binary functions $Q \in \{\wedge, \vee, \supset, \equiv, \Rightarrow, \Leftrightarrow\}$ where $Q : (\mathbf{3} \times \mathbf{3}) \to \mathbf{3}$

by the truth-tables in Figure 18.2.

LEMMA 18.2.2. *The truth-tables in Figure 18.2 are related as per the equalities in Figure 18.3, where for the purposes of these equations, p and q are considered to range over elements of* $\mathbf{3}$.

Proof. By checking truth-tables. □

REMARK 18.2.3. Definition 18.2.1 and Figure 18.2 and Lemma 18.2.2 and Figure 18.3 are elementary, but they express some useful observations:

1. Figure 18.2 presents truth-tables for propositional connectives in a three-valued paraconsistent logic [Wik24d]. There is nothing particularly unusual here within the genre of three-valued logic: we have three truth-values, and the connectives do what they do.
2. $\wedge$ and $\vee$ in Figure 18.2 are least upper bound and greatest lower bound operators on $\mathbf{3}$ considered as a simple lattice with $\mathbf{f} \leq \mathbf{b} \leq \mathbf{t}$, and $\neg$ inverts the lattice order. More on this in Subsection 18.2.2.
3. The equivalence $p{\supset}q = (\neg p){\vee}q$ from Figure 18.3 characterises $\supset$ as a **material implication** [Edg20]. Similarly, $p{\equiv}q = (p{\supset}q){\wedge}(q{\supset}q)$ is a **material equivalence.**
   The symbols $\supset$ and $\equiv$ for material implication and equivalence follow the (now standard) notation used in [WR10, page 7].[6]
4. $\mathsf{T}p$ and $\mathsf{TB}p$ are de Morgan duals, as per the equivalences $\mathsf{T}p = \neg\mathsf{TB}\neg p$ and $\mathsf{TB}p = \neg\mathsf{T}\neg p$ in Figure 18.3. Looking at the truth-tables, T acts as a modality that identifies $\mathbf{t}$, TB is a modality that identifies being $\mathbf{t}$ or $\mathbf{b}$.
5. The equivalence

$$\mathsf{M}(p \circ q) = \mathsf{M}p \circ \mathsf{M}q \quad (\mathsf{M}{\in}\{\mathsf{T}, \mathsf{TB}\},\ {\circ}{\in}\{\wedge, \vee\})$$

   is a little unexpected; it means in particular that we have these two properties:

$$\mathsf{T}(p{\vee}q) = \mathsf{T}p{\vee}\mathsf{T}q \quad \text{and} \quad \mathsf{TB}(p{\wedge}q) = \mathsf{TB}p{\wedge}\mathsf{TB}q.$$

   If we think of T as a *necessitation* (or 'box') modality and TB as a *possibility* (or 'diamond') modality, and look at these two properties in terms of Kripke structures [BRV01], then it is not hard to see that they are characteristic of Kripke frames in which each world sees precisely one world.[7]
6. We take a moment to compute the equivalence $\mathsf{TB}(p{\Rightarrow}q) = (\mathsf{TB}p){\Rightarrow}\mathsf{TB}q$ from the other equivalences:

$$\mathsf{TB}(p{\Rightarrow}q) = \mathsf{TB}(\neg p{\vee}q) = (\mathsf{TB}\neg\mathsf{TB}p){\vee}\mathsf{TB}q = (\neg\mathsf{T}\mathsf{TB}p){\vee}\mathsf{TB}q = (\neg\mathsf{TB}p){\vee}\mathsf{TB}q$$
$$(\mathsf{TB}p){\Rightarrow}\mathsf{TB}q = (\neg\mathsf{TB}\mathsf{TB}p){\vee}\mathsf{TB}q = (\neg\mathsf{TB}p){\vee}\mathsf{TB}q$$

[6] The phrase *material implication* itself goes back to Bertrand Russell in 1903 [Rus03, Section 37]. It is traditionally used to refer to $(\neg A) \vee B$ in classical logic (having two truth-values). I am not aware of a standard terminology for this in multi-valued logics, but the generalisation of 'material implication' to refer to 'not-A or B' in a multi-valued logic (for whatever 'not' and 'or' are understood to mean in that logic), seems reasonable.

[7] Our denotation for T and TB uses truth-values in $\mathbf{3}$, not Kripke structures. We are just pointing out where such axioms can and do arise in Kripke semantics for modalities.

18.2.2 Conjunction and disjunction

DEFINITION 18.2.4. Suppose $V \subseteq \mathbf{3}$ is a set of truth-values. Define $\bigwedge V$ and $\bigvee V$ to be the least upper bound and greatest lower bound of V in $\mathbf{3}$ considered as a lattice with $\mathbf{f} < \mathbf{b} < \mathbf{t}$.

As noted in Remark 18.2.3(2), the truth-tables for $\wedge$ and $\vee$ in Figure 18.2 also compute least upper bounds and greater lower bounds, so that $v \wedge v' = \bigwedge\{v, v'\}$ and $v \vee v' = \bigvee\{v, v'\}$. We may elide the difference between $\wedge$-the-binary-operator and its generalisation to $\bigwedge$ henceforth, and similarly for $\vee$ and $\bigvee$.

LEMMA 18.2.5. *Suppose $V \subseteq \mathbf{3}$ is a set of truth-values. Then:*

1. $\bigwedge V \in \{\mathbf{t}, \mathbf{b}\}$ *if and only if* $V \subseteq \{\mathbf{t}, \mathbf{b}\}$.
2. $\bigwedge V = \mathbf{f}$ *if and only if* $V \between \{\mathbf{f}\}$ *(if and only if* $\mathbf{f} \in V$*)*.
3. $\bigvee V \in \{\mathbf{t}, \mathbf{b}\}$ *if and only if* $V \between \{\mathbf{t}, \mathbf{b}\}$.
4. $\bigvee V = \mathbf{f}$ *if and only if* $V \subseteq \{\mathbf{f}\}$.

Proof. A fact of Definition 18.2.4. □

18.2.3 Implication(s)

REMARK 18.2.6 (Two implications). Figure 18.2 has two implication operators; $\supset$ (material implication) and also $\Rrightarrow$ (we underline the differences in the truth-tables in Figure 18.2).

Having multiple implication connectives in paraconsistent logic is typical: e.g. the authors of [AAZ11] note the existence of *sixteen* possible implications just in a three-valued logic [AAZ11, note 5, page 22].

We chose to include $\supset$ and $\Rrightarrow$ as primitive, from the sixteen implications available, because they will be especially useful. See — just for example — Proposition 18.4.14(3), Corollary 18.4.15, Definitions 19.1.1 and 22.4.3, and Figure 22.2.

We make some simple observations:

LEMMA 18.2.7. *For $v, v' \in \mathbf{3}$ we have:*

1. $v \supset v' = \neg v' \supset \neg v$
2. *It is not necessarily the case that* $v \Rrightarrow v' = \neg v' \Rrightarrow \neg v$.

In words: the contrapositive rule is valid for $\supset$, but not for $\Rrightarrow$.

Proof. We compute and compare truth-tables; differences are underlined:

$p \supset q$	t	b	f
t	t	b	f
b	t	b	b
f	t	t	t

$\neg q \supset \neg p$	t	b	f
t	t	b	f
b	t	b	b
f	t	t	t

$p \Rrightarrow q$	t	b	f
t	t	$\underline{\mathbf{b}}$	f
b	t	b	$\underline{\mathbf{f}}$
f	t	t	t

$\neg q \Rrightarrow \neg p$	t	b	f
t	t	$\underline{\mathbf{f}}$	f
b	t	b	$\underline{\mathbf{b}}$
f	t	t	t

□

$$\begin{aligned}\phi ::= &(v : v\in\mathbf{3}) \mid (x : x\in\mathsf{Var}) \mid (p : p\in\mathbf{P}) \\ &\mid (\mathsf{Q}\phi : \mathsf{Q}\in\{\neg, \mathsf{T}, \mathsf{TB}, \mathsf{B}\}) \\ &\mid (\phi\mathsf{Q}\phi' : \mathsf{Q}\in\{\wedge, \vee, \supset, \equiv, \Rightarrow, \Leftrightarrow\}) \\ &\mid \forall Val\,\phi \mid \exists Val\,\phi \\ &\mid \forall x.\phi \mid \exists x.\phi\end{aligned}$$

Figure 18.4: Predicate syntax

REMARK 18.2.8. We noted in Figure 18.3 and Lemma 18.2.2 that

$$p{\Rightarrow}q = (\mathsf{TB}p){\supset}q = (\neg\mathsf{TB}p){\vee}q.$$

A minimal set of propositional connectives for our logic could leave out $\Rightarrow$ — and then perhaps leave out $\supset$ in favour of $\vee$ and $\neg$, and arrive at (say) $\{\mathbf{f}, \vee, \neg, \mathsf{TB}\}$. More discussion of $\Rightarrow$ and $\supset$ is in Subsection 18.4.4.

18.3 Predicates

NOTATION 18.3.1. Fix a countably infinite set of **variable symbols** Var. We will use variable symbols to construct our syntax in Definition 18.3.2.

DEFINITION 18.3.2. Suppose $\mathbf{P}$ is any set; we will always assume this is disjoint from Var. Define the syntax $\mathsf{Pred3}(\mathbf{P})$ of **predicates over $\mathbf{P}$** by the grammar in Figure 18.4.

In this syntax:

1. We may call $p\in\mathbf{P}$ a **propositional atom** or **propositional constant**. When we give our syntax a denotation, it will get assigned a truth-value in $\mathbf{3}$.
2. Variable symbols $x\in\mathsf{Var}$ range over the propositional atoms.
3. $\forall$ and $\exists$ are quantifiers. They bind variable symbols. We treat predicates up to α-equivalence henceforth.
4. We define the **free variables of a predicate** as usual:

$$\begin{aligned} fv(v) = fv(p) &= \varnothing & fv(\mathsf{Q}\phi) &= fv(\phi) \\ fv(\phi\mathsf{Q}\phi') &= fv(\phi)\cup fv(\phi') \\ fv(\forall Val\,\phi) = fv(\phi) &= fv(\exists Val\,\phi) & fv(\forall x.\phi) &= fv(\phi)\setminus\{x\} = fv(\exists x.\phi)\end{aligned}$$

   Note that $fv(\phi)$ does not include the propositional constants in ϕ; we only include the free variables.
5. We call a predicate **closed** when it has no free variables. We write $\mathsf{ClosedPred3}(\mathbf{P})$ for the set of closed predicates in $\mathsf{Pred3}(\mathbf{P})$, in symbols:

$$\mathsf{ClosedPred3}(\mathbf{P}) = \{\phi\in\mathsf{Pred3}(\mathbf{P}) \mid fv(\phi) = \varnothing\}.$$

   Note that a closed predicate $\phi\in\mathsf{ClosedPred3}(\mathbf{P})$ may still mention propositional constants from $\mathbf{P}$.

$$
\begin{array}{rll}
[\![v]\!]_f &= v & v\in\mathbf{3}\\
[\![p]\!]_f &= f(p) & p\in\mathsf{P}\\
[\![\mathsf{Q}\phi]\!]_f &= Q\,[\![\phi]\!]_f & \mathsf{Q}\in\{\neg,\mathsf{T},\mathsf{TB},\mathsf{B}\}\\
[\![\phi\mathsf{Q}\phi']\!]_f &= [\![\phi]\!]_f\,Q\,[\![\phi']\!]_f & \mathsf{Q}\in\{\wedge,\vee,\supset,\equiv,\Rightarrow,\Leftrightarrow\}\\
\\
[\![\forall Val\ \phi]\!]_f &= \bigwedge\{[\![\phi]\!]_f \mid f:\mathsf{P}\to\mathbf{3}\}\\
[\![\exists Val\ \phi]\!]_f &= \bigvee\{[\![\phi]\!]_f \mid f:\mathsf{P}\to\mathbf{3}\}\\
[\![\forall x.\phi]\!]_f &= \bigwedge\{[\![\phi[x{:=}p]]\!]_f \mid p\in\mathsf{P}\}\\
[\![\exists x.\phi]\!]_f &= \bigvee\{[\![\phi[x{:=}p]]\!]_f \mid p\in\mathsf{P}\}
\end{array}
$$

Figure 18.5: Denotation of a predicate ϕ with respect to a valuation f

DEFINITION 18.3.3 (Denotation). Suppose $(\mathsf{P}, \mathsf{Open})$ is a semitopology on a set P and $f : \mathsf{P} \to \mathbf{3}$ is a valuation (Definition 18.1.2(2)), and suppose $\phi\in\mathsf{ClosedPred3}(\mathsf{P})$ is a closed predicate over P (Definition 18.3.2(5)).

Then define $[\![\phi]\!]_f$ inductively as in Figure 18.5, where:

1. The unary and binary connectives ($\{\neg, \mathsf{T}, \mathsf{TB}, \mathsf{B}\}$ and $\{\wedge, \vee, \supset, \equiv, \Rightarrow, \Leftrightarrow\}$) are interpreted using the tables in Figure 18.2, as discussed in Remark 18.3.4.
2. $\phi[x{:=}p]$ is obtained from ϕ by replacing (substituting) x with p throughout.

REMARK 18.3.4. For each unary and binary connective Q considered in Figure 18.2 there is a corresponding unary or binary syntax connective Q in the syntax of Figure 18.4. (This is no different from how in the usual two-valued propositional logic, the symbol '$\wedge$' can represent both a function $\mathbb{B} \times \mathbb{B} \to \mathbb{B}$ and also a predicate connective.)

Each Q is interpreted as the corresponding Q when we define the denotation in Figure 18.5.

REMARK 18.3.5.

1. The denotation of the connectives in $\{\neg, \mathsf{T}, \mathsf{TB}, \mathsf{B}\}$ and $\{\wedge, \vee, \supset, \equiv, \Rightarrow, \Leftrightarrow\}$ is straightforward: we look up the input(s) in the corresponding truth-table in Figure 18.2, and return the corresponding output; see Remark 18.3.4.

2. The modalities T, TB, and B are not continuous as functions from $\mathbf{3}$ to $\mathbf{3}$; e.g. $\lambda v.\mathsf{T}v$ is not continuous at $\mathbf{b}$. Likewise $\lambda v.v{\Rightarrow}\mathbf{f}$ is not continuous.

3. $\forall Val\ \phi$ takes a conjunction (greatest lower bound) over the truth-values of ϕ for every possible assignment of truth-values from $\mathbf{3}$ to the propositional atoms in ϕ.

   Intuitively, $\forall Val$ universally quantifies over all possible valuations; if we view valuations as possible worlds, then $\forall Val$ is a box-style modality.

4. $\exists Val\ \phi$ is the de Morgan dual: it takes a disjunction (least upper bound) over the truth-values of ϕ for every possible assignment of truth-values from $\mathbf{3}$ to the propositional atoms in ϕ.

   Intuitively, $\exists Val$ existentially quantifies over all possible valuations; if we view valuations as possible worlds, then $\exists Val$ is a diamond-style modality.

5. $\forall x.\phi$ takes a conjunction over the truth-values of $\phi[x{:=}p]$ for every $p \in \mathsf{P}$. So $\forall$ quantifies universally over propositional atoms.

6. $\exists x.\phi$ takes a disjunction over the truth-values of $\phi[x{:=}p]$ for every $p \in \mathsf{P}$. So $\exists$ quantifies existentially over propositional atoms.

18.4 Validity

18.4.1 The definition

DEFINITION 18.4.1 (Validity). Suppose $(\mathsf{P}, \mathsf{Open})$ is a semitopology and $f : \mathsf{P} \to \mathbf{3}$ is a valuation and $\phi \in \mathsf{ClosedPred3}(\mathsf{P})$ is a closed predicate.

1. Define $f \vDash \phi$ by
$$f \vDash \phi \quad \text{when} \quad [\![\phi]\!]_f \in \{\mathbf{t}, \mathbf{b}\}.$$
In this case we call ϕ **valid** in f.
2. We may call $\{\mathbf{t}, \mathbf{b}\}$ the set of **designated values**, following a terminology from paraconsistent logic — these are the *valid* truth-values.
Note that $\mathbf{t}$ is valid, as we would expect, but so is $\mathbf{b}$.
3. If f does not matter to calculating $[\![\phi]\!]_f$ (e.g. because ϕ mentions no variable symbols or predicate atoms) then we may write $[\![\phi]\!]_f$ just as $[\![\phi]\!]$, and $f \vDash \phi$ just as $\vDash \phi$.
4. Suppose $\Phi \subseteq \mathsf{ClosedPred3}(\mathsf{P})$ is a set of predicates. Write
$$f \vDash \Phi \quad \text{when} \quad \forall \phi \in \Phi. f \vDash \phi,$$
and say that Φ is **valid** in f.
5. Write $\mathsf{P} \vDash \phi$ when $f \vDash \phi$ for every valuation $f : \mathsf{P} \to \mathbf{3}$. and similarly for $\mathsf{P} \vDash \Phi$.
6. Write $\mathsf{P}, \mathsf{Open} \vDash \phi$ when $f \vDash \phi$ for every valuation $f : \mathsf{P} \to \mathbf{3}$ that is continuous on $(\mathsf{P}, \mathsf{Open})$, and similarly for $\mathsf{P} \vDash \Phi$ and $\mathsf{P}, \mathsf{Open} \vDash \Phi$.

DEFINITION 18.4.2. A **sequent** is a pair of finite sets of predicates $\Phi \vdash \Psi$.
We define a notion of **sequent validity** $\mathsf{P} \vDash (\Phi \vdash \Psi)$ by
$$\mathsf{P} \vDash (\Phi \vdash \Psi) \quad \text{when} \quad \mathsf{P} \vDash (\bigwedge \Phi) \Rightarrow (\bigvee \Psi).$$

REMARK 18.4.3. Recall from Remark 18.2.3(3) that $\supset$ is a material implication. Yet $\Rightarrow$ can also claim to be an implication operator, and unlike $\supset$ it satisfies Modus Ponens. We examine this in Propositions 18.4.12 and 18.4.14.

In Definition 18.4.1(2) we noted that $\mathbf{t}$ is valid, but so is $\mathbf{b}$. We can make this formal with a small lemma:

LEMMA 18.4.4. *Suppose* P *is a set and* $f : \mathsf{P} \to \mathbf{3}$ *is a valuation and* $\phi \in \mathsf{ClosedPred3}(\mathsf{P})$. *Then*
$$f \vDash \phi \quad \textit{if and only if} \quad f \vDash \mathbb{T}\mathbb{B}\phi.$$

Proof. Simply because $\mathbb{T}\mathbb{B}v \in \{\mathbf{t}, \mathbf{b}\}$ if and only if $v \in \{\mathbf{t}, \mathbf{b}\}$, for every $v \in \mathbf{3}$. □

18.4.2 $f^{\models}$ and $f^{\models\neg}$: the designated sets of a valuation

In this easy Subsection we continue Remark 18.1.8 and follow a basic exercise in unpacking the definitions. The only extra ingredient we have now, relative to the results surrounding Remark 18.1.8, is that we can interpret an element of $\{\mathbf{t}, \mathbf{b}\}$ as 'valid':

DEFINITION 18.4.5. Suppose P is a set and $f : \mathsf{P} \to \mathbf{3}$ is a valuation. Then define the **designated set** $f^{\models} \subseteq \mathsf{P}$ and the **neg-designated set** $f^{\models\neg} \subseteq \mathsf{P}$ of the valuation f by:

$$f^{\models} = f^{-1}\{\mathbf{t}, \mathbf{b}\}$$
$$f^{\models\neg} = f^{-1}\{\mathbf{f}, \mathbf{b}\}$$

REMARK 18.4.6.

1. We call $f^{\models}$ the designated set of f because this is the set of $p \in \mathsf{P}$ such that $f(p)$ is a *designated value* (i.e. is $\mathbf{t}$ or $\mathbf{b}$), as per Definition 18.4.1(2).
2. Similarly, $f^{\models\neg}$ is the set of $p \in \mathsf{P}$ such that $f(\neg p)$ is a *designated value* — note that this is *not the same thing* as the set of $p \in \mathsf{P}$ such that $f(p)$ is not a designated value, because $\{\mathbf{f}, \mathbf{b}\}$ is not the same thing as $\{\mathbf{f}\}$.
3. The pair $(f^{\models}, f^{\models\neg})$ characterises f in the sense that

$$char_{CC}(f) = (f^{\models}, f^{\models\neg}) \qquad \text{and} \qquad f = \delta_{C_{\mathbf{tb}}, C_{\mathbf{fb}}}(f^{\models}, f^{\models\neg}),$$

where *char* is from Definition 18.1.7, and δ is from Definition 18.1.5.

LEMMA 18.4.7. *Suppose* $(\mathsf{P}, \mathsf{Open})$ *is a semitopology and* $f : \mathsf{P} \to \mathbf{3}$ *is a continuous function on* $(\mathsf{P}, \mathsf{Open})$*. Then both* $f^{\models}$ *and* $f^{\models\neg}$ *are closed sets.*

Proof. By Definition 18.1.2(1) $\{\mathbf{t}, \mathbf{b}\}$ and $\{\mathbf{f}, \mathbf{b}\}$ are closed sets in $\mathbf{3}$, and by Corollary 5.1.12 the inverse image of a closed set under a continuous function is closed. □

The notation in Definition 18.4.5 is justified by the following very easy lemma:

LEMMA 18.4.8. *Suppose* P *is a set and* $p \in \mathsf{P}$ *and* $f : \mathsf{P} \to \mathbf{3}$ *is a valuation. Then:*

1. $p \in f^{\models}$ *if and only if* $f \models p$.
2. $p \in f^{\models\neg}$ *if and only if* $f \models \neg p$.

Proof. Direct from Figure 18.2, Definitions 18.4.5 and 18.4.1(1), and the case for $f \models p$ in Figure 18.5. □

18.4.3 Validity of conjunction and quantification

The $\mathbf{\wedge}$ and $\mathbf{\forall}$ are 'obviously' a conjunction and universal quantification respectively, but we still need to check that the truth-tables work and the notion of validity $\models$ interacts correctly with them. We do this and find that there are no surprises:

LEMMA 18.4.9. *Suppose* P *is a set and* $f : \mathsf{P} \to \mathbf{3}$ *is a valuation and* $\phi, \phi' \in \mathsf{ClosedPred3}(\mathsf{P})$. *Then:*

1. $f \models \phi \mathbf{\wedge} \phi'$ *if and only if* $f \models \phi \wedge f \models \phi'$.
2. $f \models \phi \mathbf{\vee} \phi'$ *if and only if* $f \models \phi \vee f \models \phi'$.

Proof. We reason as follows for the case of $\land$; the case of $\lor$ is exactly similar:

$$\begin{aligned} f \models \phi\land\phi' &\iff [\![\phi\land\phi']\!]_f \in \{\mathbf{t},\mathbf{b}\} && \text{Definition 18.4.1(1)} \\ &\iff [\![\phi]\!]_f \land [\![\phi']\!]_f \in \{\mathbf{t},\mathbf{b}\} && \text{Figure 18.5} \\ &\iff [\![\phi]\!]_f \in \{\mathbf{t},\mathbf{b}\} \wedge [\![\phi']\!]_f \in \{\mathbf{t},\mathbf{b}\} && \text{Lemma 18.2.5 (or Figure 18.2)} \\ &\iff f \models \phi \wedge f \models \phi' && \text{Definition 18.4.1(1)} \end{aligned}$$

□

LEMMA 18.4.10. *Suppose* P *is a set and* $f : \mathsf{P} \to \mathbf{3}$ *is a valuation and* $\forall x.\phi \in \mathsf{ClosedPred3}(\mathsf{P})$. *Then:*

1. $f \models \forall x.\phi$ *if and only if* $\forall p \in \mathsf{P}. f \models \phi[x{:=}p]$.
2. $f \models \exists x.\phi$ *if and only if* $\exists p \in \mathsf{P}. f \models \phi[x{:=}p]$.

Proof. For the case of $\forall$ we reason as follows:

$$\begin{aligned} f \models \forall x.\phi &\iff [\![\forall x.\phi]\!]_f \in \{\mathbf{t},\mathbf{b}\} && \text{Definition 18.4.1(1)} \\ &\iff \textstyle\bigwedge_{p\in\mathsf{P}} [\![\phi[x{:=}p]]\!]_f \in \{\mathbf{t},\mathbf{b}\} && \text{Figure 18.5} \\ &\iff \forall p\in\mathsf{P}.[\![\phi[x{:=}p]]\!]_f \in \{\mathbf{t},\mathbf{b}\} && \text{Lemma 18.2.5} \\ &\iff \forall p\in\mathsf{P}. f \models \phi[x{:=}p] && \text{Definition 18.4.1(1)} \end{aligned}$$

The case of $\exists$ is precisely similar. □

18.4.4 Logical implications

REMARK 18.4.11. Continuing Remarks 18.2.3(3) and 18.4.3, Figure 18.2 has truth-tables for two binary connectives expressing notions of 'implications': $\supset$ and $\Rightarrow$. We will need both:

1. $\Rightarrow$ expresses logical implication (see Lemma 18.4.13(3) and Proposition 18.4.14(3)).
2. $\supset$ helps us express the property of 'being intertwined' in (see Lemma 20.1.3 and 20.1.6).

Both implications interact nicely with $\land$:

PROPOSITION 18.4.12. *Suppose* P *is any set and* $p, q, r \in \mathsf{P}$ *and* $f : \mathsf{P} \to \mathbf{3}$ *is a valuation. Then:*

1. $[\![(p\land q)\supset r]\!]_f = [\![p\supset(q\supset r)]\!]_f$.
2. $[\![(p\land q)\Rightarrow r]\!]_f = [\![p\Rightarrow(q\Rightarrow r)]\!]_f$.

Proof. We simplify using the equivalences in Figure 18.3 (we could also just check truth-tables, of course):

$$\begin{aligned} (p\land q)\supset r &= \neg(p\land q)\lor r = \neg p\lor\neg q\lor r \\ p\supset(q\supset r) &= \neg p\lor\neg q\lor r \end{aligned}$$

$$\begin{aligned} (p\land q)\Rightarrow r &= \neg\mathsf{TB}(p\land q)\lor r = \neg(\mathsf{TB}p\land\mathsf{TB}q)\lor r = \mathsf{T}\neg p\lor\mathsf{T}\neg q\lor r \\ p\Rightarrow(q\Rightarrow r) &= \neg\mathsf{TB}p\lor\neg\mathsf{TB}q\lor r = \mathsf{T}\neg p\lor\mathsf{T}\neg q\lor r. \end{aligned}$$

□

LEMMA 18.4.13. *Suppose* $v, v' \in \mathbf{3}$. *Then:*

1. *If* $v \in \{\mathbf{t},\mathbf{b}\}$ *implies* $v' \in \{\mathbf{t},\mathbf{b}\}$, *then* $v\supset v' \in \{\mathbf{t},\mathbf{b}\}$.
2. *It is possible that* $v\supset v' \in \{\mathbf{t},\mathbf{b}\}$ *and* $v \in \{\mathbf{t},\mathbf{b}\}$, *but* $v' \notin \{\mathbf{t},\mathbf{b}\}$.

3. $v \Rightarrow v' \in \{\mathbf{t}, \mathbf{b}\}$ *if and only if* $v \in \{\mathbf{t}, \mathbf{b}\} \implies v' \in \{\mathbf{t}, \mathbf{b}\}$.

Proof. We consider each part in turn:

1. We check the truth-table for $\supset$ in Figure 18.2 and prove the contrapositive: if $v \supset v' = \mathbf{f}$, then $v = \mathbf{t} \wedge v' = \mathbf{f}$, and so $v \in \{\mathbf{t}, \mathbf{b}\}$ does not imply $v' \in \{\mathbf{t}, \mathbf{b}\}$.
2. We set $v = \mathbf{b}$ and $v' = \mathbf{f}$ and note from the truth-table for $\supset$ in Figure 18.2 that $\mathbf{b} \supset \mathbf{f} = \mathbf{b}$.
3. We prove two implications:
   - For the right-to-left implication, we prove the contrapositive. Suppose $v \Rightarrow v' = \mathbf{f}$. We check the truth-table for $\Rightarrow$ in Figure 18.2 and see that $v' = \mathbf{f}$ and $v \in \{\mathbf{t}, \mathbf{b}\}$. Thus $v \in \{\mathbf{t}, \mathbf{b}\}$ implies $v' \notin \{\mathbf{t}, \mathbf{b}\}$, and it is *not* the case that $v \in \{\mathbf{t}, \mathbf{b}\}$ implies $v' \in \{\mathbf{t}, \mathbf{b}\}$.
   - For the left-to-right implication, there are three sub-cases:
     - If $v' = \mathbf{b}$ or $v' = \mathbf{t}$ then $v' \in \{\mathbf{t}, \mathbf{b}\}$ and there is nothing to prove.
     - If $v = \mathbf{f}$ then there is nothing to prove.
     - Suppose $v' = \mathbf{f}$ and $v \in \{\mathbf{b}, \mathbf{t}\}$. We check the truth-table for $\Rightarrow$ in Figure 18.2 and see that the result holds. □

PROPOSITION 18.4.14. *Suppose* P *is a set and* $f : \mathsf{P} \to \mathbf{3}$ *is a valuation and* $\phi, \phi' \in \mathsf{ClosedPred3}(\mathsf{P})$. *Then:*

1. *If* $f \vDash \phi$ *implies* $f \vDash \phi'$, *then* $f \vDash \phi \supset \phi'$.
2. *It is possible that* $f \vDash \phi \supset \phi'$ *and* $f \vDash \phi$ *but* $f \nvDash \phi'$.
3. $f \vDash \phi \Rightarrow \phi'$ *if and only if* $f \vDash \phi \implies f \vDash \phi'$.

Proof. By Definition 18.4.1(1), $f \vDash \phi$ when $[\![\phi]\!]_f \in \{\mathbf{t}, \mathbf{b}\}$, and similarly for $f \vDash \phi \supset \phi'$, $f \vDash \phi \Rightarrow \phi'$, and $f \vDash \phi'$. By Figure 18.5 $[\![\phi \circ \phi']\!]_f = [\![\phi]\!]_f \circ [\![\phi']\!]_f$ for $\circ \in \{\supset, \Rightarrow\}$. We use Lemma 18.4.13. □

We can pull the results in this Subsection together as follows:

COROLLARY 18.4.15. *Suppose* P *is a set and* $f : \mathsf{P} \to \mathbf{3}$ *is a valuation and* $\phi, \phi' \in \mathsf{ClosedPred3}(\mathsf{P})$. *Then:*

$$\begin{aligned} f \vDash \phi \Rightarrow \phi' &\iff f \nvDash \phi \vee f \vDash \phi' \\ &\iff [\![\phi]\!]_f \in \{\mathbf{f}\} \vee [\![\phi']\!]_f \in \{\mathbf{t}, \mathbf{b}\} \\ &\iff [\![\phi]\!]_f \in \{\mathbf{t}, \mathbf{b}\} \implies [\![\phi']\!]_f \in \{\mathbf{t}, \mathbf{b}\} \end{aligned}$$

$$\begin{aligned} f \vDash \phi \supset \phi' &\iff f \vDash \neg\phi \vee f \vDash \phi' \\ &\iff [\![\phi]\!]_f \in \{\mathbf{f}, \mathbf{b}\} \vee [\![\phi']\!]_f \in \{\mathbf{t}, \mathbf{b}\} \\ &\iff [\![\phi]\!]_f \in \{\mathbf{t}\} \implies [\![\phi']\!]_f \in \{\mathbf{t}, \mathbf{b}\}. \end{aligned}$$

Proof. We reason as follows:

- That $f \vDash \phi \Rightarrow \phi'$ if and only if $f \nvDash \phi \vee f \vDash \phi'$ just rephrases Proposition 18.4.14(3).
- That $f \vDash \phi \supset \phi'$ if and only if $f \vDash \neg\phi \vee f \vDash \phi'$ follows from the characterisation $p \supset q = (\neg p) \mathbf{\vee} q$ of $\supset$ as material equivalence in Figure 18.3, and from Lemma 18.4.9(2).
- The rest of the equivalences just unpack the definition of validity from Definition 18.4.1(1).

□

18.5 Logical equivalence

Lemma 18.5.1 is closely related to Proposition 18.4.14, though it is easiest to give a direct proof:

LEMMA 18.5.1. *Suppose* P *is a set and* $f : \mathsf{P} \to \mathbf{3}$ *is a valuation and* $\phi, \phi' \in \mathsf{ClosedPred3}(\mathsf{P})$ *are closed predicates. Then*

$$f \models \phi{\Leftrightarrow}\phi' \quad \textit{if and only if} \quad f \models \phi \iff f \models \phi'.$$

Proof. A direct proof like that in Proposition 18.4.14(3) is straightforward. Or, we can note the equivalence $p{\Leftrightarrow}q = (p{\Rightarrow}q){\wedge}(q{\Rightarrow}p)$ from Figure 18.3, and use Proposition 18.4.14(3) and Lemma 18.4.9. □

COROLLARY 18.5.2. *Suppose* $(\mathsf{P}, \mathsf{Open})$ *is a semitopology and* $f : \mathsf{P} \to \mathbf{3}$ *is a valuation and* $p, p' \in \mathsf{P}$. *Then the following are equivalent:*

1. $f \models p{\Leftrightarrow}p'$.
2. $p \in f^{\models} \iff p' \in f^{\models}$.
3. $p \in f^{\text{-}1}\{\mathbf{f}\} \iff p' \in f^{\text{-}1}\{\mathbf{f}\}$.

Proof. Equivalence of parts 1 and 2 follows from Lemmas 18.5.1 and 18.4.8 (a direct calculation from the truth-table in Figure 18.2 is also straightforward).

Equivalence of parts 2 and 3 follows noting from Definition 18.4.5 that $f^{\models} = f^{\text{-}1}\{\mathbf{t}, \mathbf{b}\}$ so that $p \in f^{\models} \iff p \notin f^{\text{-}1}\{\mathbf{f}\}$ and $p' \in f^{\models} \iff p' \notin f^{\text{-}1}\{\mathbf{f}\}$. □

COROLLARY 18.5.3. *Suppose* P *is any set and* $f : \mathsf{P} \to \mathbf{3}$ *is a valuation, and suppose* $\phi, \phi' \in \mathsf{Pred3}(\mathsf{P})$ *and* $\mathit{fv}(\phi) \cup \mathit{fv}(\phi') \subseteq \{x\}$, *so that* $\forall x.\phi{\Leftrightarrow}\phi' \in \mathsf{ClosedPred3}(\mathsf{P})$. *Then the following are equivalent:*

1. $f \models \forall x.(\phi{\Leftrightarrow}\phi')$.
2. $f \models \phi[x{:=}p] \iff f \models \phi'[x{:=}p]$ *for every* $p \in \mathsf{P}$.
3. $\{p \mid f \models \phi[x{:=}p]\} = \{p \mid f \models \phi'[x{:=}p]\}$.
4. $\{p \mid [\![\phi[x{:=}p]]\!]_f \in \{\mathbf{t}, \mathbf{b}\}\} = \{p \mid [\![\phi'[x{:=}p]]\!]_f \in \{\mathbf{t}, \mathbf{b}\}\}$.
5. $\{p \mid [\![\phi[x{:=}p]]\!]_f = \mathbf{f}\} = \{p \mid [\![\phi'[x{:=}p]]\!]_f = \mathbf{f}\}$.

Proof. Equivalence of parts 1 and 2 is from Lemmas 18.5.1 and 18.4.9. Equivalence of parts 2 and 3 is a fact of sets. Equivalence of parts 3 and 4 is immediate from Definition 18.4.1(1). Equivalence of parts 4 and 5 follows since $\mathbf{3} = \{\mathbf{t}, \mathbf{b}, \mathbf{f}\}$. □

18.6 A sequent system

In this short subsection we develop a sequent system for our logic that is sound and complete with respect to validity.

The system here follows the style of the simple and elegant sequent system presented in [WM16]. There are important differences: they consider a four-valued purely propositional logic, whereas we have a three-valued logic but with far richer connectives including modalities and quantifiers. So the sequent system here can be viewed as an elaboration of a simplification of the one in [WM16].

But in a way, the details do not matter. What matters is that a legitimate notion of derivation is shown to exist, so that we can talk meaningfully about derivability and proof-search.

$$\frac{((\mathsf{a},\mathsf{b})\in\{(tb,\mathit{ff}),(fb,tt)\})}{\vdash \Sigma,\mathsf{a}:\phi,\mathsf{b}:\phi}\,(\mathbf{Ax}) \qquad \frac{\vdash \Sigma, neg(\mathsf{a}):\phi}{\vdash \Sigma,\mathsf{a}:\neg\phi}\,(\neg\mathbf{LR})$$

$$\frac{\vdash \Sigma,\mathsf{a}:\phi \quad \vdash \Sigma,\mathsf{a}:\psi \quad (\mathsf{a}\in\{tb,tt\})}{\vdash \Sigma,\mathsf{a}:\phi\wedge\psi}\,(\wedge\mathbf{R}) \qquad \frac{\vdash \Sigma,\mathsf{a}:\phi,\mathsf{a}:\psi \quad (\mathsf{a}\in\{fb,\mathit{ff}\})}{\vdash \Sigma,\mathsf{a}:\phi\wedge\psi}\,(\wedge\mathbf{L})$$

$$\frac{\vdash \Sigma,tt:\phi \quad (\mathsf{a}\in\{tb,tt\})}{\vdash \Sigma,\mathsf{a}:\mathsf{T}\phi}\,(\mathsf{T}\mathbf{R}) \qquad \frac{\vdash \Sigma,\mathit{ff}:\phi,fb:\phi \quad (\mathsf{a}\in\{fb,\mathit{ff}\})}{\vdash \Sigma,\mathsf{a}:\mathsf{T}\phi}\,(\mathsf{T}\mathbf{L})$$

$$\frac{\vdash \mathsf{a}:\phi \quad (\mathsf{a}\in\{tb,tt\})}{\vdash \Sigma,\mathsf{a}:\forall Val\,(\phi)}\,(\forall Val\,\mathbf{R}) \qquad \frac{\nvdash \mathsf{a}:\phi \quad ((\mathsf{a},\mathsf{b})\in\{(tb,\mathit{ff}),(tt,fb)\})}{\vdash \Sigma,\mathsf{b}:\forall Val\,(\phi)}\,(\forall Val\,\mathbf{L})$$

$$\frac{\vdash \Sigma,\mathsf{a}:\phi[x{:=}p] \quad (\mathsf{a}\in\{tt,tb\},\text{every }p)}{\vdash \Sigma,\mathsf{a}:\forall x.\phi}\,(\forall\mathbf{R}) \qquad \frac{\vdash \Sigma,\mathsf{a}:\phi[x{:=}p] \quad (\mathsf{a}\in\{fb,\mathit{ff}\})}{\vdash \Sigma,\mathsf{a}:\forall x.\phi}\,(\forall\mathbf{L})$$

Figure 18.6: Derivable tag-sequents

DEFINITION 18.6.1. For this Subsection, we fix some finite set **P** of propositional constant, as per Definition 18.3.2(1).

DEFINITION 18.6.2. 1. Define **tags** tb**,** ff**,** fb**, and** tt by:

$$tb = \{\mathbf{t},\mathbf{b}\} \quad \mathit{ff} = \{\mathbf{f}\} \quad fb = \{\mathbf{f},\mathbf{b}\} \quad tt = \{\mathbf{t}\}$$

We will let a and b range over tags.

2. A **tag-sequent** Σ is a finite set of pairs $\mathsf{a}:\phi$ where a is a tag and $\phi\in\mathsf{ClosedPred3}(\mathbf{P})$ is a closed predicate (Definition 18.3.2(5)).
3. Define a neg operation on tags such that $neg(t) = \{\neg x \mid x\in t\}$. Spelling this out:

$$neg(tb) = fb \quad neg(fb) = tb \quad neg(\mathit{ff}) = tt \quad neg(tt) = \mathit{ff}$$

4. The **derivable tag-sequents** are defined inductively by the sequent rules in Figure 18.6. Note that Figure 18.6 presents a subsystem of the logic, but this is fine because we can derive rules for $\mathbb{B}$, B, $\exists x$, $\vee$, $\supset$, $\equiv$, $\Rightarrow$, $\Leftrightarrow$, and $\exists Val$, as per the following equivalences:

$$\begin{array}{lll} \mathbb{B}\phi = \neg\mathsf{T}\neg\phi & \mathsf{B}\phi = \mathbb{B}\phi\wedge\neg\mathsf{T}\phi & \exists x.\phi = \neg\forall x.\neg\phi \\ \phi\supset\phi' = (\neg\phi)\vee\phi' & \phi\equiv\phi' = (\phi\supset\phi')\wedge(\phi'\supset\phi) & \phi\vee\phi' = \neg((\neg\phi)\wedge(\neg\phi')) \\ \phi\Rightarrow\phi' = (\mathbb{B}\phi)\supset\phi' & \phi\Leftrightarrow\phi' = (\phi\Rightarrow\phi')\wedge(\phi'\Rightarrow\phi) & \exists Val\,\phi = \neg\exists Val\,\neg\phi \end{array}$$

5. Write $\vdash \Sigma$ for the judgement 'Σ is derivable'.

REMARK 18.6.3. The derivation rule $(\forall Val\,\mathbf{L})$ is unusual because it has $\nvdash$ above the line. This is still a well-defined inductive definition, by induction on the syntax of the tag-sequent.

LEMMA 18.6.4. *Suppose* $\phi,\psi\in\mathsf{ClosedPred3}(\mathbf{P})$ *and suppose* $f:\mathbf{P}\to\mathbf{3}$ *is any valuation. We check properties corresponding to soundness and completeness of the rules in Figure 18.6:*

1. *The case of* $(\mathbf{Ax})$ *with* $(\mathtt{a},\mathtt{b}) = (tb, ff)$.
   $[\![\phi]\!]_f \in tb \vee [\![\phi]\!]_f \in ff$ *is a fact.*
2. *The case of* $(\mathbf{Ax})$ *with* $(\mathtt{a},\mathtt{b}) = (fb, tt)$.
   $[\![\phi]\!]_f \in fb \vee [\![\phi]\!]_f \in tt$ *is a fact.*
3. *The case of* $(\neg\mathbf{R})$.
   $[\![\phi]\!]_f \in x$ *if and only if* $[\![\neg\phi]\!]_f \in neg(x)$.
4. *The case of* $(\wedge\mathbf{R})$ *with* $\mathtt{a} = tb$.
   $[\![\phi]\!]_f \in tb \wedge [\![\psi]\!]_f \in tb$ *if and only if* $[\![\phi\wedge\psi]\!]_f \in tb$.
5. *The case of* $(\wedge\mathbf{R})$ *with* $\mathtt{a} = tt$.
   $[\![\phi]\!]_f \in tt \wedge [\![\psi]\!]_f \in tt$ *if and only if* $[\![\phi\wedge\psi]\!]_f \in tt$.
6. *The case of* $(\wedge\mathbf{L})$ *with* $\mathtt{a} = fb$.
   $[\![\phi]\!]_f \in fb \vee [\![\psi]\!]_f \in fb$ *if and only if* $[\![\phi\wedge\psi]\!]_f \in fb$.
7. *The case of* $(\wedge\mathbf{L})$ *with* $\mathtt{a} = ff$.
   $[\![\phi]\!]_f \in ff \vee [\![\psi]\!]_f \in ff$ *if and only if* $[\![\phi\wedge\psi]\!]_f \in ff$.
8. *The case of* $(\mathsf{T}\mathbf{R})$ *with* $x \in \{tb, tt\}$.
   $[\![\phi]\!]_f = \mathbf{t}$ *if and only if* $[\![\mathsf{T}\phi]\!]_f = \mathbf{t}$ *if and only if* $[\![\mathsf{T}\phi]\!]_f \in tb$ *if and only if* $[\![\mathsf{T}\phi]\!]_f \in tt$.
9. *The case of* $(\mathsf{T}\mathbf{L})$ *with* $x \in \{fb, ff\}$.
   $[\![\phi]\!]_f \in fb \vee [\![\phi]\!]_f \in ff$ *if and only if* $[\![\phi]\!]_f \neq \mathbf{t}$ *if and only if* $[\![\mathsf{T}\phi]\!]_f = \mathbf{f}$ *if and only if* $[\![\mathsf{T}\phi]\!]_f \in ff$ *if and only if* $[\![\mathsf{T}\phi]\!]_f \in fb$.
10. *The case of* $(\forall Val\,\mathbf{R})$ *with* $\mathtt{a}$.
    If $\mathtt{a} \in \{tb, tt\}$ *then* $\forall f'.[\![\phi]\!]_{f'} \in \mathtt{a}$ *if and only if* $[\![\forall Val\ \phi]\!]_f \in \mathtt{a}$.
11. *The case of* $(\forall Val\,\mathbf{L})$ *with* $\mathtt{a}$.
    If $\mathtt{a} \in \{fb, ff\}$ *then* $\exists f'.[\![\phi]\!]_{f'} \in \mathtt{a}$ *if and only if* $[\![\forall Val\ \phi]\!]_f \in \mathtt{a}$.
12. *The case of* $(\forall\mathbf{R})$ *with* $\mathtt{a} = tb$.
    $[\![\phi[x{:=}p]]\!]_f \in tb$ *for every p, if and only if* $[\![\forall x.\phi]\!]_f \in tb$.
13. *The case of* $(\forall\mathbf{R})$ *with* $\mathtt{a} = tt$.
    $[\![\phi[x{:=}p]]\!]_f = tt$ *for every p, if and only if* $[\![\forall x.\phi]\!]_f = tt$.
14. *The case of* $(\forall\mathbf{L})$ *with* $\mathtt{a} = fb$.
    $[\![\phi[x{:=}p]]\!]_f \in fb$ *for some p, if and only if* $[\![\forall x.\phi]\!]_f \in fb$.
15. *The case of* $(\forall\mathbf{L})$ *with* $\mathtt{a} = ff$.
    $[\![\phi[x{:=}p]]\!]_f = \mathbf{f}$ *for some p, if and only if* $[\![\forall x.\phi]\!]_f = \mathbf{f}$.

Proof. These are all facts of the definition of $[\![\text{-}]\!]_f$ from Figure 18.5 and the truth-tables from Figure 18.2. □

DEFINITION 18.6.5. Suppose $\phi \in \mathsf{ClosedPred3}(\mathbf{P})$ and a is a tag and Σ is a tag-sequent. Then:

1. Write $[\![\mathtt{a} : \phi]\!]_f$ when $[\![\phi]\!]_f \in \mathtt{a}$.
2. Write $\models \Sigma$ when for every valuation $f : \mathbf{P} \to \mathbf{3}$, there exists some element $\mathtt{a} : \phi$ in Σ such that $[\![\mathtt{a} : \phi]\!]_f$.
3. When $\models \Sigma$ holds, we call Σ a **valid tag-sequent**.

PROPOSITION 18.6.6. *Suppose* Σ *is a tag-sequent. Then we have:*

1. Soundness: *If* $\vdash \Sigma$ *then* $\models \Sigma$.
2. Completeness: *If* $\models \Sigma$ *then* $\vdash \Sigma$.

Thus, Σ *is derivable if and only if it is valid.*

Proof. We prove both by by a simultaneous induction on the syntax of tag-sequences:[8]

1. By unpacking Definition 18.6.5 and using the relevant clause of Lemma 18.6.4 from left-to-right.

2. By unpacking Definition 18.6.5 and using the relevant clause of Lemma 18.6.4 from right-to-left. □

[8]An induction on derivations will not work, because of the $\not\vdash$ in $(\forall \mathit{Val}\,\mathbf{L})$ observed in Remark 18.6.3. However, the derivation rules are all syntax-directed, meaning that they reduce the size of a tag-sequent when read bottom-up, so we can work by induction on syntax.

19 Axiomatisations

19.1 Theory arising from a witness function

DEFINITION 19.1.1. Suppose $wf : \mathsf{P} \to \mathcal{W}(\mathsf{P})$ is a witness function on a finite set P (Definition 8.2.2(1)).

1. Define the **axioms arising from** wf to be sets of predicates in $\mathsf{ClosedPred3}(\mathsf{P})$ as follows:

$$\begin{aligned} \mathsf{closedAx}_{wf}(p) &= \big(\textstyle\bigwedge_{w \in wf(p)} \bigvee_{q \in w} q\big) \Rightarrow p \\ \mathsf{closedAx}^{\neg}_{wf}(p) &= \big(\textstyle\bigwedge_{w \in wf(p)} \bigvee_{q \in w} \neg q\big) \Rightarrow \neg p \end{aligned}$$

$$\begin{aligned} \mathsf{closedAx}_{wf} &= \textstyle\bigwedge \{\mathsf{closedAx}_{wf}(p) \mid p \in \mathsf{P}\} \\ \mathsf{closedAx}^{\neg}_{wf} &= \textstyle\bigwedge \{\mathsf{closedAx}^{\neg}_{wf}(p) \mid p \in \mathsf{P}\} \end{aligned}$$

2. We can collect these axioms into a conjunction $\mathsf{Ax}(wf)$, which we call the **theory arising from** wf:

$$\mathsf{Ax}(wf) = \mathsf{closedAx}_{wf} \wedge \mathsf{closedAx}^{\neg}_{wf}.$$

3. We may omit the wf annotation where this is unimportant or understood, writing (for example) $\mathsf{closedAx}^{\neg}_{wf}(p)$ just as $\mathsf{closedAx}^{\neg}(p)$.

REMARK 19.1.2. We can read $\mathsf{closedAx}_{wf}$ and $\mathsf{closedAx}^{\neg}_{wf}$ above as asserting that

$$\begin{aligned} |f^{-1}\{\mathbf{t},\mathbf{b}\}| &\subseteq f^{-1}\{\mathbf{t},\mathbf{b}\} \quad \text{and} \\ |f^{-1}\{\mathbf{f},\mathbf{b}\}| &\subseteq f^{-1}\{\mathbf{f},\mathbf{b}\}. \end{aligned}$$

The proofs below implicitly reflect this intuition, and make it formal.

LEMMA 19.1.3. *Suppose that:*

- $wf : \mathsf{P} \to \mathcal{W}(\mathsf{P})$ *is a witness function on a finite set* P.
- $p \in \mathsf{P}$ *and* $B \subseteq \mathsf{P}$ *blocks* p *(Definition 8.2.4).*
- $f : \mathsf{P} \to \mathbf{3}$ *is a valuation and* $f \vDash \mathsf{Ax}(wf)$.

Then:

1. *$(\forall q \in B. f \vDash q)$ implies $f \vDash p$.*
2. *$(\forall q \in B. f \vDash \neg q)$ implies $f \vDash \neg p$.*

Proof. We consider each part in turn:

1. Recall from Definition 19.1.1 that

$$\mathsf{closedAx}_{wf}(p) = \left(\bigwedge_{w \in wf(p)} \bigvee_{q \in w} q\right) \Rightarrow p,$$

and $\mathsf{closedAx}_{wf}(p)$ appears in $\mathsf{Ax}(wf)$ by Definition 19.1.1.

Choose one $w \in wf(p)$ (by construction in Definition 8.2.2(1) at least one such exists) and consider the disjunction $\bigvee_{q \in w} q$ for $w \in wf(p)$ in $\mathsf{closedAx}_{wf}$. We assumed that B blocks p, so by assumption in Definition 8.2.4 $B \between w$. We also assumed that $f \vDash q$ for every $q \in B$, so in particular $f \vDash q$ for some $q \in w$, and thus by Lemma 18.4.9 $f \vDash \bigvee_{q \in w} q$.

Now this holds for *every* $w \in wf(p)$, so by Lemma 18.4.9 $f \vDash \bigwedge_{w \in wf(p)} \bigvee_{q \in w} q$. We assumed $f \vDash \mathsf{closedAx}_{wf}(p)$, and by Proposition 18.4.14(3) $f \vDash p$ follows.

2. The reasoning to show that $\forall q \in B. f \vDash \neg q$ implies $f \vDash \neg p$ is precisely similar to the previous case, using $\mathsf{closedAx}^{\neg}_{wf}(p) = \left(\bigwedge_{w \in wf(p)} \bigvee_{q \in w} \neg q\right) \Rightarrow \neg p$. □

REMARK 19.1.4. Note a curious thing:

- Definition 8.2.5(3) characterises closed sets in the witness semitopology using a single property $p \,\text{¡}_{wf}\, P$.
- Definition 19.1.1 contains *two* axioms for each $p \in \mathsf{P}$: $\mathsf{closedAx}_{wf}(p)$ and $\mathsf{closedAx}^{\neg}_{wf}(p)$.

We can look at the definition of $p \,\text{¡}_{wf}\, P$ in Definition 8.2.4(1), and we see that $\mathsf{closedAx}_{wf}(p)$ visibly imitates it — but what is $\mathsf{closedAx}^{\neg}_{wf}(p)$ there for?

A formal answer to this question is that the proof of Proposition 19.2.3 — which characterises those valuations $f : \mathsf{P} \to \mathbf{3}$ that are continuous on the witness semitopology $(\mathsf{P}, \mathsf{Open}(wf))$ — requires both axioms.

This formal answer raises a semi-formal question: why *should* it require both axioms? Because Definition 19.1.1 (axiomatisation of continuity) has to work harder than Definition 8.2.5(3) (definition of open/closed sets). Defining an open (or a closed) set requires two truth-values (whether a point is in or not in that set) whereas the logic has three values and correspondingly a valuation $f : \mathsf{P} \to \mathbf{3}$ can return $\mathbf{t}$, $\mathbf{f}$, or $\mathbf{b}$. This makes it a more complex entity than just an open (or closed) set.

In fact we saw in Remark 18.4.6(3) that a valuation f can be thought of as a *pair* of closed sets $char_{CC}(f) = (f^{\vDash}, f^{\vDash\neg}) = (f^{-1}\{\mathbf{t}, \mathbf{b}\}, f^{-1}\{\mathbf{f}, \mathbf{b}\})$; see Definition 18.1.7 and the surrounding discussion. In this view, axiom $\mathsf{closedAx}_{wf}(p)$ controls the behaviour of $f^{\vDash}$, and $\mathsf{closedAx}^{\neg}_{wf}(p)$ controls the behaviour of $f^{\vDash\neg}$.

19.2 Axiomatisation of continuity

Recall from Definition 19.1.1 $\mathsf{Ax}(wf)$ the theory arising from wf.

LEMMA 19.2.1. *Suppose $wf : \mathsf{P} \to \mathcal{W}(\mathsf{P})$ is a witness function on a finite set P, and suppose $f : \mathsf{P} \to \mathbf{3}$ is a valuation. Then*

- *if $f \vDash \mathsf{Ax}(wf)$ then*
- *f is continuous on the witness semitopology $(\mathsf{P}, \mathsf{Open}(wf))$ from Definition 8.2.5(3).*

Proof. Suppose $f \vDash \mathsf{Ax}(wf)$.

To prove that f is continuous, it suffices by Lemma 18.1.4 to show that $f^{\vDash} = f^{-1}\{\mathbf{t}, \mathbf{b}\}$ and $f^{\vDash\neg} = f^{-1}\{\mathbf{f}, \mathbf{b}\}$ are closed sets in $(\mathsf{P}, \mathsf{Open}(wf))$. We consider each in turn:

- *We show that $f^{\vDash}$ is closed.*[1]

  Following Definition 8.2.5(2), $f^{\vDash}$ is closed in $(\mathsf{P}, \mathsf{Open}(wf))$ when for every $p \in \mathsf{P}$ and $B \subseteq f^{\vDash}$, if B blocks p (Definition 8.2.4), then $p \in f^{\vDash}$.

  Suppose $p \in \mathsf{P}$ and suppose $B \subseteq f^{\vDash}$ for some blocking set for p, which by Lemma 18.4.8 means precisely that $f \vDash B$. By assumption $f \vDash \mathsf{Ax}(p)$, and it follows from Lemma 19.1.3 that $f \vDash p$, and so by Lemma 18.4.8 $p \in f^{\vDash}$ as required.

- *We show that $f^{-1}\{\mathbf{f}, \mathbf{b}\}$ is closed.*

  This is just like the previous case, but using $\mathsf{closedAx}^{\neg}(p)$. □

LEMMA 19.2.2. *Suppose $wf : \mathsf{P} \to \mathcal{W}(\mathsf{P})$ is a witness function on a finite set P, and suppose $f : \mathsf{P} \to \mathbf{3}$ is a valuation.*

Then if f is continuous on the witness semitopology $(\mathsf{P}, \mathsf{Open}(wf))$, then $f \vDash \mathsf{Ax}(wf)$.

Proof. Suppose $f : \mathsf{P} \to \mathbf{3}$ is continuous. By Definition 18.4.5 $f^{\vDash} = f^{-1}\{\mathbf{t}, \mathbf{b}\}$, and $f^{\vDash\neg} = f^{-1}\{\mathbf{f}, \mathbf{b}\}$ are closed.

We consider $p \in \mathsf{P}$ and show that $f \vDash \mathsf{closedAx}(p)$ and $f \vDash \mathsf{closedAx}^{\neg}(p)$. By Proposition 18.4.14(3) it would suffice to show that if $f \vDash \bigwedge_{w \in wf(p)} \bigvee_{q \in w} q$ then $f \vDash p$, and if $f \vDash \bigwedge_{w \in wf(p)} \bigvee_{q \in w} \neg q$ then $f \vDash \neg p$.

- *Suppose $f \vDash \bigwedge_{w \in wf(p)} \bigvee_{q \in w} q$.*

  Using Lemmas 18.4.9 and 18.4.8 this means precisely that there exists a blocking set B for p such that $B \subseteq f^{\vDash}$. We noted above that $f^{\vDash}$ is closed, so that by Definition 8.2.5(2) it follows from $B \subseteq f^{\vDash}$ that $p \in f^{\vDash}$, and so that $f \vDash p$ as required.

- *Suppose $f \vDash \bigwedge_{w \in wf(p)} \bigvee_{q \in w} \neg q$.*

  By reasoning on $f^{\vDash\neg}$ precisely similar to the previous case, we deduce that $f \vDash \neg p$. □

PROPOSITION 19.2.3. *Suppose $wf : \mathsf{P} \to \mathcal{W}(\mathsf{P})$ is a witness function on a finite set P, and suppose $f : \mathsf{P} \to \mathbf{3}$ is a valuation. Then the following are equivalent:*

1. *$f \vDash \mathsf{Ax}(wf)$ (Definition 19.1.1).*

[1] We cannot just use Lemma 18.4.7 to deduce this, because that Lemma requires us to know that f is continuous, but this is what we are trying to prove.

2. *f is continuous on* $(\mathsf{P}, \mathsf{Open}(wf))$.

In words we can write:

> $\mathsf{Ax}(wf)$ *axiomatises continuity over the witness semitopology.*

Proof. The top-down implication is Lemma 19.2.1. The bottom-up implication is Lemma 19.2.2. □

COROLLARY 19.2.4. *Suppose $wf : \mathsf{P} \to \mathcal{W}(\mathsf{P})$ is a witnessed function on a finite set P. Then*

$$\mathsf{P}, \mathsf{Open}(wf) \vDash \phi \quad \text{if and only if} \quad \mathsf{P} \vDash \mathsf{Ax}(wf) \Rightarrow \phi.$$

Proof. By Definition 18.4.1(6), $\mathsf{P}, \mathsf{Open}(wf) \vDash \phi$ holds when $f \vDash \phi$ for every $f : \mathsf{P} \to \mathbf{3}$ that is continuous on $(\mathsf{P}, \mathsf{Open}(wf))$. By Propositions 19.2.3 and 18.4.14(3) and Lemma 18.4.9, this is precisely what $\mathsf{P} \vDash \mathsf{Ax}(wf) \Rightarrow \phi$ expresses. □

REMARK 19.2.5 (Alternative axiomatisation). An equivalent set of axioms to $\mathsf{Ax}(wf)$ from Definition 19.1.1 is as follows:

$$\begin{aligned}
\mathsf{openAx}_{wf}(p) &= \mathsf{T}p \Rightarrow \bigvee_{w \in wf(p)} \bigwedge_{q \in w} \mathsf{T}q \\
\mathsf{openAx}^{\neg}_{wf}(p) &= \mathsf{T}\neg p \Rightarrow \bigvee_{w \in wf(p)} \bigwedge_{q \in w} \mathsf{T}\neg q \\
\mathsf{OpenAx}(wf) &= \bigwedge\{\mathsf{openAx}_{wf}(p) \mid p \in \mathsf{P}\} \wedge \bigwedge\{\mathsf{openAx}^{\neg}_{wf}(p) \mid p \in \mathsf{P}\}
\end{aligned}$$

These axioms characterise that $f^{-1}\{\mathrm{t}\}$ and $f^{-1}\{\mathrm{f}\}$ are open sets as per Definition 8.2.4(2), so by Lemma 18.1.4 this characterises continuity of f, which is also what $\mathsf{Ax}(wf)$ does: thus we obtain a lemma that

$$f \vDash \mathsf{OpenAx}(wf) \quad \text{if and only if} \quad f \vDash \mathsf{Ax}(wf).$$

We leave filling in the details as an exercise to the reader.

Note a *false* argument for *false* axioms: it is not enough to claim that $p \Rightarrow \bigvee_{w \in wf(p)} \bigwedge_{q \in w} q$ is just the contrapositive of $\mathsf{closedAx}_{wf}(p)$, and $\mathsf{closedAx}^{\neg}_{wf}(p)$ is just the contrapositive of $\neg p \Rightarrow \bigvee_{w \in wf(p)} \bigwedge_{q \in w} \neg q$, because by Lemma 18.2.7 we know that $\Rightarrow$ does not satisfy the contrapositive property.

We use $\mathsf{Ax}(wf)$ because it is simpler than $\mathsf{OpenAx}(wf)$, in the sense that the latter requires modalities. This is because Definition 18.4.1(1) uses the set $\{\mathrm{t}, \mathrm{b}\}$ as designated values, and $\{\mathrm{t}, \mathrm{b}\}$ is a *closed* set in the semitopology on $\mathbf{3}$ from Definition 18.1.2(1). Thus, it is slightly more natural to use our logic to express topological properties in terms of closed sets.

19.3 Quantifying over continuous valuations, within the logic

19.3.1 Basic properties

NOTATION 19.3.1. Suppose $wf : \mathsf{P} \to \mathcal{W}(\mathsf{P})$ is a witness function on a finite set P, and suppose $\phi \in \mathsf{Pred3}(\mathsf{P})$. Define $\forall Val_{wf}\, \phi$ and $\exists Val_{wf}\, \phi$ by:

$$\begin{aligned}
\forall Val_{wf}\, \phi &= \forall Val\, (\mathsf{Ax}(wf) \Rightarrow \phi) \\
\exists Val_{wf}\, \phi &= \exists Val\, (\mathsf{Ax}(wf) \wedge \phi)
\end{aligned}$$

PROPOSITION 19.3.2. *Suppose* $wf : \mathsf{P} \to \mathcal{W}(\mathsf{P})$ *is a witness function on a finite set* P, *and suppose* $\phi \in \mathsf{ClosedPred3}(\mathsf{P})$ *(Definition 18.3.2(5)).*[2] *Suppose further that* $f' : \mathsf{P} \to \mathbf{3}$ *is any valuation. Then:*

1. $f' \vDash \forall Val_{wf}\ \phi$ *if and only if* $f \vDash \phi$ *for every continuous* $f : (\mathsf{P}, \mathsf{Open}(wf)) \to \mathbf{3}$.
2. $f' \vDash \forall Val_{wf}\ \phi$ *if and only if* $\mathsf{P}, \mathsf{Open}(wf) \vDash \phi$.
3. $f' \vDash \exists Val_{wf}\ \phi$ *if and only if* $f \vDash \phi$ *for some continuous* $f : (\mathsf{P}, \mathsf{Open}(wf)) \to \mathbf{3}$.

Proof. We consider each part in turn:

1. Following Definition 18.4.1(1), Figure 18.5, and Notation 19.3.1, it is routine to check that $f' \vDash \forall Val_{wf}\ \phi$ precisely when

$$\forall f : \mathsf{P} \to \mathbf{3}. f \vDash \mathsf{Ax}(wf) \Rightarrow \phi.$$

   By Propositions 18.4.14(3) and 19.2.3 this is equivalent to insisting that $f \vDash \phi$ for every continuous $f : (\mathsf{P}, \mathsf{Open}(wf)) \to \mathbf{3}$, as required.

2. From part 1 of this result, using Definition 18.4.1(6).

3. By reasoning exactly similar to that used in part 1. □

Lemma 19.3.3 will be helpful later; see for instance Lemma 20.1.5:

LEMMA 19.3.3. *Suppose* $wf : \mathsf{P} \to \mathcal{W}(\mathsf{P})$ *is a witness function on a finite set* P, *and suppose* $\phi \in \mathsf{ClosedPred3}(\mathsf{P})$. *Suppose* $f, f' : \mathsf{P} \to \mathbf{3}$ *are valuations. Then*

$$f \vDash \forall Val_{wf}\ \phi \iff f' \vDash \forall Val_{wf}\ \phi, \quad \text{and}$$
$$f \vDash \exists Val_{wf}\ \phi \iff f' \vDash \exists Val_{wf}\ \phi.$$

In words: the valuation in which we evaluate $\forall Val_{wf}$ (-) *and* $\exists Val_{wf}$ (-) *is not relevant.*

Proof. Routine from Proposition 19.3.2. □

19.3.2 Valuation quantification treated as a modality

Fix a witness function $wf : \mathsf{P} \to \mathcal{W}(\mathsf{P})$. We saw in Proposition 19.2.3 and Corollary 19.2.4 that continuity on the witness semitopology $(\mathsf{P}, \mathsf{Open}(wf))$ can be captured in our logic as an axiom $\mathsf{Ax}(wf)$ (Definition 8.5.2(2)).

We can view $\forall Val_{wf}$ and $\exists Val_{wf}$ as a pair of modalities, where

- $\forall Val_{wf}$ behaves like a box modality, quantifying over all possible worlds, where possible worlds are valuations $f : \mathsf{P} \to \mathbf{3}$ that are continuous over $(\mathsf{P}, \mathsf{Open}(wf))$, and
- $\exists Val_{wf}$ behaves like a diamond modality, quantifying over the existence of some possible world.

[2] 'Closed' means no free variables. ϕ might still mention propositional atoms $p \in \mathsf{P}$.

Just for this Subsection, we write

$$\forall Val_{wf} \text{ as } \Box \qquad \text{and} \qquad \exists Val_{wf} \text{ as } \Diamond$$

and we fix a valuation $f : \mathsf{P} \to \mathbf{3}$ that is continuous on $(\mathsf{P}, \mathsf{Open}(wf))$.

It is routine to check that $\Box$ and $\Diamond$ satisfy standard modal axioms, including:

1. Necessitation: If $\mathsf{P}, \mathsf{Open}(wf) \vDash \phi$ then $\mathsf{P}, \mathsf{Open}(wf) \vDash \Box\phi$.
2. Distribution / K: $f \vDash \Box(\phi \Rightarrow \phi') \Rightarrow \Box\phi \Rightarrow \Box\phi'$.
3. Reflexivity / T: $f \vDash (\Box\phi) \Rightarrow \phi$.
4. 4: $f \vDash (\Box\phi) \Rightarrow \Box\Box\phi$.
5. 5: $f \vDash \Diamond\phi \Rightarrow \Box\Diamond\phi$.
6. B: $f \vDash \phi \Rightarrow \Box\Diamond\phi$

We recognise these as axioms for the well-known modal logic S5, which is equal to Necessitation + K + T + 5.[3]

S5 is characteristic of Kripke structures in which the accessibility relation is an equivalence relation. The proofs are easy, and it is nice to find S5 in our logic.

What may be interesting to highlight here is that although our logical syntax from Figure 18.4 assumes $\forall Val$ and $\exists Val$ (without wf), what actually interests us in here is to use this syntax to construct a family of modalities $\forall Val_{wf}$ and $\exists Val_{wf}$; one for each witness function $wf : \mathsf{P} \to \mathcal{W}(\mathsf{P})$ and corresponding judgement-form $\mathsf{P}, \mathsf{Open}(wf) \vDash$ (-).

So, it is the wf-indexed modal structure $\forall Val_{wf}$ and $\exists Val_{wf}$ obtained by quantifying over all continuous valuations on $(\mathsf{P}, \mathsf{Open}(wf))$, rather than the raw syntax of $\forall Val$ and $\exists Val$, that — for now, at least — we are using.

[3]In the presence of the other axioms, axiom 5 is equivalent to axioms 4 and B.

Intertwined-ness & regularity via three-valued logic 20

We now set about characterising semitopological properties — specifically, properties having to do with *sets intersections*, like being intertwined — inside our logic. This is a first step to building a framework within which a participant can reason about network properties (or at least, about those parts that the participant can access information about).

20.1 Logical characterisation of being intertwined

REMARK 20.1.1. In this Subsection we proceed in three steps:

1. Lemma 20.1.2 gives characterisations of points being interwined using closed sets, rather than the open sets used in Definition 3.6.1(1).
2. Lemma 20.1.3 uses this characterisation to express $p \between p'$ as a logical judgement $\mathsf{P}, \mathsf{Open} \models p{\equiv}p'$.
3. Notation 20.1.4 and Proposition 20.1.6 use a $\forall Val$ binding to internalise this judgement as a predicate $\forall Val_{wf}\ (p{\equiv}p')$.

LEMMA 20.1.2. *Suppose* $(\mathsf{P}, \mathsf{Open})$ *is a semitopology and* $p, p' \in \mathsf{P}$. *Then the following are equivalent:*

1. $p \between p'$.
2. *For every* $O, O' \in \mathsf{Open}$, *if* $p \in O$ *and* $p' \in O'$ *then* $O \between O'$ *(as per Definition 3.6.1(1)).*
3. *For every* $C, C' \in \mathsf{Closed}$, *if* $p \notin C$ *and* $p' \notin C'$ *then* $C \cup C' \neq \mathsf{P}$.
4. *For every* $C, C' \in \mathsf{Closed}$, *if* $C \cup C' = \mathsf{P}$ *then* $p \in C$ *or* $p' \in C'$.
5. *For every* $C, C' \in \mathsf{Closed}$, *if* $C \cup C' = \mathsf{P}$ *then* $p, p' \in C$ *or* $p, p' \in C'$.

Proof. We check the assertions are equivalent by routine manipulations.

Equivalence of parts 1 and 2 was observed already in Definition 3.6.1(1).

By Lemma 5.1.9 part 3 just restates part 2 using closed sets instead of open sets. Part 4 is then just the contrapositive of part 3.

We now prove equivalence of parts 4 and 5 (the more interesting part is to show that part 4 implies part 5):

- *Assume that for every* $C, C' \in \mathsf{Closed}$, *if* $C \cup C' = \mathsf{P}$ *then* $p \in C \vee p' \in C'$.

  Suppose $C, C' \in \mathsf{Closed}$ and $C \cup C' = \mathsf{P}$. By assumption $p \in C \vee p' \in C'$; suppose $p \in C$ (the case of $p' \in C'$ is symmetric).

  Now $C \cup C' = \mathsf{P}$ so $p' \in C$ or $p' \in C'$. If $p' \in C$ then we have $p, p' \in C$ and we are done. Otherwise, we have $p' \notin C$ and thus $p' \in C'$ — and we have $p' \notin C$ and thus by our assumption (since $C' \cup C = \mathsf{P}$ and $p' \notin C$) also $p \in C'$.

  Thus we have $p, p' \in C'$ and we are done.

- *Assume for every* $C, C' \in \mathsf{Closed}$, *if* $C \cup C' = \mathsf{P}$ *then* $p, p' \in C \vee p, p' \in C'$.

  Suppose $C, C' \in \mathsf{Closed}$ and $C \cup C' = \mathsf{P}$. Clearly $p, p' \in C$ implies $p \in C$ and $p, p' \in C'$ implies $p' \in C'$, so $p \in C \vee p' \in C'$ is immediate. □

LEMMA 20.1.3. *Suppose* $(\mathsf{P}, \mathsf{Open})$ *is a semitopology and* $p, p' \in \mathsf{P}$. *Then the following are equivalent:*

1. $p \between p'$ *in* $(\mathsf{P}, \mathsf{Open})$.
2. $\mathsf{P}, \mathsf{Open} \vDash p \equiv p'$.
3. $\mathsf{P}, \mathsf{Open} \vDash (p \wedge p') \vee (\neg p \wedge \neg p')$.
4. $\mathsf{P}, \mathsf{Open} \vDash (p \vee p') \supset (p \wedge p')$.

Proof. Equivalence of parts 2, 3, and 4 is routine from the equivalences in Figure 18.3 and Lemma 18.2.2 (or just by checking truth-tables from Figure 18.2).

Equivalence of parts 1 and 3 is just by unfolding definitions and checking truth-tables, but we spell out the details. We consider two implications:

- *Suppose* $p \between p'$ *in* $(\mathsf{P}, \mathsf{Open})$. We wish to show that $\mathsf{P}, \mathsf{Open} \vDash (p \wedge p') \vee (\neg p \wedge \neg p')$.

  Consider a valuation $f : \mathsf{P} \to \mathbf{3}$ that is continuous on $(\mathsf{P}, \mathsf{Open})$, and consider $f^{\vDash} = f^{-1}\{\mathbf{t}, \mathbf{b}\}$ and $f^{\vDash\neg} = f^{-1}\{\mathbf{f}, \mathbf{b}\}$ from Definition 18.4.5. These are closed sets in $(\mathsf{P}, \mathsf{Open})$ by Lemma 18.4.7.

  We note that $f^{\vDash} \cup f^{\vDash\neg} = \mathsf{P}$, and so by Lemma 20.1.2(1&5) (since $p \between p'$) $p, p' \in f^{\vDash}$ or $p, p' \in f^{\vDash\neg}$. It follows using Lemma 18.4.9 that $f \vDash (p \wedge p') \vee (\neg p' \wedge \neg p)$.

  Now f was an arbitrary continuous valuation on $(\mathsf{P}, \mathsf{Open})$, so the result follows by Definition 18.4.1(6).

- Suppose $\mathsf{P}, \mathsf{Open} \vDash (p \wedge p') \vee (\neg p \wedge \neg p')$.

  Consider $C, C' \in \mathsf{Closed}$ such that $C \cup C' = \mathsf{P}$; by Lemma 20.1.2(1&5) it would suffice to show that $p, p' \in C$ or $p, p' \in C'$.

We define $f = \delta_{C_{\mathbf{tb}},C_{\mathbf{fb}}}(C,C') : \mathsf{P} \to \mathbf{3}$ from Definition 18.1.5, so $f(\mathsf{P}\setminus C) = \mathbf{t}$ and $f(\mathsf{P}\setminus C') = \mathbf{f}$ and $f(C \cap C') = \mathbf{b}$, and $C = f^{\models}$ and $C' = f^{\models\neg}$. This is well-defined since $C \cup C' = \mathsf{P}$, and continuous by Lemma 18.1.6. Thus, $f \models (p \wedge p') \vee (\neg p \wedge \neg p')$.

By Lemma 18.4.9, we see that at least one of $f \models p \wedge f \models p'$ or $f \models \neg p \wedge f \models \neg p'$ must hold. By Lemma 18.4.8, either $p, p' \in C$ or $p, p' \in C'$, as required. ☐

Notation 20.1.4. Suppose $wf : \mathsf{P} \to \mathcal{W}(\mathsf{P})$ is a witness function on a finite set P, and suppose $p, p' \in \mathsf{P}$. Then define a predicate $p \between_{wf} p'$ in $\mathsf{ClosedPred3}(\mathsf{P})$ by

$$p \between_{wf} p' \;=\; \forall Val_{wf}\,(p{\equiv}p').$$

We make an elementary observation:

Lemma 20.1.5. *Suppose $wf : \mathsf{P} \to \mathcal{W}(\mathsf{P})$ is a witness function on a finite set P and $p, p' \in \mathsf{P}$, and suppose $f, f' : \mathsf{P} \to \mathbf{3}$ are any valuations. Then*

$$f \models p \between_{wf} p' \quad \textit{if and only if} \quad f' \models p \between_{wf} p'.$$

In words: $f \models p \between_{wf} p'$ — the validity of $p \between_{wf} p'$ — depends on p, p', and wf. It does not depend on f.

Proof. Direct from Lemma 19.3.3. ☐

Proposition 20.1.6. *Suppose $wf : \mathsf{P} \to \mathcal{W}(\mathsf{P})$ is a witness function on a finite set P, and suppose $f' : \mathsf{P} \to \mathbf{3}$ is any valuation. Then the following are equivalent:*

1. *$p \between p'$ in $(\mathsf{P}, \mathsf{Open}(wf))$ (Definition 3.6.1(1)).*
2. *$\mathsf{P}, \mathsf{Open}(wf) \models p{\equiv}p'$.*
3. *$f' \models p \between_{wf} p'$ (f' does not matter, as per Lemma 20.1.5).*

Proof. We prove two equivalences:

- Equivalence of parts 1 and 2 is just Lemma 20.1.3(1&2), for the witness semitopology $(\mathsf{P}, \mathsf{Open}(wf))$.
- To prove equivalence of parts 1 and 3, we unpack Notation 20.1.4 and use Proposition 19.3.2(1) to see that $f' \models p \between_{wf} p'$ when $f \models p{\equiv}p'$ for every $f : \mathsf{P} \to \mathbf{3}$ that is continuous on $(\mathsf{P}, \mathsf{Open}(wf))$. We use Lemma 20.1.3. ☐

Remark 20.1.7. By Lemmas 19.3.3 and 20.1.5, the valuation f' is not used to evaluate $f' \models p \between_{wf} p'$ in Proposition 20.1.6. The valuation f' is there because the validity judgement '$f' \models p \between_{wf} p'$' from Definition 18.4.1(1) requires us to provide a valuation.[1]

The reader might ask why in the statement of Proposition 20.2.3(3) we do not replace

- the judgement $f' \models p \between_{wf} p'$ (Definition 18.4.1(1)) with
- the judgement $\mathsf{P} \models p \between_{wf} p'$ (Definition 18.4.1(5)).

Because this is a different result. It is the difference between proving $\forall f'.(A \Leftrightarrow B)$ — which permits us to replace A with B, given some f' — and $A \Leftrightarrow \forall f'.B$, which does not.

[1]This happens in first-order logic too, for example with $f \models x{=}y \Leftrightarrow y{=}x$.

Corollary 20.1.8 will be useful later. It shows how to characterise more sophisticated inclusions and equalities between $f^{\vDash}$ and $p_{\between}$ ($f^{\vDash}$ is from Definition 18.4.5; $p \between_{wf} x$ is from Definition 3.6.1(2)):

COROLLARY 20.1.8. *Suppose $wf : \mathsf{P} \to \mathcal{W}(\mathsf{P})$ is a witness function on a finite set P, and suppose $p \in \mathsf{P}$ and $f : \mathsf{P} \to \mathbf{3}$ is a valuation. Then:*

1. *$f \vDash \forall x.x \Leftrightarrow (p \between_{wf} x)$ if and only if $f^{\vDash} = p_{\between}$.*
2. *$f \vDash \forall x.x \Rightarrow (p \between_{wf} x)$ if and only if $f^{\vDash} \subseteq p_{\between}$.*
3. *$f \vDash \forall x.(p \between_{wf} x) \Rightarrow x$ if and only if $p_{\between} \subseteq f^{\vDash}$.*

Proof. We consider each part in turn:

1. Using Corollary 18.5.3, $f \vDash \forall x.x \Leftrightarrow (p \between_{wf} x)$ holds when
$$\{p' \in \mathsf{P} \mid f \vDash p'\} = \{p' \in \mathsf{P} \mid f \vDash p \between_{wf} p'\}.$$
By Definition 18.4.5, the left-hand side of this equality is just $f^{\vDash}$.
By Proposition 20.1.6 $f \vDash p \between_{wf} p'$ if and only if $p \between p'$, so the right-hand side of this equality is $\{p' \in \mathsf{P} \mid p \between p'\} = p_{\between}$ (Definition 3.6.1(2)).
The result follows.

2. Using Proposition 18.4.14(3) and Definition 18.4.5, $f \vDash \forall x.x \Rightarrow (p \between_{wf} x)$ holds when
$$f^{\vDash} \subseteq \{p' \in \mathsf{P} \mid f \vDash p \between_{wf} p'\}.$$
We use Proposition 20.1.6 just as we did in part 1 of this result.

3. Precisely as for part 2 of this result. □

20.2 Logical characterisation of topological indistinguishability

By Proposition 20.1.6, $\forall Val_{wf}\ (p \equiv p')$ characterises being intertwined in $(\mathsf{P}, \mathsf{Open}(wf))$. But recall that Figure 18.2 features two connectives for logical equivalence: $\equiv$ and $\Leftrightarrow$. In this Subsection we briefly consider the natural question: what does $\forall Val_{wf}\ (p \Leftrightarrow p')$ characterise? The answer is very simple: this is topological indistinguishability.

Recall from Definition 13.4.6 the notion of *topological indistinguishability* $p \doteq p'$. This has a beautiful logical characterisation:

LEMMA 20.2.1. *Suppose $(\mathsf{P}, \mathsf{Open})$ is a semitopology and $p, p' \in \mathsf{P}$. Then the following are equivalent:*

1. *$p \doteq p'$.*
2. *$\mathsf{P}, \mathsf{Open} \vDash p \Leftrightarrow p'$.*

Proof. This result looks like it should have a trivial proof. Not so: the proof is simple, but not trivial. We prove two implications:

- Suppose $p \doteq p'$ and consider a continuous $f : \mathsf{P} \to \mathbf{3}$. By assumption $p \in f^{-1}\{\mathbf{f}\} \iff p' \in f^{-1}\{\mathbf{f}\}$, so by Corollary 18.5.2(1&3) $f \vDash p \Leftrightarrow p'$ as required.

- Suppose $\mathsf{P}, \mathsf{Open} \vDash p \Leftrightarrow p'$ and consider some $O \in \mathsf{Open}$. We set $f = \delta_{O_\mathbf{t},O_\mathbf{f}}(\varnothing, O)$ — using $\delta_{O_\mathbf{t},O_\mathbf{f}}(O, \varnothing)$ will not give us what we need — so f maps a point to $\mathbf{f}$ if it is in O, and to $\mathbf{b}$ otherwise. By Corollary 18.5.2(1&3), $p \in O \iff p' \in O$ as required. □

Definition 20.2.2. Suppose $wf : \mathsf{P} \to \mathcal{W}(\mathsf{P})$ is a witness function on a finite set P, and suppose $p, p' \in \mathsf{P}$. Then define a predicate $\mathsf{TopIndis}_{wf}(p, p')$ in $\mathsf{ClosedPred3}(\mathsf{P})$ by:

$$\mathsf{TopIndis}_{wf}(p, p') \;=\; \forall Val_{wf}\,(p \Leftrightarrow p').$$

Proposition 20.2.3. *Suppose $wf : \mathsf{P} \to \mathcal{W}(\mathsf{P})$ is a witness function on a finite set P and $p, p' \in \mathsf{P}$, and suppose $f' : \mathsf{P} \to \mathbf{3}$ is any valuation. Then the following are equivalent:*

1. *$p \mathrel{\mathring{=}} p'$ in $(\mathsf{P}, \mathsf{Open}(wf))$.*
2. *$\mathsf{P}, \mathsf{Open}(wf) \vDash p \Leftrightarrow p'$.*
3. *$f' \vDash \mathsf{TopIndis}_{wf}(p, p')$*
   (By Lemma 19.3.3, f' is not used to evaluate $f' \vDash \mathsf{TopIndis}_{wf}(p, p')$. See Remark 20.1.7.)

Proof. Equivalence of parts 1 and 2 is just Lemma 20.2.1. Equivalence of parts 2 and 3 is direct from Proposition 19.3.2(2). □

20.3 Logical characterisation of $interior(f^{\vDash}) \neq \varnothing$

To characterise (weak/quasi-)regularity, which we do below, we will need to logically characterise when f takes value $\mathbf{t}$ at some point p, and when $f^{\vDash}$ has a nonempty open interior and so is a closed neighbourhood. These properties are closely related, and we study them in this Subsection:

Lemma 20.3.1. *Suppose $(\mathsf{P}, \mathsf{Open})$ is a semitopology and $f : \mathsf{P} \to \mathbf{3}$ is continuous and $p \in \mathsf{P}$. Then:*

1. *$f \vDash \mathsf{T}p$ if and only if $p \in f^{-1}\{\mathbf{t}\}$.*
2. *If $f \vDash \mathsf{T}p$ then $p \in interior(f^{\vDash})$.*
3. *If $f \vDash \mathsf{T}p$ then $f^{\vDash}$ is a closed neighbourhood of $p \in \mathsf{P}$ (a closed set with p in its open interior; Definition 5.4.1(2)).*
4. *The converse implication to part 3 need not hold: it is possible that $p \in interior(f^{\vDash})$ yet $f \nvDash \mathsf{T}p$ — later on in Corollary 22.2.7 we will identify those valuations for which the reverse implication does hold, as an interesting class of* extremal *valuations.*

Proof. We consider each part in turn:

1. Following Definition 18.4.1(1) and the rules in Figures 18.5 and 18.2, $f \vDash \mathsf{T}p$ when $f(p) = \mathbf{t}$. It is a fact that $f(p) = \mathbf{t}$ when $p \in f^{-1}\{\mathbf{t}\}$.

2. Suppose $f \vDash \mathsf{T}p$. Then $p \in f^{-1}\{\mathbf{t}\}$ by part 1 of this result and $f^{-1}\{\mathbf{t}\} \in \mathsf{Open}$ by Corollary 18.1.11(1). By construction $f^{-1}\{\mathbf{t}\} \subseteq f^{\vDash} = f^{-1}\{\mathbf{t}, \mathbf{b}\}$, so $f^{-1}\{\mathbf{t}\} \subseteq interior(f^{\vDash})$. Thus we have
$$p \in f^{-1}\{\mathbf{t}\} \subseteq interior(f^{\vDash})$$
as required.

3. By Lemma 18.4.7 $f^{\vDash}$ is a closed set. The result now just follows from part 2 of this result.

4. Consider the **singleton semitopology** $\mathsf{One} = \{*\}$ with $\mathsf{Open} = \{\varnothing, \mathsf{One}\}$. Note that $\{*\}$ is a closed neighbourhood, with open interior $\{*\}$. Define $f : \mathsf{One} \to \mathbf{3}$ by $f(*) = \mathbf{b}$. Then f is continuous and $f^{\vDash} = \{*\}$, but $f \nvDash \mathsf{T}*$. □

PROPOSITION 20.3.2. *Suppose* $(\mathsf{P}, \mathsf{Open})$ *is a semitopology and* $f : \mathsf{P} \to \mathbf{3}$ *is a valuation that is closed on* $(\mathsf{P}, \mathsf{Open})$. *Then:*

1. $f \vDash \exists x.\mathsf{T}x$ *if and only if* $f^{\text{-}1}\{\mathbf{t}\} \neq \varnothing$.
2. *If* $f \vDash \exists x.\mathsf{T}x$ *then* $f^{\vDash}$ *is a closed neighbourhood.*
3. *It is possible for* $f^{\vDash}$ *to be a closed neighbourhood, yet* $f \nvDash \exists x.\mathsf{T}x$.

Proof. We consider each part in turn:

1. Suppose $f \vDash \exists x.\mathsf{T}x$. By Lemma 18.4.10 this is if and only if there exists some $p \in \mathsf{P}$ such that $f \vDash \mathsf{T}p$. By Lemma 20.3.1(1) $p \in f^{\text{-}1}\{\mathbf{t}\}$, so that $f^{\text{-}1}\{\mathbf{t}\} \neq \varnothing$.

   The reverse implication follows just reversing this reasoning.

2. Suppose $f \vDash \exists x.\mathsf{T}x$. By Lemma 18.4.7 $f^{\vDash} = f^{\text{-}1}\{\mathbf{t}, \mathbf{b}\}$ is a closed set, and by part 1 of this result it has a nonempty open interior.

3. See the counterexample in Lemma 20.3.1(4). □

COROLLARY 20.3.3. *Suppose* $(\mathsf{P}, \mathsf{Open})$ *is a semitopology and* $C \subseteq \mathsf{Closed}$ *is a closed set. Define*

$$f = \delta_{O_{\mathbf{t}}, C_{\mathbf{tb}}}(\mathit{interior}(C), C)$$

and note from the construction of $\delta_{O_{\mathbf{t}}, C_{\mathbf{tb}}}(\mathit{interior}(C), C)$ *in Definition 18.1.5) that* $f^{\vDash} = C$. *Then the following are equivalent:*

1. *C is a closed neighbourhood (closed set with nonempty open interior; Definition 5.4.1(3)).*
2. $\delta_{O_{\mathbf{t}}, C_{\mathbf{tb}}}(O, C) \vDash \exists x.\mathsf{T}x$.

Proof. By an easy argument from the results so far:

- Suppose C is a closed neighbourhood, meaning by Definition 5.4.1(3) that $\mathit{interior}(C) \neq \varnothing$. Choose $p \in \mathit{interior}(C)$ and note that $f(p) = \mathbf{t}$. By Proposition 20.3.2(1), $f \vDash \exists x.\mathsf{T}x$ as required.

- Suppose $f \vDash \exists x.\mathsf{T}x$. By Proposition 20.3.2(2), $f^{\vDash}$ is a closed neighbourhood. □

20.4 Logical characterisation of being quasiregular

By Definition 4.1.4(5), a point is quasiregular when its community is nonempty. We express this as follows:

DEFINITION 20.4.1. Suppose $wf : \mathsf{P} \to \mathcal{W}(\mathsf{P})$ is a witness function on a finite set P, and suppose $p \in \mathsf{P}$. We define a predicate $\mathsf{QuasiRegular}_{wf}(p)$ by:

$$\mathsf{QuasiRegular}_{wf}(p) \;=\; \exists Val_{wf}\,(\exists x.\mathsf{T}x \wedge \forall x.x {\Leftrightarrow} p \between_{wf} x).$$

PROPOSITION 20.4.2. *Suppose* $wf : \mathsf{P} \to \mathcal{W}(\mathsf{P})$ *is a witness function on a finite set* P *and* $p \in \mathsf{P}$, *and suppose* $f' : \mathsf{P} \to \mathbf{3}$ *is any valuation. Then the following are equivalent:*

1. p *is quasiregular in* $(\mathsf{P}, \mathsf{Open}(wf))$.
2. *There exists a valuation* $f : \mathsf{P} \to \mathbf{3}$ *such that*

$$f \vDash \mathsf{Ax}(wf) \quad \textit{and} \quad f \vDash \exists x.\mathrm{T}x \quad \textit{and} \quad f \vDash \forall x.x \Leftrightarrow p \between_{wf} x.$$

3. $f' \vDash \mathsf{Quasiregular}_{wf}(p)$.
   (By Lemma 19.3.3, f' *is not used to evaluate* $f' \vDash \mathsf{QuasiRegular}_{wf}(p)$. *See Remark 20.1.7.)*

Proof. Equivalence of parts 2 and 3 is just by unpacking Definition 20.4.1 and using Proposition 19.3.2(3) and Lemma 18.4.9.

Note from Proposition 5.4.9 that p is quasiregular when $p_{\between}$ is a closed neighbourhood (Definition 5.4.1(3)). We prove equivalence of parts 1 and 2 by proving two implications:

- *Suppose* p *is quasiregular, so that* $p_{\between}$ *is a closed neighbourhood and* $K(p) \neq \varnothing$.

  We set $f = \delta_{O_{\mathbf{t}}, C_{\mathbf{tb}}}(K(p), p_{\between}) : \mathsf{P} \to \mathbf{3}$; so f returns $\mathbf{t}$ on $K(p) = interior(p_{\between})$ — which is nonempty since we assumed that $p_{\between}$ is a closed neighbourhood — and $\mathbf{b}$ on $boundary(p_{\between}) = p_{\between} \setminus K(p)$, and $\mathbf{f}$ on $\mathsf{P} \setminus p_{\between}$. Then:

  - By Lemma 18.1.6 f is continuous on $(\mathsf{P}, \mathsf{Open}(wf))$, so by Proposition 19.2.3 $f \vDash \mathsf{Ax}(wf)$.
  - By Proposition 20.3.2(1) (right-to-left direction, since $f^{-1}\{\mathbf{t}\} = K(p) \neq \varnothing$) $f \vDash \exists x.\mathrm{T}x$.
  - By Corollary 20.1.8(1) $f \vDash \forall x.x \Leftrightarrow p \between_{wf} x$ holds when $f^{\vDash} = p_{\between}$. But we defined $f = \delta_{O_{\mathbf{t}}, C_{\mathbf{tb}}}(K(p), p_{\between})$, so this is a fact.

- *Suppose* $f \vDash \mathsf{Ax}(wf)$ *and* $f \vDash \exists x.\mathrm{T}x$ *and* $f \vDash \forall x.x \Leftrightarrow p \between_{wf} x$.

  By Proposition 19.2.3 f is continuous from $(\mathsf{P}, \mathsf{Open}(wf))$ to $\mathbf{3}$. Write $C = f^{\vDash} = f^{-1}\{\mathbf{t}, \mathbf{b}\}$. By Proposition 20.3.2(1) (left-to-right direction) C is a closed neighbourhood. By Corollary 20.1.8 $C = p_{\between}$. Thus p is quasiregular. □

20.5 Logical characterisation of being weakly regular

By Definition 4.1.4(4), a point is weakly regular when it is an element of its own community. We express this as follows:

DEFINITION 20.5.1. Suppose $wf : \mathsf{P} \to \mathcal{W}(\mathsf{P})$ is a witness function on a finite set P, and suppose $p \in \mathsf{P}$. We define a predicate $\mathsf{WeaklyRegular}_{wf}(p)$ by:

$$\mathsf{WeaklyRegular}_{wf}(p) \;=\; \exists Val_{wf}\,(\mathrm{T}p \wedge \forall x.x \Leftrightarrow (p \between_{wf} x)).$$

PROPOSITION 20.5.2. *Suppose* $wf : \mathsf{P} \to \mathcal{W}(\mathsf{P})$ *is a witness function on a finite set* P *and* $p \in \mathsf{P}$, *and suppose* $f' : \mathsf{P} \to \mathbf{3}$ *is any valuation. Then the following are equivalent:*

1. p *is weakly regular in* $(\mathsf{P}, \mathsf{Open}(wf))$.

2. $f' \vDash \mathsf{WeaklyRegular}_{wf}(p)$.
   (By Lemma 19.3.3, f' is not used to evaluate $f' \vDash \mathsf{WeaklyRegular}_{wf}(p)$. See Remark 20.1.7.)

Proof. We unpack what $f' \vDash \mathsf{WeaklyRegular}_{wf}(p)$ means. By Proposition 19.3.2(3) this means that there exists a valuation $f : \mathbf{P} \to wf$ that is continuous on $(\mathbf{P}, \mathsf{Open}(wf))$ such that $f^{\vDash} = f^{-1}\{\mathbf{t}, \mathbf{b}\}$ is a closed neighbourhood of p (by Lemma 20.3.1(3)), and $f^{\vDash}$ is equal to $p_{\between}$ (by Corollary 20.1.8). But by Proposition 5.4.10(2) this precisely means that p is weakly regular; the equivalence follows. □

20.6 Logical characterisation of being unconflicted

By Definition 6.1.1(2), a point is unconflicted when the 'intertwined with' relation is transitive at that point. We express this as follows:

DEFINITION 20.6.1. Suppose $wf : \mathbf{P} \to \mathcal{W}(\mathbf{P})$ is a witness function on a finite set $\mathbf{P}$, and suppose $p \in \mathbf{P}$. We define a predicate $\mathsf{Unconflicted}_{wf}(p)$ by

$$\mathsf{Unconflicted}_{wf}(p) \;=\; \forall x', x''.(x \between_{wf} p \wedge p \between_{wf} x'') \Longrightarrow x' \between_{wf} x''.$$

PROPOSITION 20.6.2. *Suppose $wf : \mathbf{P} \to \mathcal{W}(\mathbf{P})$ is a witness function on a finite set $\mathbf{P}$ and $p \in \mathbf{P}$, and suppose $f' : \mathbf{P} \to \mathbf{3}$ is any valuation. Then the following are equivalent:*

1. *p is unconflicted in the witness semitopology $(\mathbf{P}, \mathsf{Open}(wf))$.*
2. $f' \vDash \mathsf{Unconflicted}_{wf}(p)$.
   (By Lemma 19.3.3, f' is not used to evaluate $f' \vDash \mathsf{Unconflicted}_{wf}(p)$. See Remark 20.1.7.)

Proof. We unpack what $f' \vDash \mathsf{Unconflicted}_{wf}(p)$ means. Using Lemma 18.4.10, Proposition 18.4.14(3), and Lemma 18.4.9, $f' \vDash \mathsf{Unconflicted}_{wf}(p)$ if and only if

$$\forall p', p'' \in \mathbf{P}.(f \vDash p' \between_{wf} p \wedge f \vDash p \between_{wf} p'') \Longrightarrow f \vDash p' \between_{wf} p''.$$

We simplify with Proposition 20.1.6 to obtain

$$\forall p', p'' \in \mathbf{P}.(p' \between p \wedge p \between p'') \Longrightarrow p' \between p''.$$

But as per Definition 6.1.1(2), this is precisely what being unconflicted means. □

20.7 Logical characterisation of being regular

REMARK 20.7.1. We will consider two characterisations of being regular in Definition 20.7.2:

1. By Theorem 6.2.2 a point is regular when it is weakly regular and unconflicted. This is expressed by $\mathsf{Regular}_{wf}$ in Definition 20.7.2.
2. By Theorem 5.6.2 a point p is regular when it is weakly regular and furthermore $p_{\between}$ is a minimal closed neighbourhood. This is expressed by $\mathsf{Regular}'_{wf}$ in Definition 20.7.2.

We translate both into our logic.

DEFINITION 20.7.2. Suppose $wf : \mathsf{P} \to \mathcal{W}(\mathsf{P})$ is a witness function on a finite set P, and suppose $p \in \mathsf{P}$. We define predicates $\mathsf{Regular}_{wf}(p)$ and $\mathsf{Regular}'_{wf}(p)$ by:

$$\begin{aligned}\mathsf{Regular}_{wf}(p) &= \mathsf{WeaklyRegular}_{wf}(p) \wedge \mathsf{Unconflicted}_{wf}(p)\\ \mathsf{Regular}'_{wf}(p) &= \mathsf{WeaklyRegular}_{wf}(p) \wedge\\ &\quad \forall Val_{wf}\ (\exists x.\mathrm{T}x) \Rightarrow (\forall x'.x' \Rightarrow (p \between_{wf} x')) \Rightarrow \forall x'.((p \between_{wf} x') \Rightarrow x')\end{aligned}$$

PROPOSITION 20.7.3. *Suppose* $wf : \mathsf{P} \to \mathcal{W}(\mathsf{P})$ *is a witness function on a finite set* P *and* $p \in \mathsf{P}$, *and suppose* $f' : \mathsf{P} \to \mathbf{3}$ *is any valuation. Then the following are equivalent:*

1. *p is regular in* $(\mathsf{P}, \mathsf{Open}(wf))$.
2. $f' \vDash \mathsf{Regular}_{wf}(p)$.
   (By Lemma 19.3.3, f' *is not used to evaluate* $f' \vDash \mathsf{Regular}_{wf}(p)$. *See Remark 20.1.7.)*

Proof. The argument is routine from the machinery we have build:

1. By Theorem 6.2.2, p is regular when it is weakly regular and unconflicted.
2. By Lemma 18.4.9 and Propositions 20.5.2 and 20.6.2, $f' \vDash \mathsf{Regular}_{wf}(p)$ holds when p is weakly regular and unconflicted.

The result follows. □

PROPOSITION 20.7.4. *Suppose* $wf : \mathsf{P} \to \mathcal{W}(\mathsf{P})$ *is a witness function on a finite set* P *and* $p \in \mathsf{P}$, *and suppose* $f' : \mathsf{P} \to \mathbf{3}$ *is any valuation. Then the following are equivalent:*

1. *p is regular in* $(\mathsf{P}, \mathsf{Open}(wf))$.
2. *$p \in K(p)$ and $p_{\between}$ is a minimal closed neighbourhood.*
3. $f' \vDash \mathsf{Regular}'_{wf}(p)$.
   (By Lemma 19.3.3, f' *is not used to evaluate* $f' \vDash \mathsf{Regular}'_{wf}(p)$. *See Remark 20.1.7.)*

Proof. Equivalence of parts 1 and 2 just repeats Theorem 5.6.2. We now consider equivalence of parts 2 and 3. We prove two implications:

- *Suppose* $f' \vDash \mathsf{Regular}'_{wf}(p)$.

  Unpacking Definition 20.7.2 and using Lemma 18.4.9, we have:

  $$\begin{aligned}&f' \vDash \mathsf{WeaklyRegular}_{wf}(p) \quad \text{and}\\ &f' \vDash \forall Val_{wf}\ (\exists x.\mathrm{T}x) \Rightarrow (\forall x'.x' \Rightarrow (p \between_{wf} x')) \Rightarrow \forall x'.((p \between_{wf} x') \Rightarrow x')\end{aligned}$$

  By Proposition 20.5.2 p is weakly regular, which by Proposition 5.4.10(3) means that $p_{\between}$ is a minimal closed neighbourhood of p.

  We will now show that $p_{\between}$ is a minimal closed neighbourhood (not just of a minimal closed neighbourhood of p). Suppose $C \subseteq p_{\between}$ is some closed neighbourhood, and write $O = interior(C) \neq \varnothing$. Let $f = \delta_{O_{\mathbf{t}}, C_{\mathbf{tb}}}(O, C)$ from Definition 18.1.5, so $f(p) = \mathbf{t}$ when $p \in O$, and $f(p) = \mathbf{b}$ when $p \in C \setminus O$, and $f(p) = \mathbf{f}$ when $p \in \mathsf{P} \setminus C$. By Lemma 18.1.6 f is continuous on $(\mathsf{P}, \mathsf{Open}(wf))$. It follows from Proposition 19.3.2(1) that

  $$f \vDash (\exists x.\mathrm{T}x) \Rightarrow (\forall x'.x' \Rightarrow (p \between_{wf} x')) \Rightarrow \forall x'.((p \between_{wf} x') \Rightarrow x').$$

Now C is a closed neighbourhood so by Corollary 20.3.3 (since $f = \delta_{O_{\mathtt{t}},C_{\mathtt{tb}}}(O,C)$) $f \vDash \exists x.\mathrm{T}x$. By assumption $C \subseteq p_{\between}$ so by Corollary 20.1.8(2) $f \vDash \forall x'.x' \Rightarrow (p \between_{wf} x')$. Then by Proposition 18.4.14(3) it follows that $f \vDash \forall x'.(p \between_{wf} x') \Rightarrow x'$, and by Corollary 20.1.8(3) $p_{\between} \subseteq C$.

Thus $C \subseteq p_{\between}$ implies $p_{\between} \subseteq C$, and so $p_{\between}$ is a minimal closed neighbourhood as required.

- *Suppose $p \in K(p)$ and $p_{\between}$ is a minimal closed neighbourhood.*

  The reasoning is essentially by reversing the argument of the previous case, but we spell out the details.

  By Definition 4.1.4(4) $p \in K(p)$ means precisely that p is weakly regular, and by Proposition 20.5.2 it follows that $f' \vDash \mathsf{WeaklyRegular}_{wf}(p)$.

  By Propositions 19.3.2(1) and 18.4.14(3) and Corollary 20.1.8(2), to prove

$$\forall Val_{wf}\ (\exists x.\mathrm{T}x) \Rightarrow (\forall x'.x' \Rightarrow (p \between_{wf} x')) \Rightarrow \forall x'.((p \between_{wf} x') \Rightarrow x')$$

  it suffices to show for every $f : \mathbf{P} \to \mathbf{3}$ that is continuous on $(\mathbf{P}, \mathsf{Open}(wf))$, if $f \vDash \exists x.\mathrm{T}x$ and $f^{\vDash} \subseteq p_{\between}$, then $p_{\between} \subseteq f^{\vDash}$.

  So suppose $f \vDash \exists x.\mathrm{T}x$, so that $f^{\vDash}$ is a closed neighbourhood by Proposition 20.3.2(2), and suppose $f^{\vDash} \subseteq p_{\between}$. By minimality of $p_{\between}$ it follows that $p_{\between} \subseteq f^{\vDash}$.

  We use Lemma 18.4.9 to deduce

$$\begin{aligned} f' \vDash\ & \mathsf{WeaklyRegular}_{wf}(p)\ \wedge \\ & \forall Val_{wf}\ (\exists x.\mathrm{T}x) \Rightarrow (\forall x'.x' \Rightarrow (p \between_{wf} x')) \Rightarrow \forall x'.((p \between_{wf} x') \Rightarrow x') \end{aligned}$$

  as required. □

Computational complexity & logic programming 21

We now consider some computational aspects of semitopologies, specifically the following two questions:

1. What is the computational complexity of deciding whether two points are intertwined?
2. What is a suitable notion of logic programming over **3**?

Many excellent treatments of logic programming over two-valued logic are available [MNPS91, Lif08, Lif19]. There is some literature on the complexity of checking satisfiability of many-valued propositional logics; for example [HÖ3, Han11] consider propositional many-valued logics, and [Vid21] considers the modal case. I am not aware of any comprehensive surveys.

21.1 Translation of SAT to intertwinedness problem

Definition 21.1.1. Suppose ψ is a Boolean logic proposition over some set of **propositional atoms Q**. So ψ is built from propositional atoms $q \in \mathbf{Q}$, with connectives $\bot$, $\top$, $\neg$, $\wedge$, and $\vee$. Suppose without loss of generality that ψ is in **conjunctive normal form**, so

$$\psi = \bigwedge_{i\in I}\bigvee_{j\in i} a_{ij}q_{ij}$$

where

1. I is an indexing set and each $i \in I$ is itself an indexing set of $j \in i$,
2. each $q_{ij} \in \mathbf{Q}$ is a propositional atom, and
3. a_{ij} indicates an *arity*, which is either **positive arity** (meaning that q_{ij} is unnegated) or **negative arity** (meaning that q_{ij} is negated).

NOTATION 21.1.2. We may call the combination $a_{ij}q_{ij}$ of an arity and a propositional atom a **literal**. We may call the literal **positive** when its arity is positive (so $a_{ij}q_{ij}$ is just a propositional atom $q \in \mathbf{Q}$), and **negative** when its arity is negative (so $a_{ij}q_{ij}$ is a negated proposition atom $\neg q$).

REMARK 21.1.3. Definition 21.1.4 below is a little fiddly to write out. The reader might like to look at it together with Proposition 21.1.5, since that Proposition motivates the design of this Definition. Intuitively in Definition 21.1.4:

1. Points on the right are intended to be assigned 'true', and points on the left are intended to be assigned 'false'.
2. A conjunctive restriction is represented by a point with one witness-set, which may have several elements.
3. A disjunctive restriction is represented by a point with several witness-sets, each of which contains just one element.

Details of how this works are in Proposition 21.1.5.

DEFINITION 21.1.4. We fix some data to help us map a proposition ψ from Definition 21.1.1 to a witness function in the sense of Definition 8.2.2(1), as follows:

1. Fix symbols *left* and *right*.
2. For each $i \in I$, fix a symbol $right_i$.
3. For each $q \in \mathbf{Q}$, fix symbols +q and -q, and fix symbols $left_q$ and $right_q$.
4. Define a set of **points**

$$\mathsf{P} = \{left, right\} \cup \{right_i \mid i \in I\} \cup \bigcup_{q \in \mathbf{Q}} \{\texttt{+q}, \texttt{-q}, left_q, right_q\}.$$

5. Define a function m from literals (unnegated or negated propositional atoms, as per Notation 21.1.2) to points P, such that

$$m(q) = \texttt{+q} \quad \text{and} \quad m(\neg q) = \texttt{-q}.$$

6. Define a witness function wf on P as follows:
   a) For each $q \in \mathbf{Q}$, let +q and -q have witness-sets

   $$wf(\texttt{+q}) = \{\,\{\texttt{+q}\}\,\} \quad \text{and} \quad wf(\texttt{-q}) = \{\,\{\texttt{-q}\}\,\}.$$

   So each of +q and -q has just one witness-set, which is singleton itself.
   b) For each $q \in \mathbf{Q}$, let $left_q$ and $right_q$ have witness-sets

   $$wf(left_q) = \{\{\texttt{+q}\}, \{\texttt{-q}\}\} = wf(right_q).$$

   So each $left_q$ has two singleton witness-sets, and so does $right_q$.
   c) Let *left* have just one witness-set, given by

   $$wf(left) = \{\ \{left_q \mid q \in \mathbf{Q}\}\ \}.$$

   d) Let $right \in \mathsf{P}$ have just one witness-set, given by

   $$wf(right) = \{\ \{right_q \mid q \in \mathbf{Q}\} \cup \{right_i \mid i \in I\}\ \}.$$

e) For each $i \in I$, let $right_i$ have witness-sets given by

$$wf(right_i) = \{\ \{m(a_{ij}q_{ij})\} \mid j \in i\}.$$

So for each $j \in i$, the singleton set $\{m(a_{ij}q_{ij})\}$ is a witness-set of $right_i$.

PROPOSITION 21.1.5. *Suppose that:*

- *ψ is a Boolean logic proposition over propositional atoms* Q, *that is in conjunctive normal form in the sense of Definition 21.1.1.*
- *$wf : \mathsf{P} \to \mathcal{W}(\mathsf{P})$ is the witness function derived from ψ as per Definition 21.1.4.*

Then the following are equivalent:

1. *ψ is 2-satisfiable in (the usual) two-valued Boolean logic, as per Definition 21.2.2(1).*
2. *$left \not\between right$.*

Proof. We prove two implications.

- Suppose $left \not\between right$, so there exist a disjoint pair of open neighbourhoods

$$left \in L \in \mathsf{Open} \quad \text{and} \quad right \in R \in \mathsf{Open}.$$

Now by the construction of the witness semitopology in Definition 8.2.5(3) and by the design of wf in Definition 21.1.4, we can make observations as follows:

1. By Definition 21.1.4(6d) $right \in R$ has only one witness-set, so by Definition 8.2.5(3) that witness-set must be a subset of R. It follows that $right_q \in R$ for every $q \in \mathsf{Q}$, and $right_i \in R$ for every $i \in I$.
2. By Definition 21.1.4(6c) $left \in L$ has only one witness-set, so by Definition 8.2.5(3) that witness-set must be a subset of L. It follows that $left_q \in L$ for every $q \in \mathsf{Q}$.
3. By Definition 21.1.4(6b) each $right_q \in R$ has just two witness-sets, so (as for the cases above) at least one of those witness-sets must be a subset of R. Therefore

$$\texttt{+q} \in R \vee \texttt{-q} \in R.$$

By similar reasoning on the $left_q$ we have

$$\texttt{+q} \in L \vee \texttt{-q} \in L.$$

But we assumed L and R are disjoint, so we conclude that

$$(\texttt{+q} \in R \wedge \texttt{-q} \in L \wedge \texttt{+q} \notin L \wedge \texttt{-q} \notin R) \ \vee \ (\texttt{+q} \in L \wedge \texttt{-q} \in R \wedge \texttt{+q} \notin R \wedge \texttt{-q} \notin L).$$

4. Finally, by Definitions 21.1.4(6e) and 8.2.5(3), for $i \in I$ it follows from $right_i \in R$ that there exists some $j \in i$ such that $\{m(a_{ij}q_{ij})\} \subseteq R$, thus that $m(a_{ij}q_{ij}) \in R$.

We can now take our satisfying assignment f to be such that $f(q) = \mathbf{f}$ when $\texttt{+q} \in L$, and $f(q) = \mathbf{t}$ when $\texttt{+q} \in R$.

Intuitively, this interprets q as true when its positive literal +q is in R, and it interprets q as false when its positive literal +q is in L.

- Suppose we have a satisfying assignment $f : \mathbf{Q} \to \{\mathtt{f}, \mathtt{t}\}$. Then we set

$$L = \{\texttt{+q} \mid q \in f^{-1}\{\mathtt{f}\}\} \cup \{\texttt{-q} \mid q \in f^{-1}\{\mathtt{t}\}\} \cup \{\mathit{left}_q \mid q \in \mathbf{Q}\} \cup \{\mathit{left}\} \quad \text{and}$$
$$R = \{\texttt{+q} \mid q \in f^{-1}\{\mathtt{t}\}\} \cup \{\texttt{-q} \mid q \in f^{-1}\{\mathtt{f}\}\} \cup \{\mathit{right}_q \mid q \in \mathbf{Q}\} \cup \{\mathit{right}_i \mid i \in I\} \cup \{\mathit{right}\}.$$

Intuitively, this puts the positive and negative literals that are false on the left, and the positive and negative literals that are true on the right. By the same reasoning as above, it is routine to check that L and R are a disjoint pair of open sets. □

THEOREM 21.1.6. *The problem of determining whether two points are intertwined in a finite semitopology, is NP complete in general.*

Proof. Proposition 21.1.5 exhibits a reduction of SAT into an intertwinedness problem, and the reduction clearly runs in polynomial time (indeed, it is linear). □

REMARK 21.1.7. Theorem 21.1.6 shows that checking intertwinedness is NP complete in general. However, this is not the last word about how difficult intertwinedness is to check in practice, since there may be practical use cases where the problem is easier.

For example, suppose we have a semitopology $(\mathsf{P}, \mathsf{Open})$ of which we know that it contains a strongly transitive open set K. This may (for example) arise as a kernel atom (Definition 10.1.2(1)) of core participants who for historical reasons are considered reliable, whose witness functions have been relatively stable, and of which we can prove (using our sequent system from Subsection 18.6, or by model-checking the witness function restricted just to those participants) that they are intertwined, consensus-equivalent, and hypertwined with each other (Definition 22.6.6).[1]

Then to check whether the space P is intertwined (meaning by Notation 3.6.5 that *all* points in P are pairwise intertwined), it suffices to compute $|K|$. By Lemma 5.3.5, $|K| = \mathsf{P}$ if and only if P is intertwined.

By the algorithm given in Remark 8.4.14, the closure $|K|$ can be computed efficiently.

21.2 Horn satisfiability over Bool and 3

21.2.1 The (standard) HORNSAT algorithm for Boolean logic

We start with a brief but precise introduction to Horn clause theories and an algorithm to check their satisfiability in Boolean (two-valued) logic:

NOTATION 21.2.1. A **literal** is a possibly negated propositional atom: p or $\neg p$. We call p a **positive literal** and $\neg p$ a **negative literal**. A **Horn clause** is a possibly empty disjunct of positive and negative literals, that contains at most one positive literal. For example:

- $\neg p \lor \neg p' \lor \neg p'' \lor q$ and $\neg p \lor \neg p' \lor \neg p''$ are Horn clauses.

[1] English has many terms for such a K: clique, cabal, eminences, founding committee, in-crowd, and so on. This is a fairly typical situation in real life. For example, a secret to getting along in an organisation is to find out which small clique of well-connected members actually makes the decisions and gets things done — this may or may not line up with any formal hierarchy of the organisation. And, such a group tends in practice to be stable over time (sometimes infuriatingly so). This is the magic of mathematics: we may not be able to cure bureaucracy, but at least we can build a topological model of it!

- $p \vee q$ is not a Horn clause, because it contains two positive literals.
- $\varnothing$ the empty disjunct is a Horn clause. We may call this **empty Horn clause**, and (slightly abusing notation) we may write this as $\bot$.
- p and $\neg p$ — which are clauses consisting of a single literal — are Horn clauses. We may call a Horn clause consisting of a single literal a **unit clause**.

Call a finite set of Horn clauses a **Horn clause theory**.

DEFINITION 21.2.2.

1. Call a propositional theory (which need not consist only of Horn clauses) **2-satisfiable** when there is a valuation $f : \mathsf{P} \to \{\mathtt{t}, \mathtt{f}\}$ that makes all the clauses true in two-valued propositional logic.
2. Call a propositional theory **3-satisfiable** when there is a valuation $f : \mathsf{P} \to \mathbf{3}$ that makes all the clauses either b or t in **3**.

HORNSAT is the computational problem of deciding whether a Horn clause theory over two-valued logic is 2-satisfiable. This can be solved in linear time [DG84]:

REMARK 21.2.3 (HORNSAT algorithm). An overview of the algorithm is as follows:

1. If the set contains a positive unit clause p, then:
   a) delete every clause that contains that literal in positive form, except for p itself, and
   b) delete $\neg p$ from any of the remaining clauses.
2. Repeat the previous step until no further changes to the theory occur.
3. If the resulting theory contains an empty Horn clause then fail; otherwise succeed.

If we succeed, then we can read a satisfying assignment from the result by mapping p to t for every unit clause p that remains, and mapping every other p' to f.

21.2.2 A proposal for HORNSAT over 3

We now sketch a proposal for Horn clause programming in **3**, adapting Notation 21.1.2 (Horn clause theories) and Remark 21.2.3 (the HORNSAT algorithm) to the three-valued case.

The syntax and algorithm in Definition 21.2.4 and Remark 21.2.6 are just pencil-and-paper prototypes, but they are simple and elegant and uncover no obvious difficulties, which suggests that they could be implemented and that more advanced systems could also be practical, like those discussed in [MNPS91, Lif08, Lif19] but based on **3** instead of Bool.

DEFINITION 21.2.4 (Horn clause theories and HORNSAT over **3**). Fix a set **P**.

1. a) A **boxed positive literal** is a proposition of the form $\mathsf{T}p$, for $p \in \mathsf{P}$.
   b) An **unboxed positive literal** is a proposition of the form p, for $p \in \mathsf{P}$.
   c) A **positive literal** is a boxed or unboxed positive literal.
2. a) A **boxed negative literal** is a proposition of the form $\mathsf{T}\neg p$, for $p \in \mathsf{P}$.
   b) An **unboxed negative literal** is a proposition of the form $\neg p$, for $p \in \mathsf{P}$.
   c) A **negative literal** is a boxed or unboxed negative literal.
3. A **literal** is a positive or negative literal.
4. A **3Horn clause** is a possibly empty disjunct of literals, at most one of which is positive:

- When a 3Horn clause contains a positive literal, we call this the **head** of the Horn clause.
- When a 3Horn clause is empty, we call it **unsatisfiable 3Horn clause**.
- When a 3Horn clause is a singleton, we call it a **unit clause**.

5. Echoing Notation 21.2.1, call a finite set of 3Horn clauses a **3Horn clause theory**.

REMARK 21.2.5. Useful classes of propositions fit into the 3Horn clause syntax of Definition 21.2.4. In particular, it follows from the equivalences in Figure 18.3 that:

- $p \supset q$ is equivalent to $\neg p \lor q$, which is a Horn clause.
- $p \equiv q$ is equivalent to $(\neg p \lor q) \land (\neg q \lor p)$, which we can express as a Horn clause theory $\{\neg p \lor q,\ p \lor \neg q\}$.
- We can express the invalidity $\not\models p \supset q$ as $\{\mathsf{T}p, \mathsf{T}\neg q\}$, because (checking Figure 18.2) $[\![p \supset q]\!]_f = \mathbf{f}$ precisely when $f(p) = \mathbf{t}$ and $f(q) = \mathbf{f}$.
- $p \Rightarrow q$ is equivalent to $\mathsf{T}\neg p \lor q$.
- $(p_1 \land \ldots \land p_n) \Rightarrow p$ is equivalent to $\mathsf{T}\neg p_1 \lor \ldots \mathsf{T}\neg p_n \lor p$.

Note from Definition 18.4.1(1) and Figure 18.5 that: $f \models \mathsf{T}p$ precisely when $f(p) = \mathbf{t}$; $f \models p$ precisely when $f(p) \in \{\mathbf{t}, \mathbf{b}\}$; $f \models \mathsf{T}\neg p$ precisely when $f(p) = \mathbf{f}$; and $f \models \neg p$ precisely when $f(p) \in \{\mathbf{f}, \mathbf{b}\}$.

REMARK 21.2.6 (Three-valued HORNSAT). We now propose an algorithm to check satisfiability of Horn clause theories in $\mathbf{3}$ — which as per Definition 21.2.2(2) is the problem of deciding whether there exists a *three-valued* valuation $f : \mathsf{P} \to \mathbf{3}$ that makes all the clauses valid (have truth-value equal to $\mathbf{b}$ or $\mathbf{t}$) in $\mathbf{3}$ as per the three-valued truth-tables in Figure 18.2.

It is based on the algorithm for two-valued logic from Remark 21.2.3. Rules are applied in decreasing order of priority, and are repeated as often as they continue to act to change the theory:

1. Suppose the theory contains a unit boxed positive literal $\mathsf{T}p$ (so intuitively, a satisfying assignment must satisfy $f(p) = \mathbf{t}$). Then:
   a) We remove all other clauses that contain p or $\mathsf{T}p$, because they are now satisfied.
   b) We delete all literals of the form $\mathsf{T}\neg p$, because these literals cannot be satisfied.
   c) We delete all literals of the form $\neg p$, because these literals cannot be satisfied.
2. Suppose the theory contains a unit positive literal p (so for a satisfying assignment, $f(p) \neq \mathbf{f}$). Then:
   a) We delete all other clauses that contain a literal p (but *not* those that contain $\mathsf{T}p$), because these clauses are now satisfied.
   b) We delete all literals of the form $\mathsf{T}\neg p$ from all clauses, because these literals cannot be satisfied.
3. If the theory contains an unsatisfiable (empty) Horn clause, then fail; otherwise succeed.

If we succeed, then we can read a satisfying assignment from the result by

- mapping p to $\mathbf{t}$ if a unit clause $\mathsf{T}p$ remains,
- mapping p to $\mathbf{b}$ if a unit clause p remains (in this case, from the form of the rules a unit clause $\mathsf{T}p$ cannot remain), and
- mapping p to $\mathbf{f}$ otherwise.

22 Extremal valuations

22.1 Definition of an extremal valuation

REMARK 22.1.1. Intuitively, a valuation is *extremal* when it is continuous and it returns as many $\mathbf{t}$ and $\mathbf{f}$ values as possible (is as definite as possible), and conversely when it returns as few $\mathbf{b}$ values as possible (is no more ambivalent than necessary); the precise definition is in Definition 22.1.2.

For example, there are four extremal valuations from $\mathbf{3}$ to itself:

$$\lambda v.\mathbf{t}, \quad \lambda v.\mathbf{f}, \quad \lambda v.v, \quad \text{and} \quad \lambda v.\neg v.$$

Conversely, $\lambda v.\mathsf{T}v$ is not extremal because it is not continuous. $\lambda v.\mathbf{b}$ is continuous but not extremal. The function mapping $\mathbf{t}$ to $\mathbf{t}$, $\mathbf{b}$ to $\mathbf{b}$, and $\mathbf{f}$ to $\mathbf{b}$ is continuous but not extremal.

It turns out that extremal valuations are rather useful. If we think in terms of agreement, an extremal valuation represents a system state where algorithms have run and succeeded as much as they can; if a participant is still returning the ambivalent truth-value $\mathbf{b}$ then this is because they must, and not just because they have not yet made up their mind.

DEFINITION 22.1.2. Suppose $(\mathsf{P}, \mathsf{Open})$ is a semitopology and $p \in \mathsf{P}$ and $f, f' : \mathsf{P} \to \mathbf{3}$ are continuous valuations.

1. Call the elements $\mathbf{t}, \mathbf{f} \in \mathbf{3}$ **definite**, and call $\mathbf{b} \in \mathbf{3}$ **ambivalent** (because $\mathbf{b}$ is short for 'both').
2. If $f(p) \in \mathbf{3}$ is ambivalent / definite then call f **ambivalent / definite** at p.
   It is easy to check from Definition 18.4.1(1) that f is definite at p when $f(p) \not\models \mathsf{B}p$ and when $f \models \mathsf{T}p \mathsf{V} \mathsf{T}\neg p$.
3. Define $\mathit{definite}(f) \subseteq \mathsf{P}$ by

$$\mathit{definite}(f) = f^{-1}\{\mathbf{t}, \mathbf{f}\}.$$

   Thus, $\mathit{definite}(f)$ is the set of points at which f is definite. Note a nice characterisation of this using Definition 18.4.5 as

$$\mathit{definite}(f) = \mathsf{P} \setminus (f^{\models} \cap f^{\models\neg}) = (\mathsf{P} \setminus f^{\models}) \cup (\mathsf{P} \setminus f^{\models\neg}).$$

4. Define a partial order $f \leq f'$ on valuations $f, f' : \mathsf{P} \to \mathbf{3}$ by

$$f \leq f' \quad \text{when} \quad f|_{definite(f)} = f'|_{definite(f)}.$$

In words: $f \leq f'$ when f' agrees with f whenever f has a definite value (however, f' may be definite at other values at which f is ambivalent).
5. Call f a $(\mathsf{P}, \mathsf{Open})$**-extremal valuation** when
   a) f is continuous on $(\mathsf{P}, \mathsf{Open})$ and
   b) f is $\leq$-maximal amongst valuations that are continuous on $(\mathsf{P}, \mathsf{Open})$.

   If the semitopology is clear from context, we may just call f an **extremal valuation**.
6. We will write

$$\mathsf{P}, \mathsf{Open} \overset{\mathsf{x}}{\vDash} \phi$$

when $f \vDash \phi$ for every extremal valuation.

Remark 22.1.3. There is a slight wrinkle to the terminology here:

- In Definition 18.1.2(2) we let a valuation on P be any function $f : \mathsf{P} \to \mathbf{3}$; so if we need it to be continuous then we have to say 'f is a *continuous* valuation'.
- In Definition 22.1.2(5) we let an extremal valuation be a $(\mathsf{P}, \mathsf{Open})$-continuous valuation that is $\leq$-maximal.

We never need to say 'f is an extremal *continuous* valuation', because if f is extremal then it is assumed continuous. The notion of being extremal assumes continuity; since without it being extremal just means being a function in $\mathsf{P} \to \{\mathtt{t}, \mathtt{f}\}$.

22.2 Topological characterisation of extremal valuations

In this Subsection we characterise extremal valuations in terms of regular open/closed sets. The key technical observation powering this Subsection is Lemma 22.2.4.

We note a standard fact from topology relating inverse images, closures, and interiors [Eng89, Proposition 1.4.1(v'&vi)]. It is also valid in semitopologies:

Lemma 22.2.1. *Suppose* $(\mathsf{P}, \mathsf{Open})$ *and* $(\mathsf{P}', \mathsf{Open}')$ *are semitopologies and* $f : \mathsf{P} \to \mathsf{P}'$ *is continuous and* $P' \subseteq \mathsf{P}'$. *Then:*

1. $|f^{-1}(P')| \subseteq f^{-1}|P'|$.
2. $interior(f^{-1}(P')) \supseteq f^{-1}(interior(P'))$.

Proof. For part 1, write $P = f^{-1}(P')$ and suppose $p \in |P|$, so that by Lemma 6.5.3 $nbhd(p) \between P$. Now consider $O' \in nbhd(f(p))$; it would suffice to show that $O' \between P'$. Write $O = f^{-1}(O')$. Note that $O \in \mathsf{Open}$ by continuity and $O \in nbhd(p)$ by construction. It follows that $O \between P$, and so that $O' \between P'$ as required.

For part 2, just use part 1 and take complements, noting that $f^{-1}(\mathsf{P}' \setminus P') = \mathsf{P} \setminus f^{-1}(P')$ and by Lemma 5.2.1 that $interior(\mathsf{P}' \setminus P') = \mathsf{P}' \setminus |P'|$ and $interior(\mathsf{P}' \setminus P') = \mathsf{P}' \setminus |P'|$. □

Corollary 22.2.2. *Suppose* $(\mathsf{P}, \mathsf{Open})$ *is a semitopology and* $f : \mathsf{P} \to \mathbf{3}$ *is a continuous valuation. Then:*

1. $|f^{-1}\{\mathbf{t}\}| \subseteq f^{-1}\{\mathbf{t},\mathbf{b}\}$ *and* $|f^{-1}\{\mathbf{f}\}| \subseteq f^{-1}\{\mathbf{f},\mathbf{b}\}$.
2. *The inclusions may be strict.*

Proof. It is a fact that $|\{\mathbf{t}\}| = \{\mathbf{t},\mathbf{b}\}$ and $|\{\mathbf{f}\}| = \{\mathbf{f},\mathbf{b}\}$. The inclusions follow from Lemma 22.2.1.

To show that the inclusions may be strict, it suffices to provide an example. Consider the semitopology in Figure 3.1 (top-right example) with $f(0) = \mathbf{f}$, $f(1) = \mathbf{b}$, and $f(2) = \mathbf{t}$. The reader can check that this is continuous and $f^{-1}\{\mathbf{f}\} \subsetneq f^{-1}\{\mathbf{f},\mathbf{b}\}$ and $f^{-1}\{\mathbf{t}\} \subsetneq f^{-1}\{\mathbf{t},\mathbf{b}\}$. □

Remarkably, extremal valuations are characterised *precisely* by the property that the inclusions of Corollary 22.2.2 are equalities, as we shall see in the next few results, and in Proposition 22.2.5 in particular.

Lemma 22.2.3. *Suppose* $(\mathsf{P}, \mathsf{Open})$ *is a semitopology and* $f : \mathsf{P} \to \mathbf{3}$ *is a continuous extremal valuation. Then*

$$|f^{-1}\{\mathbf{t}\}| = f^{-1}\{\mathbf{t},\mathbf{b}\} \quad \textit{and} \quad |f^{-1}\{\mathbf{f}\}| = f^{-1}\{\mathbf{f},\mathbf{b}\}.$$

Proof. By Corollary 22.2.2 $|f^{-1}\{\mathbf{t}\}| \subseteq f^{-1}\{\mathbf{t},\mathbf{b}\}$ and $|f^{-1}\{\mathbf{f}\}| \subseteq f^{-1}\{\mathbf{f},\mathbf{b}\}$. Now suppose $p \notin |f^{-1}\{\mathbf{t}\}|$ (the case of $\mathbf{f}$ is precisely similar). So there exists $O \in nbhd(p)$ such that $O \not\between f^{-1}\{\mathbf{t}\}$. There are now two subcases:

- *Suppose* $O \subseteq f^{-1}\{\mathbf{f}\}$.
  Then $p \notin f^{-1}\{\mathbf{t},\mathbf{b}\}$ and we are done.
- *Suppose* $O \not\subseteq f^{-1}\{\mathbf{f}\}$.
  Then we obtain a strictly more definite continuous valuation as $\delta_{O_\mathbf{t},O_\mathbf{f}}(f^{-1}\{\mathbf{t}\}, f^{-1}\{\mathbf{f}\} \cup O)$. By extremality of f, this is impossible. □

The nontrivial technical content of this Subsection is here:

Lemma 22.2.4. *Suppose* $(\mathsf{P}, \mathsf{Open})$ *is a semitopology and* $f : \mathsf{P} \to \mathbf{3}$ *is a continuous valuation. Then*

$$|f^{-1}\{\mathbf{t}\}| = f^{-1}\{\mathbf{t},\mathbf{b}\} \wedge |f^{-1}\{\mathbf{f}\}| = f^{-1}\{\mathbf{f},\mathbf{b}\} \quad \textit{implies} \quad f \textit{ is extremal.}$$

Proof. We prove the contrapositive. Suppose f is not extremal, so that there exists a continuous valuation $f' : \mathsf{P} \to \mathbf{3}$ such that $f \lneq f'$. So there is a $p' \in \mathsf{P}$ at which f' is definite and f is not; suppose $f'(p') = \mathbf{t}$ (the case that $f'(p') = \mathbf{f}$ is exactly similar).

Since f' is continuous, by Definition 5.4.4 and Remark 5.4.5 there exists an $O' \in nbhd(p')$ such that $O' \subseteq (f')^{-1}\{\mathbf{t}\}$. From this it follows using Definition 5.1.2(1) that $p' \notin |f^{-1}\{\mathbf{f}\}|$. Now we assumed $f(p') = \mathbf{b}$, so that $p' \in f^{-1}\{\mathbf{f},\mathbf{b}\}$ is a fact. Thus, $|f^{-1}\{\mathbf{f}\}| \neq f^{-1}\{\mathbf{f},\mathbf{b}\}$ as required. □

Proposition 22.2.5. *Suppose* $(\mathsf{P}, \mathsf{Open})$ *is a semitopology and* $f : \mathsf{P} \to \mathbf{3}$ *is a continuous valuation. Then the following are equivalent:*

1. *f is extremal.*
2. $|f^{-1}\{\mathbf{t}\}| = f^{-1}\{\mathbf{t},\mathbf{b}\}$ *and* $|f^{-1}\{\mathbf{f}\}| = f^{-1}\{\mathbf{f},\mathbf{b}\}$.
3. $f^{-1}\{\mathbf{t},\mathbf{b}\} \subseteq |f^{-1}\{\mathbf{t}\}|$ *and* $f^{-1}\{\mathbf{f},\mathbf{b}\} \subseteq |f^{-1}\{\mathbf{f}\}|$.

It is a fact that $|\{\mathbf{t}\}| = \{\mathbf{t},\mathbf{b}\}$ *and* $|\{\mathbf{f}\}| = \{\mathbf{f},\mathbf{b}\}$ *in* $\mathbf{3}$, *so we can say:*

> *A continuous valuation $f : \mathsf{P} \to \mathbf{3}$ is extremal precisely when inverse images commute with closures.*

Proof. Equivalence of parts 1 and 2 is just Lemmas 22.2.3 and 22.2.4. Part 2 certainly implies part 3, and part 3 combined with Lemma 22.2.1 implies part 2. □

It is easy and natural to dualise Proposition 22.2.5:

COROLLARY 22.2.6. *Suppose* $(\mathsf{P}, \mathsf{Open})$ *is a semitopology and* $f : \mathsf{P} \to \mathbf{3}$ *is a continuous valuation. Then the following are equivalent:*

1. *f is extremal.*
2. $\mathit{interior}(f^{-1}\{\mathbf{f}, \mathbf{b}\}) = f^{-1}\{\mathbf{f}\}$ *and* $\mathit{interior}(f^{-1}\{\mathbf{t}, \mathbf{b}\}) = f^{-1}\{\mathbf{t}\}$.
3. $f^{-1}\{\mathbf{f}\} \supseteq \mathit{interior}(f^{-1}\{\mathbf{t}, \mathbf{b}\})$ *and* $f^{-1}\{\mathbf{t}\} \supseteq \mathit{interior}(f^{-1}\{\mathbf{f}, \mathbf{b}\})$.

It is a fact that $\mathit{interior}(\{\mathbf{f}, \mathbf{b}\}) = \{\mathbf{f}\}$ *and* $\mathit{interior}(\{\mathbf{t}, \mathbf{b}\}) = \{\mathbf{t}\}$ *in* $\mathbf{3}$, *so we can say:*

> *A continuous valuation $f : \mathsf{P} \to \mathbf{3}$ is extremal precisely when inverse images commute with interiors.*

Proof. We take complements in Proposition 22.2.5 and dualise using Lemma 5.2.1 (just as we did to derive part 2 of Lemma 22.2.1 from part 1). □

In Lemma 20.3.1(4) we noted that $f \vDash \mathsf{T}p$ implies $p \in \mathit{interior}(f^{\vDash})$, but the reverse implication need not hold. Rather nicely, it turns out that extremal valuations are characterised *precisely* as those valuations such that the reverse implications hold for $\mathsf{T}p$ and $\mathsf{T}\neg p$:

COROLLARY 22.2.7. *Suppose* $(\mathsf{P}, \mathsf{Open})$ *is a semitopology and* $f : \mathsf{P} \to \mathbf{3}$ *is a continuous valuation. Then the following are equivalent:*

1. *f is extremal.*
2. $p \in \mathit{interior}(f^{\vDash}) \Longleftrightarrow f \vDash \mathsf{T}p$ *and* $p \in \mathit{interior}(f^{\vDash\neg}) \Longleftrightarrow f \vDash \mathsf{T}\neg p$, *for every* $p \in \mathsf{P}$.
3. $p \in \mathit{interior}(f^{\vDash}) \Longrightarrow f \vDash \mathsf{T}p$ *and* $p \in \mathit{interior}(f^{\vDash\neg}) \Longrightarrow f \vDash \mathsf{T}\neg p$, *for every* $p \in \mathsf{P}$.

Proof. This just rephrases Corollary 22.2.6 using Lemma 20.3.1(1) and Definition 18.4.5. □

22.3 Maximal disjoint pairs of open sets

We now develop a different view of extremal valuations, based on maximal elements in the poset of disjoint pairs of open sets. We start by introducing some notation and recalling an easy sets bijection:

NOTATION 22.3.1. Suppose P and P' are sets and suppose $\mathcal{P} \subseteq \mathit{pow}(\mathsf{P})$ and $\mathcal{P}' \subseteq \mathit{pow}(\mathsf{P}')$ are sets of subsets. Then write $\mathcal{P} \otimes \mathcal{P}'$ for the set of disjoint pairs in $\mathcal{P} \times \mathcal{P}'$. In symbols:

$$\mathcal{P} \otimes \mathcal{P}' = \{(P, P') \mid P \in \mathcal{P},\ P' \in \mathcal{P}',\ P \not\between P'\}.$$

In particular, if $(\mathsf{P}, \mathsf{Open})$ is a semitopology then $\mathsf{Open} \otimes \mathsf{Open}$ is the set of disjoint pairs of open sets.

REMARK 22.3.2. Suppose $(\mathsf{P}, \mathsf{Open})$ is a semitopology. Recall the *indicator functions* $\delta_{O_{\mathsf{t}},O_{\mathsf{f}}}$ from Definition 18.1.5 and the *characteristic sets* $char_{OO}$ from Definition 18.1.7, and recall from Proposition 18.1.10 that

1. $char_{OO}$ maps a continuous valuation $f : \mathsf{P} \to \mathbf{3}$ to a disjoint pair

$$char_{OO}(f) = (f^{-1}\{\mathsf{t}\}, f^{-1}\{\mathsf{f}\}) \in \mathsf{Open} \otimes \mathsf{Open}$$

   and
2. $\delta_{O_{\mathsf{t}},O_{\mathsf{f}}}$ maps $(O, O') \in \mathsf{Open} \otimes \mathsf{Open}$ to the continuous valuation that maps $p \in \mathsf{P}$ to t if $p \in O$, and to $\mathbf{f}$ if $p \in O'$, and to $\mathbf{b}$ if $p \in \mathsf{P} \setminus (O \cup O')$, and
3. $char_{OO}$ and $\delta_{O_{\mathsf{t}},O_{\mathsf{f}}}$ are inverse and biject between valuations and pairs of disjoint open sets.

We now note that $char$ and δ can also be viewed as maps of posets:

DEFINITION 22.3.3. Suppose $(\mathsf{P}, \mathsf{Open})$ is a semitopology. Then:

1. Extend the subset inclusion ordering $\subseteq$ to disjoint pairs of open sets in $\mathsf{Open} \otimes \mathsf{Open}$ **componentwise** as follows:

$$(O_1, O_1') \leq (O_2, O_2') \quad \text{when} \quad O_1 \subseteq O_1' \ \wedge \ O_2 \subseteq O_2'$$

   for $(O_1, O_1'), (O_2, O_2') \in \mathsf{Open} \otimes \mathsf{Open}$.
2. Call $(O, O') \in \mathsf{Open} \otimes \mathsf{Open}$ **maximal disjoint** when it is $\leq$-maximal amongst pairs of disjoint open sets.

LEMMA 22.3.4. *Suppose* $(\mathsf{P}, \mathsf{Open})$ *is a semitopology. Then:*

1. *If* $f, f' : \mathsf{P} \to \mathbf{3}$ *are continuous valuations then the following are equivalent:*
   - $f \leq f'$ *in the sense of Definition 22.1.2(4).*
   - $char_{OO}(f) \leq char_{OO}(f')$ *in the sense of Definition 22.3.3(1).*
2. *The following are equivalent for* $(O_1, O_1'), (O_2, O_2') \in \mathsf{Open} \otimes \mathsf{Open}$*:*
   - $(O_1, O_1') \leq (O_2, O_2')$ *in the sense of Definition 22.3.3(1).*
   - $\delta_{O_{\mathsf{t}},O_{\mathsf{f}}}(O_1, O_1') \leq \delta_{O_{\mathsf{t}},O_{\mathsf{f}}}(O_2, O_2')$ *in the sense of Definition 22.1.2(4).*

Proof. By routine calculations from the definitions. □

COROLLARY 22.3.5. *Suppose* $(\mathsf{P}, \mathsf{Open})$ *is a semitopology. Then* $\delta_{O_{\mathsf{t}},O_{\mathsf{f}}}$ *and* $char_{OO}$ *determine a poset isomorphism between*

- *continuous valuations ordered by* $\leq$ *(Definition 22.1.2(4)) and*
- $\mathsf{Open} \otimes \mathsf{Open}$ *ordered by* $\leq$ *(Definition 22.3.3(1)).*

Proof. Direct from Lemma 22.3.4. □

The maximal elements of $\mathsf{Open} \otimes \mathsf{Open}$ have some useful characterisations:

PROPOSITION 22.3.6. *Suppose* $(\mathsf{P}, \mathsf{Open})$ *is a semitopology and* $(O, O') \in \mathsf{Open} \otimes \mathsf{Open}$. *Then the following are equivalent:*

1. *(O, O') is maximal disjoint.*
2. *$O = interior(\mathsf{P} \setminus O')$ and $O' = interior(\mathsf{P} \setminus O)$.*
3. *O is regular (Definition 5.7.2) and $O' = interior(\mathsf{P} \setminus O)$.*
4. *O' is regular and $O = interior(\mathsf{P} \setminus O')$.*
5. *O and O' are regular and $O = interior(\mathsf{P} \setminus O')$ and $O' = interior(\mathsf{P} \setminus O)$.*

Proof. Equivalence of parts 1 and 2 is just an easy corollary of Lemma 4.1.2. We spell out the details:

- *Suppose (O, O') is maximal disjoint.*

  Then O' is a greatest open set that is disjoint from O, meaning equivalently that O' is a greatest open set contained in $\mathsf{P} \setminus O$, so by Lemma 4.1.2 $O' = interior(\mathsf{P} \setminus O)$. Similarly, $O = interior(\mathsf{P} \setminus O')$.

- *Suppose $O' = interior(\mathsf{P} \setminus O)$ and $O = interior(\mathsf{P} \setminus O')$.*

  Consider a disjoint pair $(O'', O''') \geq (O, O')$, meaning by Definition 22.3.3(1) that $O'' \supseteq O$ and $O''' \supseteq O'$. Then $O'' \subseteq \mathsf{P} \setminus O''' \subseteq \mathsf{P} \setminus O'$ so that $O'' \subseteq interior(\mathsf{P} \setminus O') = O$, so that $O'' = O$. Similarly, $O''' = O'$. Since (O'', O''') was arbitrary, (O, O') is a maximal disjoint pair of open sets.

To prove equivalence of parts 2 and 3 we reason as follows:

- *Suppose $O = interior(\mathsf{P} \setminus O')$ and $O' = interior(\mathsf{P} \setminus O)$.*

  We reason as follows using Lemma 5.2.1(3):

$$\begin{aligned} O &= interior(\mathsf{P} \setminus O') && \text{Assumption} \\ &= interior(\mathsf{P} \setminus interior(\mathsf{P} \setminus O)) && O' = interior(\mathsf{P} \setminus O) \\ &= interior(\mathsf{P} \setminus (\mathsf{P} \setminus |O|)) && \text{Lemma 5.2.1(3)} \\ &= interior(|O|) && \text{Fact of sets} \end{aligned}$$

- *Suppose O is regular (so $O = interior(|O|)$) and $O' = interior(\mathsf{P} \setminus O)$.*

  We just reverse the reasoning of the previous case.

To prove equivalence of parts 3 and 4 follows by the same reasoning, on O'. Equivalence of parts 1 and 5 then follows easily. □

REMARK 22.3.7. Putting Proposition 22.3.6 and Corollary 22.3.5 together, we see that the following items of data are in natural correspondence:

1. *Extremal valuations $f : \mathsf{P} \to \mathbf{3}$.* An extremal f corresponds to the by Corollary 22.3.5 maximal disjoint pair $(f^{-1}\{\mathtt{t}\}, f^{-1}\{\mathtt{f}\})$, and to the by Proposition 22.3.6 regular open set $f^{-1}\{\mathtt{t}\}$.
2. *Maximal pairs $(O, O') \in \mathsf{Open} \otimes \mathsf{Open}$.* A maximal (O, O') corresponds to the by Corollary 22.3.5 extremal valuation $\delta_{O_\mathtt{t}, O_\mathtt{f}}(O, O') : \mathsf{P} \to \mathbf{3}$, and to the by Proposition 22.3.6 regular open O.
3. *Regular open sets $O \in \mathsf{Open}_{reg}$.* A regular open set O corresponds to the by Proposition 22.3.6 maximal disjoint $(O, interior(\mathsf{P} \setminus O))$ and to the by Corollary 22.3.5 extremal $\delta_{O_\mathtt{t}, O_\mathtt{f}}(O, interior(\mathsf{P} \setminus O))$.

$f : (\mathsf{P}, \mathsf{Open}) \to \mathbf{3}$	$(f^{-1}\{\mathtt{t}\}, f^{-1}\{\mathtt{f}\})$	$f^{-1}\{\mathtt{t}\}$
$\delta_{O_\mathtt{t},O_\mathtt{f}}(O, O')$	$(O, O') \in \mathsf{Open} \otimes \mathsf{Open}$	O
$\delta_{O_\mathtt{t},O_\mathtt{f}}(O, \mathit{interior}(\mathsf{P} \setminus O))$	$(O, \mathit{interior}(\mathsf{P} \setminus O))$	$O \in \mathsf{Open}_{reg}$

In row 1 $f : (\mathsf{P}, \mathsf{Open}) \to \mathbf{3}$ *is an extremal valuation; in row 2* (O, O') *is a maximal disjoint pair of open sets; in row 3* O *is a regular open set.*

Figure 22.1: Correspondence between extremal valuations, maximal disjoint pairs of open sets, and regular open sets

We sum this up in a small table in Figure 22.1.

We conclude with a small lemma noting that extremal valuations always exist. A standard argument using Zorn's lemma is possible, but we give a more direct argument using closures and regular open sets:

LEMMA 22.3.8. *Suppose* $(\mathsf{P}, \mathsf{Open})$ *is a semitopology and* $f : \mathsf{P} \to \mathbf{3}$ *is a continuous valuation. Then there exists an extremal valuation* $f' : \mathsf{P} \to \mathbf{3}$ *such that* $f \leq f'$.

Proof. Write $O = f^{-1}\{\mathtt{t}\}$ and $O' = f^{-1}\{\mathtt{f}\}$. Note from Remark 22.3.2 that O and O' are open and disjoint. Write $R = \mathit{interior}(|O|)$. Note by construction that $O \subseteq R$, and from Lemmas 5.7.3 and 5.7.8 that R is open, regular, and disjoint from O'.

Finally, write $R' = \mathit{interior}(\mathsf{P} \setminus R)$. By construction $O' \subseteq R'$ and using Proposition 22.3.6, (R, R') is maximal disjoint above (O, O') in $\mathsf{Open} \otimes \mathsf{Open}$. We set $f' = \delta_{O_\mathtt{t} O_\mathtt{f}}(R, R')$. By Corollary 22.3.5 f' is extremal and (since $O \subseteq R$ and $O' \subseteq R'$) $f \leq f'$. □

22.4 Logical characterisation of extremal valuations

REMARK 22.4.1. Suppose $\mathit{wf} : \mathsf{P} \to \mathcal{W}(\mathsf{P})$ is a witness function on a finite set P, and recall from Subsection 19.2 and Proposition 19.2.3 that a valuation $f : \mathsf{P} \to \mathbf{3}$ is continuous in the witness semitopology from Definition 8.2.5(3) if and only if $f \vDash \mathsf{Ax}(\mathit{wf})$, for $\mathsf{Ax}(\mathit{wf})$ the theory arising from wf in Definition 19.1.1. Recall that $\mathsf{Ax}(\mathit{wf})$ has for each $p \in \mathsf{P}$ the following pair of axioms:

$$\mathsf{closedAx}_{\mathit{wf}}(p) = \big(\textstyle\bigwedge_{w \in \mathit{wf}(p)} \bigvee_{q \in w} q\big) \Rightarrow p \quad \text{and}$$
$$\mathsf{closedAx}^{\neg}_{\mathit{wf}}(p) = \big(\textstyle\bigwedge_{w \in \mathit{wf}(p)} \bigvee_{q \in w} \neg q\big) \Rightarrow \neg p.$$

REMARK 22.4.2. Recall from Definition 9.5.2(3) that $\mathit{Covers}(p) = \{O \in \mathit{nbhd}(p) \mid p \lessdot O\}$ is the set of *open covers* of p, which are the minimal open neighbourhoods of p. Recall from Theorem 9.4.1 that witness semitopologies are chain-complete, so that for the case of a witness semitopology we get Corollary 9.5.5, that $p \in |O'|$ if and only if $O' \between \mathit{Covers}(p)$.

DEFINITION 22.4.3. Suppose $\mathit{wf} : \mathsf{P} \to \mathcal{W}(\mathsf{P})$ is a witness function on a finite set P. Continuing Definition 19.1.1:

1. Define the **extremal axioms arising from** wf to be sets of predicates in $\mathsf{ClosedPred3}(\mathsf{P})$

as follows:

$$\begin{aligned}\mathsf{closedAxEx}_{wf}(p) &= p \Rightarrow \big(\bigwedge_{Q\in Covers(p)} \bigvee_{q\in Q} \mathsf{T}q\big)\\ \mathsf{closedAxEx}^{\neg}_{wf}(p) &= \neg p \Rightarrow \big(\bigwedge_{Q\in Covers(p)} \bigvee_{q\in Q} \mathsf{T}\neg q\big)\end{aligned}$$

$$\begin{aligned}\mathsf{closedAxEx}_{wf} &= \bigwedge\{\mathsf{closedAxEx}_{wf}(p) \mid p\in \mathsf{P}\}\\ \mathsf{closedAxEx}^{\neg}_{wf} &= \bigwedge\{\mathsf{closedAxEx}^{\neg}_{wf}(p) \mid p\in \mathsf{P}\}\end{aligned}$$

2. We collect these axioms into a large conjunction $\mathsf{AxEx}(wf)$, which we call the **extremal theory arising from** wf:

$$\mathsf{AxEx}(wf) = \mathsf{Ax}(wf) \wedge \mathsf{closedAxEx}_{wf} \wedge \mathsf{closedAxEx}^{\neg}_{wf}.$$

We may omit the wf annotation where this is unimportant or understood, writing (for example) $\mathsf{closedAxEx}^{\neg}_{wf}(p)$ just as $\mathsf{closedAxEx}^{\neg}(p)$.

Proposition 22.4.4. *Suppose* $(\mathsf{P}, \mathsf{Open})$ *is a semitopology and* $f : \mathsf{P} \to \mathbf{3}$ *is a valuation. Then the following conditions are equivalent:*

1. *f is an extremal valuation.*
2. $f \vDash \mathsf{AxEx}(wf)$.

Proof. We prove two implications:

- *Suppose $f : \mathsf{P} \to \mathbf{3}$ is an extremal valuation.*

  By assumption f is continuous, so by Proposition 19.2.3 $f \vDash \mathsf{Ax}(wf)$. It remains to show that $f \vDash \mathsf{AxEx}(wf)$. From Proposition 22.2.5 we have that $f^{-1}\{\mathbf{t},\mathbf{b}\} \subseteq |f^{-1}\{\mathbf{t}\}|$ and $f^{-1}\{\mathbf{f},\mathbf{b}\} \subseteq |f^{-1}\{\mathbf{f}\}|$. But by Corollary 9.5.5 this is precisely what $\mathsf{closedAxEx}_{wf}$ and $\mathsf{closedAxEx}^{\neg}_{wf}$ express, so we are done.

- *Suppose $f \vDash \mathsf{AxEx}(wf)$.*

  Then $f \vDash \mathsf{Ax}(wf)$ so that by Proposition 19.2.3 f is continuous. Also, by Corollary 9.5.5 $f \vDash \mathsf{closedAxEx}_{wf} \wedge f \vDash \mathsf{closedAxEx}^{\neg}_{wf}$ expresses precisely that $f^{-1}\{\mathbf{t},\mathbf{b}\} \subseteq |f^{-1}\{\mathbf{t}\}|$ and $f^{-1}\{\mathbf{f},\mathbf{b}\} \subseteq |f^{-1}\{\mathbf{f}\}|$, so by Proposition 22.2.5 f is extremal as required. □

Remark 22.4.5. We note in passing that:

1. In Definition 22.4.3 we could equivalently set

$$\begin{aligned}\mathsf{closedAxEx}_{wf}(p) &= p \Rightarrow \big(\bigwedge_{Q\in nbhd(p)} \bigvee_{q\in Q} \mathsf{T}q\big) \quad \text{and}\\ \mathsf{closedAxEx}^{\neg}_{wf}(p) &= \neg p \Rightarrow \big(\bigwedge_{Q\in nbhd(p)} \bigvee_{q\in Q} \mathsf{T}\neg q\big).\end{aligned}$$

   This would not be wrong — but in the case of a witness semitopology, which is chain-bounded, by Corollary 9.5.5 we might as well restrict to $Covers(p) \subseteq nbhd(p)$.

2. We can use the equivalences in Figure 18.3 to rewrite the extremal axioms from Definition 22.4.3 into a more clausal form as follows:

   - $p \Rightarrow \bigwedge_i \bigvee_j \mathsf{T}q_{ij}$ becomes $\mathsf{T}\neg p \vee (\bigwedge_i \bigvee_j \mathsf{T}q_{ij})$.
   - $\neg p \Rightarrow \bigwedge_i \bigvee_j \neg q_{ij}$ becomes $\mathsf{T}p \vee (\bigwedge_i \bigvee_j \mathsf{T}\neg q_{ij})$.

22.5 Extremal valuations and regular points

We can think of Proposition 22.5.1 (which considers continuous functions to **3**) as generalising Theorem 3.2.2 (which regards partially continuous functions, but to a discrete space):

PROPOSITION 22.5.1. *Suppose* $(\mathsf{P}, \mathsf{Open})$ *is a semitopology and* $f : \mathsf{P} \to \mathbf{3}$ *is an extremal valuation and* $T \in \mathsf{Topen}$ *and* $p, p' \in T$. *Then:*

1. $f(p) = f(p')$.
2. $f(p) \neq \mathbf{b}$.

Proof. Write $(O, O') = \mathit{char}_{OO}(f) = (f^{-1}\{\mathbf{t}\}, f^{-1}\{\mathbf{f}\})$, and note from Remark 22.3.7(1) that this is a maximal pair of disjoint open sets. We consider each part in turn:

1. If $p \in O$ then $T \between O$ and by Proposition 5.3.2 $T \subseteq O$, so that $p' \in O$. By the same reasoning $p' \in O$ implies $p \in O$, and similarly $p \in O'$ if and only if $p' \in O'$. Thus

   $$p \in O \iff p' \in O \quad \text{and} \quad p \in O' \iff p' \in O'.$$

   It follows that $f(p) = f(p')$ as required.

2. Suppose $f(p) = \mathbf{b}$; we will arrive at a contradiction. By assumption p has a topen neighbourhood T and it follows using Theorem 4.2.6(5&2) that $p \in K(p) \in \mathsf{Topen}$.

   Using part 1 of this result (for the topen $K(p) \in \mathsf{Topen}$) we know that $f(p'') = \mathbf{b}$ for every $p'' \in K(p)$. But then $K(p)$ is an open set that is disjoint from both O and O', contradicting their maximality. □

LEMMA 22.5.2. *Suppose* $(\mathsf{P}, \mathsf{Open})$ *is a semitopology and* $p \in \mathsf{P}$. *Then:*

1. *If p is regular then every extremal valuation is definite at p.*
2. *The converse implication need not hold: it is possible for every extremal valuation to be definite at p, yet p is not regular.*

Proof. For part 1, suppose p is regular, so that by Definition 4.1.4(3) $p \in K(p) \in \mathsf{Topen}$, and suppose $f : \mathsf{P} \to \mathbf{3}$ is an extremal valuation. By Proposition 22.5.1(2) (taking $p' = p$) $f(p) \neq \mathbf{b}$, so by Definition 22.1.2(2) f is definite at p.

For part 2, it suffices to provide a counterexample. Consider the semitopology in Figure 5.2, so:

- $\mathsf{P} = \{0, 1, 2, 3\}$ and
- Open is generated by $\{\{0, 1\}, \{1, 2\}, \{2, 3\}, \{3, 4\}\}$.

The reader can check that extremal valuations are constant and definite on $\{0, 1\}$ and $\{2, 3\}$, or on $\{0, 3\}$ and $\{1, 2\}$ — for instance we can map $\{0, 1\}$ to $\mathbf{t}$ and $\{2, 3\}$ to $\mathbf{f}$. However, by Lemma 5.6.7 no points in this space are regular (see the discussion in Example 5.6.8(3)). □

22.6 Characterisations of intertwinedness properties

22.6.1 Studying $p \between p'$ and $p \equiv p'$

LEMMA 22.6.1. *For a semitopology* $(\mathsf{P}, \mathsf{Open})$ *and* $p, p' \in \mathsf{P}$, *the following are equivalent:*

1. $\forall O, O' \in \mathsf{Open}.(p \in O \wedge p' \in O') \Longrightarrow O \between O'$ *(meaning by Definition 3.6.1(1) that* $p \between p'$*).*
2. $\forall O, O' \in \mathsf{Open}_{reg}.(p \in O \wedge p' \in O') \Longrightarrow O \between O'$.

Proof. This just repackages Corollary 5.7.10. □

Lemma 22.6.2 gives yet another view of being intertwined; we use it in Definition 22.6.6:

LEMMA 22.6.2. *For a semitopology* $(\mathsf{P}, \mathsf{Open})$ *and* $p, p' \in \mathsf{P}$, *the following are equivalent:*

1. $p \between p'$.
2. $\forall O \in \mathsf{Open}. p, p' \in |O| \vee p, p' \in \mathsf{P} \setminus O$.
3. $\forall O \in \mathsf{Open}_{reg}. p, p' \in |O| \vee p, p' \in \mathsf{P} \setminus O$.
4. $\forall C \in \mathsf{Closed}. p, p' \in C \vee p, p' \in |\mathsf{P} \setminus C|$.
5. $\forall C \in \mathsf{Closed}_{reg}. p, p' \in C \vee p, p' \in |\mathsf{P} \setminus C|$.

Proof. For the equivalence of parts 1 and 2 we prove two implications:

- *Suppose* $p \between p'$.

  Write $C = |O|$ and $C' = \mathsf{P} \setminus O$. By Lemma 5.1.6 C is closed, and by Lemma 5.1.9 so is $\mathsf{P} \setminus O$. By construction and Lemma 5.1.3(2) $C \cup C' = \mathsf{P}$, so by Lemma 20.1.2(5) $p, p' \in C \vee p, p' \in C'$.

- *Suppose* $p \not\between p'$, so there exist $p \in O \in \mathsf{Open}$ and $p' \in O' \in \mathsf{Open}$ such that $O \not\between O'$.

  Because $p' \in O' \not\between O$, we know from Definition 5.1.2(1) that $p' \notin |O|$. Also, $p \notin \mathsf{P} \setminus O$ (because $p \in O$). It follows that $p, p' \in |O|$ is impossible, and $p, p' \in \mathsf{P} \setminus O$ is impossible.

For the equivalence of parts 1 and 3 it suffices to show that part 2 implies part 3, and that the negation of Lemma 22.6.1(2) implies the negation of part 3:

- *Suppose* $\forall O \in \mathsf{Open}. p, p' \in |O| \vee p, p' \in \mathsf{P} \setminus O$.

  Then certainly $\forall O \in \mathsf{Open}_{reg}. p, p' \in |O| \vee p, p' \in \mathsf{P} \setminus O$, because every regular open set is also an open set.

- *Suppose we have* $O, O' \in \mathsf{Open}_{reg}$ *such that* $p \in O \wedge p' \in O'$ *and* $O \not\between O'$.

  Since $p \in O$, we have by Lemma 5.1.3(2) that $p \in |O|$, and by construction that $p \notin \mathsf{P} \setminus O$. Since $p' \in O' \not\between O$, we have that $p' \in \mathsf{P} \setminus O$, and by Definition 5.1.2(1) that $p' \notin |O|$.

We prove equivalence of parts 1 and 4 as follows:

- *Suppose* $p \between p'$.

  Suppose $C \in \mathsf{Closed}$. Write $C' = |\mathsf{P} \setminus C|$. By Lemma 5.1.6 C' is closed, and by Lemma 5.1.3(2) $C \cup C' = \mathsf{P}$. It follows by Lemma 20.1.2(5) that $p, p' \in C \vee p, p' \in C'$, so we are done.

- *Suppose* $p \not\between p'$, so there exist $p \in O \in \mathsf{Open}$ and $p' \in O' \in \mathsf{Open}$ such that $O \not\between O'$.

  Write $C = |O|$ and $C' = \mathbf{P} \setminus O$. Because $p' \in O' \not\between O$, we know from Definition 5.1.2(1) that $p' \notin C$. Also, $p \notin \mathbf{P} \setminus O$ (because $p \in O$). Now using Lemma 5.2.1(3&2) we have that

  $$|\mathbf{P} \setminus |O|| \overset{L5.2.1(3)}{=} |interior(\mathbf{P} \setminus O)| = |interior(C')| \overset{L5.2.1(2)}{\subseteq} C'.$$

  So if $p \notin C'$ then also $p \notin |\mathbf{P} \setminus C|$. It follows that $p, p' \in C$ is impossible, and $p, p' \in \mathbf{P} \setminus C$ is impossible, so we are done.

Equivalence of parts 3 and 5 follows by a routine argument from regularity (Definition 5.7.2) using Corollary 5.7.6 and Lemma 5.2.1. □

Proposition 22.6.3. *For a semitopology* $(\mathbf{P}, \mathsf{Open})$ *and* $p, p' \in \mathbf{P}$, *the following are equivalent:*

1. $p \between p'$
2. $\mathbf{P}, \mathsf{Open} \models p \equiv p'$
3. $\mathbf{P}, \mathsf{Open} \overset{\mathsf{x}}{\models} p \equiv p'$ *(Definition 22.1.2(6))*

Proof. Parts 1 and 2 just repeat Lemma 20.1.3(1&2). For equivalence of parts 2 and 3 we prove two implications:

- Suppose $\mathbf{P}, \mathsf{Open} \models p \equiv p'$. Then certainly $\mathbf{P}, \mathsf{Open} \overset{\mathsf{x}}{\models} p \equiv p'$, since every extremal valuation is a continuous valuation.

- Suppose $\mathbf{P}, \mathsf{Open} \not\models p \equiv p'$. By the equivalence of parts 1 and 2 of this result and by Lemma 22.6.2, there exists an $O \in \mathsf{Open}_{reg}$ such that $p \notin \mathbf{P} \setminus O$ — so $p \in O$ — and $p' \notin |O|$ — so $p' \in interior(\mathbf{P} \setminus O)$ — or $p' \in O$ and $p \in \mathbf{P} \setminus O$. We consider the former case; the proof for the latter is precisely similar.

  We set $f = \delta_{O_\mathbf{t} O_\mathbf{f}}(O, interior(\mathbf{P} \setminus O))$. By Remark 22.3.7 this is extremal, and by construction $f(p) = \mathbf{t}$ and $f(p') = \mathbf{f}$ so that $f \not\models p \equiv p'$. □

Remark 22.6.4. Lemma 22.6.2 and Proposition 22.6.3 are interesting not just for what they are, but for the design space that they suggest:

1. Given that we have
   - two turnstiles ($\models$ and $\overset{\mathsf{x}}{\models}$),
   - two notions of logical equivalence ($\equiv$ and $\Leftrightarrow$), and also
   - the T modality,

   when Proposition 22.6.3 notes that these are equivalent

   $$p \between p' \quad \Longleftrightarrow \quad \mathbf{P}, \mathsf{Open} \models p \equiv p' \quad \Longleftrightarrow \quad \mathbf{P}, \mathsf{Open} \overset{\mathsf{x}}{\models} p \equiv p',$$

   this only draws attention to the rest of design space in which they are embedded (below, we omit $\mathbf{P}, \mathsf{Open}$ for compactness):[1]

   $$\models p \equiv p' \quad \overset{\mathsf{x}}{\models} p \equiv p' \quad \models p \Leftrightarrow p' \quad \overset{\mathsf{x}}{\models} p \Leftrightarrow p'$$

   $$\models \mathsf{T}(p \equiv p') \quad \overset{\mathsf{x}}{\models} \mathsf{T}(p \equiv p') \quad \models \mathsf{T}(p \Leftrightarrow p') \quad \overset{\mathsf{x}}{\models} \mathsf{T}(p \Leftrightarrow p')$$

[1] The design space is even larger than this, of course. We have multiple modalities, and Remark 18.2.6 notes that there are actually sixteen different implications in three-valued logic. But we have to start somewhere.

Intertwined $\between$	$\models p{\equiv}p'$	$\overset{\times}{\models} p{\equiv}p'$	See P.22.6.3
Consensus equivalent $\between^{=}$	$\overset{\times}{\models} p{\Leftrightarrow}p'$		See L.22.6.14
Hypertwined $\between^{\mathsf{T}}$	$\overset{\times}{\models} \mathsf{T}(p{\equiv}p')$	$\overset{\times}{\models} \mathsf{T}(p{\Leftrightarrow}p')$	See L.22.6.16
Topologically indistinguishable $\mathrel{\overset{\circ}{=}}$	$\models p{\Leftrightarrow}p'$		See L.20.2.1
Always false $\varnothing$	$\models \mathsf{T}(p{\equiv}p')$	$\models \mathsf{T}(p{\Leftrightarrow}p')$	See L.22.6.5

Figure 22.2: Summary of antiseparation properties

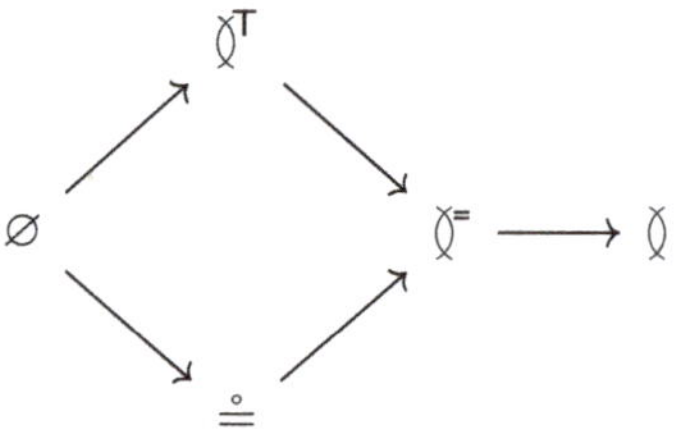

Figure 22.3: Summary of Lemmas 22.6.12 & 22.6.13

2. The equivalences in Lemma 22.6.2 lack the tidy conciseness of the logical presentation, but they invite the same question:

   What is the design space of possible antiseparation properties between points?

We sum our answer up in a short table in Figure 22.2, and we can use Lemmas 22.6.12 and 22.6.13 (proved below) to reformat Figure 22.2 diagrammatically as per Figure 22.3, where arrows indicate implication / relation inclusion.

22.6.2 Topological indistinguishability $\mathrel{\overset{\circ}{=}}$, and the empty relation $\varnothing$

In Lemma 20.2.1 we noted that in a semitopology $(\mathbf{P}, \mathsf{Open})$,

$$p \mathrel{\overset{\circ}{=}} p' \quad \text{holds precisely when} \quad \mathbf{P}, \mathsf{Open} \models p{\Leftrightarrow}p'.$$

We now show that $\mathbf{P}, \mathsf{Open} \models \mathsf{T}(p{\equiv}p')$ and $\mathbf{P}, \mathsf{Open} \models \mathsf{T}(p{\Leftrightarrow}p')$ yield the empty relation:

Lemma 22.6.5. *Suppose* $(\mathbf{P}, \mathsf{Open})$ *is a semitopology and* $p, p' \in \mathbf{P}$.[2] *Then*

$$\mathbf{P}, \mathsf{Open} \not\models \mathsf{T}(p{\equiv}p') \quad \textit{and} \quad \mathbf{P}, \mathsf{Open} \not\models \mathsf{T}(p{\Leftrightarrow}p')$$

always (even if $p = p'$*).*

Proof. We consider the valuation $\lambda x.\mathbf{b} : \mathbf{P} \to \mathbf{3}$. The reader can check that this is continuous, and from Figures 18.2 and 18.5

$$[\![\mathsf{T}(p{\equiv}p')]\!]_f = \mathbf{f} \quad \text{and} \quad [\![\mathsf{T}(p{\Leftrightarrow}p')]\!]_f = \mathbf{f}.$$

The result follows from Definition 18.4.1(6&1). □

[2] If the semitopology is empty then there is no $p \in \mathbf{P}$ to choose.

22.6.3 Consensus equivalence $\between^{=}$, hypertwined $\between^{\mathsf{T}}$, and hyperdefinite

Definition 22.6.6. Suppose $(\mathsf{P}, \mathsf{Open})$ is a semitopology and $p, p' \in \mathsf{P}$. Then:

1. By Lemma 22.6.2(3), p and p' are intertwined when for every $O \in \mathsf{Open}_{reg}$,

$$p, p' \in |O| \ \lor \ p, p' \in \mathsf{P} \setminus O.$$

Then we can define:

2. Call p and p' **consensus equivalent** and write $p \between^{=} p'$ when for every $O \in \mathsf{Open}_{reg}$,

$$p, p' \in O \ \lor \ p, p' \in interior(\mathsf{P} \setminus O) \ \lor \ p, p' \notin O \cup interior(\mathsf{P} \setminus O).$$

By elementary propositional manipulations we can write this equivalently as

$$(p \in O \iff p' \in O) \ \land \ (p \in interior(\mathsf{P} \setminus O) \iff p' \in interior(\mathsf{P} \setminus O)).$$

We may use these two forms synonymously without comment.

3. Call p and p' **hypertwined** and write $p \between^{\mathsf{T}} p'$ when for every $O \in \mathsf{Open}_{reg}$,

$$p, p' \in O \ \lor \ p, p' \in interior(\mathsf{P} \setminus O).$$

4. Recall from Definition 22.1.2 the notion of *definite* truth-values. Expanding on this, call p **hyperdefinite** when for every $O \in \mathsf{Open}_{reg}$,

$$p \in O \ \lor \ p \in interior(\mathsf{P} \setminus O).$$

5. We may write $p_{\between^{=}}$ and $p_{\between^{\mathsf{T}}}$ for the set of points that are consensus equivalent with and hypertwined with p respectively:

$$p_{\between^{=}} = \{p' \in \mathsf{P} \mid p \between^{=} p'\} \qquad p_{\between^{\mathsf{T}}} = \{p' \in \mathsf{P} \mid p \between^{\mathsf{T}} p'\}.$$

Remark 22.6.7. Definition 22.6.6 uses quantifications over regular open sets, but recall from Corollary 5.7.6 that taking closures and taking open interiors yield bijections between Open_{reg} and Closed_{reg}. Thus, reasonable alternative forms of the Definition are easy to write out with regular closed sets.

Example 22.6.8. Figure 22.4 illustrates a space with two points, 1 and 2, that are unconflicted, hyperregular, and hypertwined with each other; but they are not even quasiregular.

This means that they have good consensus behaviour in the sense that they always agree on an unambivalent (non-$\mathbf{b}$) truth-value for any extremal valuation, but they are not intertwined with any topen set of points. The 'problem' (if we wish to call it a problem) is that 1 and 2 will agree with each other and will agree with at least one of 0 or 3, but (intuitively) 1 and 2 can choose *which* of 0 or 3 to agree with when they decide their state.

We make a few more small observations:

- $p_{\between^{\mathsf{T}}}$ and $p_{\between^{=}}$ are not necessarily open. Consider $p = 0$ in the semitopology illustrated in Figure 5.2; then $p_{\between^{\mathsf{T}}} = \{0\} = p_{\between^{=}}$.
- $p_{\between^{\mathsf{T}}}$ and $p_{\between^{=}}$ are not necessarily a singleton. Consider $p = 1$ in the semitopology illustrated in Figure 22.4; $p_{\between^{\mathsf{T}}} = \{1, 2\} = p_{\between^{=}}$.

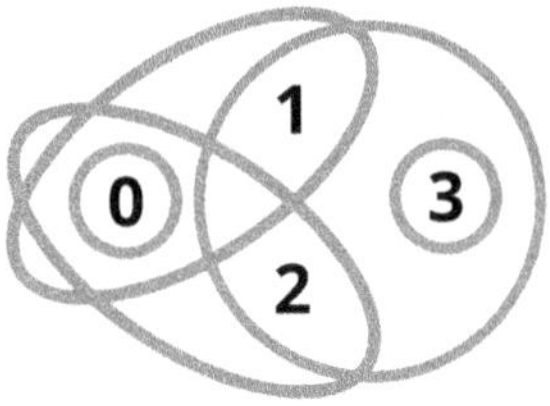

Figure 22.4: 1 and 2 are hypertwined and hyperdefinite, but they are not even quasiregular (Example 22.6.8)

- $p_{\between^{\top}}$ and $p_{\between^{=}}$ do not necessarily have a nonempty open interior. Again, consider $p = 1$ in the semitopology illustrated in Figure 22.4; $interior(p_{\between^{\top}}) = \varnothing = p_{\between^{=}}$.

If p is regular then things are simpler: see Proposition 22.6.19 and Remark 22.6.20.

LEMMA 22.6.9. *For a semitopology* $(\mathsf{P}, \mathsf{Open})$ *and* $p, p' \in \mathsf{P}$, *the following are equivalent:*

1. *$p \between^{\top} p'$ (Definition 22.6.6(3)).*
2. *$p \between^{=} p'$ and at least one of p and p' is hyperdefinite.*
3. *$p \between^{=} p'$ and both p and p' are hyperdefinite.*

We can write:

hypertwined = consensus equivalent + hyperdefinite.

Proof. Routine from Definition 22.6.6(2&3). □

By Definition 22.6.6(4) we call p *hyperdefinite* when for every regular $O \in \mathsf{Open}_{reg}$, $p \in O$ or $p \in interior(\mathsf{P} \setminus O)$. By Definition 6.5.4 we call p *hypertransitive* when for every $O, O' \in \mathsf{Open}$, if $p \in |O| \cap |O'|$ then $O \between O'$. Remarkably, these two conditions are equivalent:

PROPOSITION 22.6.10. *For a semitopology* $(\mathsf{P}, \mathsf{Open})$ *and* $p, p' \in \mathsf{P}$, *the following are equivalent:*

1. *p is hyperdefinite in the sense of Definition 22.6.6(4) ($p \in O$ or $p \in interior(\mathsf{P} \setminus O)$ for every $O \in \mathsf{Open}_{reg}$).*
2. *p is hypertransitive in the sense of Definition 6.5.4.*

We can write:

Hyperdefinite = hypertransitive.

Proof. We prove two implications:

- Suppose p is hypertransitive and consider $O \in \mathsf{Open}_{reg}$. Write $O' = interior(\mathsf{P} \setminus O)$; note by Lemma 5.7.7 that $\mathsf{P} \setminus O$ is also regular, so that $|O'| = \mathsf{P} \setminus O$ and $O \cup O' = \mathsf{P}$. By construction $O \not\between O'$, and it follows by the contrapositive of Lemma 6.5.5(3) that $p \in O$ or $p \in O'$ as required.

- Suppose p is hyperdefinite and consider $O, O' \in \mathsf{Open}_{reg}$ such that $p \in |O| \cap |O'|$. Set $R' = interior(\mathsf{P} \setminus O)$. By assumption (since p is hyperdefinite) $p \in O$ or $p \in R'$.

  If $p \in R'$ then by Definition 5.1.2(1) (because $p \in |O|$) also $R' \between O$, which is impossible by construction, so $p \notin R'$.

  Therefore it must be that $p \in O$. Then by Definition 5.1.2(1) (because $p \in |O'|$) we have $O \between O'$ as required. □

COROLLARY 22.6.11. *For a semitopology* $(\mathsf{P}, \mathsf{Open})$ *and* $p \in \mathsf{P}$, *the following are equivalent:*

1. *p is hyperdefinite.*
2. *p is hypertransitive.*
3. $p \between^{\mathsf{T}} p$.
4. $p \between^{\mathsf{T}} p'$ *for any* $p' \in \mathsf{P}$.

Proof. Routine from Definition 22.6.6(2&3) and Proposition 22.6.10. □

LEMMA 22.6.12. *Suppose* $(\mathsf{P}, \mathsf{Open})$ *is a semitopology and* $p, p' \in \mathsf{P}$. *Then:*

1. *If* $p \between^{\mathsf{T}} p'$ *then* $p \between^{=} p'$. *The converse implication need not hold.*
2. *If* $p \between^{=} p'$ *then* $p \between p'$. *The converse implication need not hold.*

In symbols, we can write:

$$p_{\between^{\mathsf{T}}} \subseteq p_{\between^{=}} \subseteq p_{\between}.$$

Proof. The implications are routine from Definition 22.6.6, using Lemma 5.1.3(2) and arguments on sets.

For the non-implications it suffices to provide counterexamples.

1. Consider the semitopology **3** from Definition 18.1.2 Set $p = p' = \mathbf{b}$. The reader can check that $p \between^{=} p'$ but $\neg(p \between^{\mathsf{T}} p')$.[3]
2. Consider the semitopology **3** again, but now set $p = \mathbf{t}$ and $p' = \mathbf{b}$. The reader can check that $p \between p'$ but $\neg(p \between^{=} p')$. □

Topological indistinguishability is stronger than consensus equivalence, but it is incomparable with being hypertwined:

LEMMA 22.6.13. *Suppose* $(\mathsf{P}, \mathsf{Open})$ *is a semitopology and* $p, p' \in \mathsf{P}$. *Then:*

1. *If* $p \circeq p'$ *then* $p \between^{=} p'$. *The converse implication need not hold.*
2. *The properties* $p \circeq p'$ *and* $p \between^{\mathsf{T}} p'$ *are incomparable. That is:*
   a) *p and p' can be topologically indistinguishable but not hypertwined.*
   b) *p and p' can be hypertwined but not topologically indistinguishable.*

Proof. Suppose p and p' are topologically indistinguishable. Then it is routine to check Definition 22.6.6(2) and see that they are consensus equivalent.

For the non-implications, it suffices to provide counterexamples.

[3]If the reader finds it a little confusing to think about continuous valuations from **3** to itself, use the semitopology illustrated in Figure 3.1, top-left diagram, instead. This is isomorphic, but instead of **f**, **b**, **t** we have 0, 1, and 2.

- Consider the semitopology **3** from Definition 18.1.2. Set $p = p' = \mathbf{b}$. Clearly, p and p' are topologically indistinguishable; however $\neg(p \between^{\mathsf{T}} p')$.
- Consider the Sierpiński space $\mathsf{P} = \{0,1\}$ and $\mathsf{Open} = \{\varnothing, \{1\}, \{0,1\}\}$ (cf. Example 5.8.4 and Remark 18.1.1).
  Set $p = 0$ and $p' = 1$; these are topologically distinguishable since $p' \in \{1\}$ and $p \notin \{1\}$. But, there are only two extremal valuations — $\lambda x.0$ and $\lambda x.1$ — and it follows that $p \between^{=} p'$ and also $p \between^{\mathsf{T}} p'$. $\square$

Lemma 22.6.14. *For a semitopology* $(\mathsf{P}, \mathsf{Open})$ *and* $p, p' \in \mathsf{P}$, *the following are equivalent:*

1. $p \between^{=} p'$ *(Definition 22.6.6(2)).*
2. $\mathsf{P}, \mathsf{Open} \overset{\mathsf{X}}{\models} p \Leftrightarrow p'$ *(Definition 22.1.2(6)).*
3. $f(p) = f(p')$ *for every extremal* $f : \mathsf{P} \to \mathbf{3}$.

Proof. For equivalence of parts 1 and 2 we prove two implications:

- *Suppose* $p \between^{=} p'$.
  Suppose $f : \mathsf{P} \to \mathbf{3}$ is an extremal valuation. By Definition 22.6.6(2), $p \in f^{-1}\{\mathbf{f}\} \iff p' \in f^{-1}\{\mathbf{f}\}$. By Corollary 18.5.2 $f \models p \Leftrightarrow p'$. Since f was arbitrary, by Definition 22.1.2(6) $\mathsf{P}, \mathsf{Open} \overset{\mathsf{X}}{\models} p \Leftrightarrow p'$ as required.
- *Suppose* $\mathsf{P}, \mathsf{Open} \overset{\mathsf{X}}{\models} p \Leftrightarrow p'$.
  Consider some regular open set $O \in \mathsf{Open}_{reg}$. Set
  $$f = \delta_{O_{\mathbf{t}}, O_{\mathbf{f}}}(O, interior(\mathsf{P} \setminus O)) \quad \text{and} \quad f' = \delta_{O_{\mathbf{t}}, O_{\mathbf{f}}}(interior(\mathsf{P} \setminus O), O).$$
  By Proposition 22.3.6(1&3) and Corollary 22.3.5 these are both extremal valuations, and from Corollary 18.5.2 it follows that $p \in O \iff p' \in O$ and $p \in interior(\mathsf{P} \setminus O) \iff p' \in interior(\mathsf{P} \setminus O)$. By Definition 22.6.6(2) $p \between^{=} p'$ as required.

Equivalence of parts 1 and 3 is by routine calculations from Definition 22.6.6(2) and Remark 22.3.2. $\square$

Lemma 22.6.15. *For a semitopology* $(\mathsf{P}, \mathsf{Open})$ *and* $p \in \mathsf{P}$, *the following are equivalent:*

1. *p is hyperdefinite (Definition 22.6.6(4)).*
2. $\mathsf{P}, \mathsf{Open} \overset{\mathsf{X}}{\models} \mathsf{T}p \vee \mathsf{T}\neg p$.
3. $\mathsf{P}, \mathsf{Open} \not\models^{X} \mathsf{B}p$.
4. $f(p) \in \{\mathbf{t}, \mathbf{b}\}$ *for every extremal* $f : \mathsf{P} \to \mathbf{3}$.

Proof. By routine calculations using Remark 22.3.2 and using the truth-tables in Figure 18.2. $\square$

Lemma 22.6.16. *For a semitopology* $(\mathsf{P}, \mathsf{Open})$ *and* $p, p' \in \mathsf{P}$, *the following are equivalent:*

1. $p \between^{\mathsf{T}} p'$ *(Definition 22.6.6(3)).*
2. $\mathsf{P}, \mathsf{Open} \overset{\mathsf{X}}{\models} \mathsf{T}(p \Leftrightarrow p')$.
3. $\mathsf{P}, \mathsf{Open} \overset{\mathsf{X}}{\models} \mathsf{T}(p \equiv p')$.

4. $f(p) = f(p') \in \{\mathrm{t}, \mathrm{b}\}$ *for every extremal* $f : \mathsf{P} \to \mathbf{3}$.

Proof. By routine calculations combining Lemmas 22.6.15, 22.6.14, and 22.6.9 and using the truth-tables in Figure 18.2. □

REMARK 22.6.17. It may be helpful to sum up the high points of the results above (this summary is for intuition; references to the precise results are included):

1. Consensus equivalence $\between^{=}$ is when two points return the same truth-value in $\mathbf{3}$ for all extremal valuations. This is an equivalence relation, because equality is an equivalence relation.
2. Being *hypertwined* $\between^{\mathsf{T}}$ is the partial equivalence relation obtained by restricting consensus equivalence to the *hyperdefinite points* (those that return t or $\mathbf{f}$ in all extremal valuations; never b).
3. Being hyperdefinite is the same as being hypertransitive, and also the same as being hypertwined with yourself or with any other point, and the same as validating $\mathsf{P}, \mathsf{Open} \models \mathsf{T}p \vee \mathsf{T}\neg p$ (Corollary 22.6.11).
4. So in answer to our question of Remark 22.6.4, the design space is populated by four distinct entities:

   a) *Topological indistinguishability* $\circeq$. By Lemma 20.2.1 this lives naturally in the world of possibly non-extremal valuations as $\mathsf{P}, \mathsf{Open} \models p \Leftrightarrow p'$; see Figure 22.2. $p \circeq p'$ when $f(p) = f(p')$ for every continuous valuation. This relation is, of course, very familiar from topology; a space is T_0 precisely when $\circeq$ coincides with $=$.
   b) Being *intertwined* $\between$. This lives both in the world of extremal and possibly non-extremal valuations, as $\mathsf{P}, \mathsf{Open} \overset{\mathsf{X}}{\models} p \equiv p'$ and $\mathsf{P}, \mathsf{Open} \models p \equiv p'$ respectively, as per Figure 22.2. We have studied $\between$ very closely, but there is also another one:
   c) Being *consensus equivalent* $\between^{=}$. This lives naturally in the world of extremal valuations as $\mathsf{P}, \mathsf{Open} \overset{\mathsf{X}}{\models} p \Leftrightarrow p'$, as per Figure 22.2. $p \between^{=} p'$ holds when $f(p) = f(p')$ for every *extremal* valuation f (i.e. like topological indistinguishability, but only for extremal valuations). This is a very natural notion.
   d) Being *hyperdefinite / hypertransitive*. Remarkably, and like being intertwined, this lives naturally in both worlds, as per Proposition 22.6.10.

REMARK 22.6.18. Remark 22.6.17 invites the question of whether other relevant regularity or intertwinedness properties exist?[4] Yes — and in fact we can note examples of both.

A regularity property. In Theorem 5.6.2(3) we showed that p is regular when p is weakly regular and $p_{\between}$ is a minimal closed neighbourhood. If we write $MCN(p)$ for the property that $p_{\between}$ is a minimal closed neighbourhood, then we can write Theorem 5.6.2 as

'regular = weakly regular + MCN'

following the style of other results of this genre such as Theorem 6.2.2 for 'regular = weakly regular + unconflicted' and Theorem 6.5.8 for 'regular = quasiregular + hypertransitive'. Now note that MCN does not coincide with being unconflicted (the points in Figure 5.2 are unconflicted but not MCN). And, MCN does not coincide with being hypertransitive (the points 1

[4] The list so far: (quasi/weak/indirect)-regularity; intertwined, consensus equivalent, hypertwined; and being unconflicted, hyperdefinite, hypertransitive, and strongly compatible.

and 2 in Figure 22.4 are hypertwined but not MCN). So MCN is ... another well-behavedness property. It is an open problem how MCN fits in with the well-behavedness properties that we have considered so far.[5]

An intertwinedness property. In Lemma 3.6.4 we noted that 'being intertwined' $\between$ is symmetric and reflexive, but not necessarily transitive, and it became clear that special cases where $\between$ *is* transitive are of particular interest. But, we can also directly study $\between^*$, the transitive closure of the intertwined-with relation $\between$: that is, call p and p' **transitively intertwined** and write $p \between^* p'$ when there exist points $p_1, \ldots, p_n$ such that $p \between p_1 \between \cdots \between p_n \between p'$. This transitive closure is another intertwinedness property with its own properties [Gab26].

These are just some, of presumably many, such questions that remain to be considered.

We conclude with a brief investigation of the connections between $\between^{\mathsf{T}}$, regularity, and $K(p)$. This is just a routine application of the tools we have already built:

PROPOSITION 22.6.19. *Suppose* $(\mathbf{P}, \mathsf{Open})$ *is a semitopology and* $p, p' \in \mathbf{P}$. *Then:*

1. *If there exists a topen* $T \in \mathsf{Topen}$ *such that* $p, p' \in T$, *then* $p \between^{\mathsf{T}} p'$.
2. p *is regular if and only if* $K(p) = p_{\between^{\mathsf{T}}} \neq \varnothing$.
3. p *is regular if and only if* $K(p) \neq \varnothing$ *and* $p_{\between^{\mathsf{T}}} \neq \varnothing$.

Proof. We consider each part in turn:

1. Suppose there exists a topen $T \in \mathsf{Topen}$ such that $p, p' \in T$. By Proposition 22.5.1(1&2) $f(p) = f(p') \in \{\mathbf{t}, \mathbf{f}\}$. We use Lemma 22.6.16(4).

2. Suppose p is regular, meaning by Definition 4.1.4(3) that $p \in K(p) \in \mathsf{Topen}$. We prove two subset inclusions:

   - By part 1 of this result $K(p) \subseteq p_{\between^{\mathsf{T}}}$.
   - Set $f = \delta_{O_{\mathbf{t}} C_{\mathbf{tb}}}(K(p), p_{\between})$. By Corollary 5.7.4 $K(p) \in \mathsf{Open}_{reg}$, and from Remark 22.3.7(3) this is an extremal valuation. It follows that $p_{\between^{\mathsf{T}}} \subseteq K(p)$.

   Suppose $K(p) = p_{\between^{\mathsf{T}}} \neq \varnothing$. By Definition 4.1.4(5) p is quasiregular, and by Corollary 22.6.11(4) p is hypertransitive. We use Theorem 6.5.8.

3. Suppose p is regular. Then we use part 2 of this result and we are done.

   Suppose $K(p) \neq \varnothing$ and $p_{\between^{\mathsf{T}}} \neq \varnothing$. By Definition 4.1.4(5) p is quasiregular, and by Corollary 22.6.11(4) p is hypertransitive. We just use Theorem 6.5.8. □

REMARK 22.6.20. By Proposition 22.6.19(2) and Lemma 22.6.12, if p is regular then

$$K(p) = p_{\between^{=}} = p_{\between^{\mathsf{T}}}.$$

If p is not regular then the equalities need not hold. The semitopology illustrated in Figure 22.4 gives an example, where $K(1) = \varnothing$ but $1_{\between^{\mathsf{T}}} = \{1, 2\}$.

[5] Beyond a few simple comments, e.g. it can be proved that MCN implies indirect regularity (Definition 9.3.2).

Part IV

Conclusions

Conclusions 23

We started by noticing that a notion of 'actionable coalition' as discussed in the Introduction, suggests a topology-like structure which we call *semitopologies*.

We simplified and purified our motivating examples to two mathematical questions:

1. understand antiseparation properties, and
2. understand the implications of these for value assignments.[1]

We have surveyed the implications of these ideas and seen that they are mathematically rich and varied. Point-set semitopologies have an interesting theory, and a family of results which resemble those of point-set topology, but which are different enough to have their own distinct character. We build and study a dual category of semiframes; and we provide a many-valued modal logic for describing them.

Generalising topology to drop the condition that intersections of open sets must be open, brings a wealth of new and interesting structure. We have considered many results, but we also hope that putting this story together will serve as a stimulus to considering semitopologies as a new field of research.

23.1 Topology vs. semitopology

We briefly compare and contrast topology and semitopology:

1. *Topology:* Topology considers a wealth of separation properties, but we are not aware of a taxonomy of anti-separation properties in the topological literature.[2]

   Semitopology: We consider a taxonomy of antiseparation properties, including: points being intertwined (see Definition 3.6.1 and Remark 3.6.10) and hypertwined and consensus

[1] A value assignment is just a not-necessarily-continuous map from a semitopology to a discrete space (Definition 2.1.3(2)).

[2] The Wikipedia page on separation axioms [Wik24e] includes an excellent overview with over a dozen separation axioms; no anti-separation axioms are proposed. Important non-Hausdorff spaces do exist; e.g. the *Zariski topology* [Hul03, Subsection 1.1.1].

equivalent (see the overviews in Figures 22.2 and 22.3); and points being quasiregular, indirectly regular, weakly regular, and regular (Definitions 4.1.4 and 9.3.2), (un)conflicted (Definition 6.1.1(2)), and hypertransitive (Definition 6.5.4).[3]

2. *Topology:* If a minimal open neighbourhood of a point exists then it is least, because we can intersect two minimal neighbourhoods to get a smaller one which by minimality is equal to both.

   Yet, in topology the existence of a least open neighbourhood is not guaranteed (e.g. $0 \in \mathbb{R}$ has no least open neighbourhood).

   Semitopology: A point may have multiple minimal open neighbourhoods — examples are very easy to generate, see e.g. the top-right example in Figure 3.1. Furthermore, in the useful special case of a chain-complete semitopology, every open neighbourhood of p contains a(t least one) minimal open neighbourhood of p (Corollary 9.5.4) so that existence of minimal open neighbourhoods is assured.

3. *Topology:* Every finite T_0 topology is sober. A topology is sober if and only if every nonempty irreducible closed set is the closure of a unique point.

   Semitopology: Neither property holds. See Lemma 14.3.8.

4. *Topology:* We are typically interested in functions on topologies that are continuous (or mostly so, e.g. $f(x) = 1/x$). Thus for example, the definition of Top the category of topological spaces takes continuous functions as morphisms, essentially building in assumptions that continuous functions are of most interest and that finding them is enough of a solved problem that we can restrict to continuous functions in the definition.

   Semitopology: For our intended application to consensus, we are explicitly interested in functions that may be discontinuous. This models initial and intermediate states where local consensus has not yet been achieved, or final states on semitopologies that include disjoint topens and non-regular points (e.g. conflicted points), as well as adversarial or failing behaviour. Thus, having continuity is neither a solved problem, nor even necessarily desirable.

5. Sometimes, definitions from topology transfer to semitopology but split into multiple distinct notions when they do. For example: topology has one notion of *dense subset of* and, as discussed in Remark 11.5.1, when we transfer this to semitopologies it splits into two notions — *weakly* dense and *strongly* dense (Definition 11.1.2) — both of which turn out to be important.

[3] An extra word on the converse of this: Our theory of semitopologies admits spaces whose points partition into distinct communities, as discussed in Theorem 3.5.4 and Remark 3.5.5. To a professional blockchain engineer it might seem terrible if two points points are *not* intertwined, since this means they might not be in consensus in a final state. Should this not be excluded by the definition of semitopology, as is done in the literature on quorum systems, where it is typically definitionally assumed that all quorums in a quorum system intersect? No! Separation is a fact of life which we permit not only so that we can mathematically analyse it (and we do), but also because we may need it for certain *normal situations*. For example, most blockchains have a *mainnet* and several *testnets* and it is understood that each should be coherent within itself, but different nets *need not* be in consensus with one another. Indeed, if the mainnet had to agree with a testnet then this would likely be a bug, not a feature. So the idea of having multiple partitions is nothing new *per se*. It is a familiar idea, which semitopologies put in a powerfully general mathematical context.

Sometimes, ideas that come from semitopology project down to topology but may lose impact in doing so; they make mathematical sense, but become less interesting, or at least lose finesse. For example: our theory of semitopologies considers notions of *topen set* and *strongly topen set* (Definitions 3.2.1 and 3.7.5). In topology these are equivalent to one another, and to a known and simpler topological property of being *hyperconnected* (Definition 3.7.12).[4] Something similar happens with the semitopological notion of *strongly dense for*; see the discussion in Remark 11.4.7.

6. A natural space of functions for describing topologies is continuous maps to the Sierpiński space. The natural space of functions for describing semitopologies seems to be possibly non-continuous maps to $\mathbf{3}$.

7. Semitopological questions such as *'is this a topen set'* or *'are these two points intertwined'* or *'does this point have a topen neighbourhood'* — and many other definitions, such as our taxonomy of points into *regular*, *weakly regular*, *indirectly regular*, *quasiregular*, *unconflicted*, and *hypertransitive*; or the notions of *witnessed set*, and *kernel* ... and so on — appear to be novel.

   Also in the background is that we are particularly interested in properties and algorithms that work well using local and possibly incomplete or even partially incorrect information.

   Thus semitopologies have their own distinct character: because they are mathematically distinct, and because modern applications having to do with actionable coalitions motivate us to ask questions that have not necessarily been considered before.

23.2 Related work

Union sets, closure spaces, and minimal structures There is a thread of research into *union-closed families*; these are subsets of a finite powerset closed under unions, so that a union-closed family is precisely just a finite semitopology. The motivation is to study the combinatorics of finite subsemilattices of a powerset. Some progress has been made in this [Poo92]; the canonical reference for the relevant combinatorial conjectures is the 'problem session' on page 525 (conjectures 1.9, 1.9', and 1.9") of [Riv85]. See also recent progress in a conjecture about union-closed families.[5] There is no direct connection to semitopologies, and certainly no consideration of duality results. Perhaps the duality that we present may be of some interest in that community.

A *closure space* is a subset of a powerset that is closed under intersections [Ern09, page 173], so that the set of complements of a closure space is just a semitopology (and likewise a finite closure space is, up to taking sets complements, just a union-closed family). The motivation for closure spaces is, as the name suggests, to study closure operations in a topology-flavoured style, so closure spaces (unlike union-closed sets) share a topological flavour with semitopologies.

It is not clear that semitopologies are 'just' closure spaces, union sets, or complete semilattices (see also Remark 2.1.7). This is because of how the theory of semitopologies (e.g. the basic notion of intertwined points) uses sets intersection $\between$. Semiframes (the dual to semitopologies)

[4] ...but (strong) topens are their own thing. Analogy: a projection from $\mathbb{C}$ to $\mathbb{R}$ maps $a + bi$ to a; this is not evidence that i is equivalent to 0!

[5] https://web.archive.org/web/20230330170701/https://en.wikipedia.org/wiki/Union-closed_sets_conjecture#Partial_results.

make this explicit with the compatibility relation $*$ (Definitions 12.2.1 and 12.3.1), demonstrating that semitopologies are in fact *compatible* complete sublattices of a powerset. This seems to be a new idea. And of course: the applications are new.

A *minimal structure* on a set X is a subset of $pow(X)$ that contains $\varnothing$ and X. Thus a semitopology is a minimal structure that is also closed under arbitrary unions. There is a thread of research into minimal structures, studying how notions familiar from topology (such as continuity) fare in weak (minimal) settings [PN01] and how this changes as axioms (such as closure under unions) are added or removed. An accessible discussion is in [Szá07], and see the brief but comprehensive references in Remark 3.7 of that paper. Of course our focus is on properties of semitopologies which are not considered in that literature; but we share an observation with minimal structures that it is useful to study topology-like constructs, in the absence of closure under intersections.

Gradecast converges on a topen Many consensus algorithms have the property that once consensus is established in a quorum O, it propagates to $|O|$. For example, in the Grade-Cast algorithm [FM88], participants assign a confidence grade of 0, 1 or 2 to their output and must ensure that if any participant outputs v with grade 2 then all must output v with grade at least 1. If all the quorums of a participant intersect some set S that unanimously supports value v, then the participant assigns grade at least 1 to v.

From the view of our paper, this is just taking a closure, which suggests that, to convince a topen to agree on a value, it would suffice to first convince an open neighbourhood that intersects the topen, and then use Grade-Cast to convince the whole topen. More on this in Proposition 5.3.2 and Remark 5.3.3.

Algebraic topology as applied to distributed computing tasks The reader may know that solvability results about distributed computing tasks have been obtained from algebraic topology, starting with the impossibility of wait-free k-set consensus and the Asynchronous Computability Theorem [HS93, BG93a, SZ93] in 1993. See [HKR13] for numerous such results.

The basic observation is that the set of states of a distributed algorithm forms a simplicial complex, called its *protocol complex*, and topological properties of this complex, like connectivity, are constrained by the underlying communication and fault model. These topological properties in turn can determine what tasks are solvable. For example: every algorithm in the wait-free model with atomic read-write registers has a connected protocol complex, and because the consensus task's output complex is disconnected, consensus in this model is not solvable [HKR13, Chapter 4].

This work is also topological, but in a different way: we use (semi)topologies to study consensus in and of itself, rather than the solvability of consensus or other tasks in particular computation models. Put another way: the papers cited above use topology to study the solvability of distributed tasks, but this work shows how the very idea of 'distribution' can be viewed as having a semitopological foundation.

Of course we can imagine that these might be combined — that in future work we may find interesting and useful things to say about the topologies of distributed algorithms when viewed as algorithms *on* and *in* a semitopology. See also the discussion of 'algebraic semitopology' in Remark 23.3.10.

Fail-prone systems and quorum systems Given a set of processes P, a *fail-prone* system [MR98] (or *adversary structure* [HM00]) is a set of *fail-prone sets* $\mathcal{F} = \{F_1, ..., F_n\}$ where, for every $1 \leq i \leq n$, $F_i \subseteq \mathsf{P}$. $\mathcal{F}$ denotes the assumptions that the set of processes that will fail (potentially maliciously) is a subset of one of the fail-prone sets. A *dissemination quorum system* for $\mathcal{F}$ is a set $\{Q_1, ..., Q_m\}$ of quorums where, for every $1 \leq i \leq m$, $Q_i \subseteq \mathsf{P}$, and such that

- for every two quorums Q and Q' and for every fail-prone set F, $(Q \cap Q') \setminus F \neq \emptyset$ and
- for every fail-prone set F, there exists a quorum disjoint from F.

Several distributed algorithms, such as Bracha Broadcast [Bra87] and PBFT [CL02], rely on a quorum system for a fail-prone system $\mathcal{F}$ in order to solve problems such as reliable broadcast and consensus assuming (at least) that the assumptions denoted by $\mathcal{F}$ are satisfied.

Several recent works generalise the fail-prone system model to heterogeneous systems. Under the failure assumptions of a traditional fail-prone system, Bezerra et al. [BKK22] study reliable broadcast when participants each have their own set of quorums. Asymmetric Fail-Prone Systems [ACTZ24] generalise fail-prone systems to allow participants to make different failure assumptions and have different quorums. In Permissionless Fail-Prone Systems [CLZ23], participants not only make assumptions about failures, but also make assumptions about the assumptions of other processes; the resulting structure seems closely related to witness semi-topologies, but the exact relationship still needs to be elucidated.

In Federated Byzantine Agreement Systems [Maz15], participants declare quorum slices and quorums emerge out of the collective quorum slices of their members. Quorum slices are a special case of the notion of witness-set in Definition 8.2.2(2). García-Pérez and Gotsman [GPG18] rigorously prove the correctness of broadcast abstractions in Stellar's Federated Byzantine Agreement model and investigate the model's relationship to dissemination quorum systems. The Personal Byzantine Quorum System model [LGM19] is an abstraction of Stellar's Federated Byzantine Agreement System model and accounts for the existence of disjoint consensus clusters (in the terminology of the paper) which can each stay in agreement internally but may disagree between each other. Consensus clusters are closely related to the notion of topen in Definition 3.2.1(2).

Sheff et al. study heterogeneous consensus in a model called Learner Graphs [SWRM21] and propose a consensus algorithm called Heterogeneous Paxos.

Cobalt, the Stellar Consensus Protocol, Heterogeneous Paxos, and the Ripple Consensus Algorithm [Mac18, Maz15, SWRM21, SYB14] are consensus algorithms that rely on heterogeneous quorums or variants thereof. The Stellar network [LLM+19] and the XRP Ledger [SYB14] are two global payment networks that use heterogeneous quorums to achieve consensus among an open set of participants; the Stellar network is an instance of a witness semitopology.

Quorum systems and semitopologies are not the same thing. Quorum systems are typically taken to be such that all quorums intersect (in our terminology: they are *intertwined*), whereas semitopologies do not require this. On the other hand, quorums are not always taken to be closed under arbitrary unions, whereas semitopologies are (see the discussion in Example 2.1.4(7)).

The literature on fail-prone systems and quorum systems is most interested in synchronisation algorithms for distributed systems and has been less concerned with their deeper mathematical structure. Some work [LGM19] gets as far as proving an analogue to Proposition 3.5.2 (though it seems fair to say that the semitopological presentation is simpler and clearer), but it fails to notice the connection with topology and there is no consideration of algebra.

Dualities We discussed duality results in detail in Remark 15.4.3. The reader may know that there are a great many such results, starting with Stone's classic duality between Boolean algebras and compact Hausdorff spaces with a basis of clopen sets [Sto36, Joh86]. A nice representation result for semilattices (not the compatible semilattices we consider here) is in [Bre84]. The duality between frames and topologies is described in [MM92, page 479, Corollary 4]. See also the encyclopedic treatment in [Car11], with an overview in Example 2.9 on page 17. See also a recent accessible text with clear exposition in [GvG24].

Our duality between semiframes and semitopologies fits into this canon.

(Semi)lattices with extra structure We are not aware of semiframes having been studied in the literature, but they are in excellent company, in the sense that things have been studied that are structurally similar. We mention two examples to give a flavour of this extensive literature:

1. A **quantale** is a complete lattice $(\mathsf{Q}, \bigvee)$ with an associative *multiplication* operation $* : (\mathbf{Q} \times \mathbf{Q}) \to \mathbf{Q}$ that distributes over $\bigvee$ in both arguments [Ros90]. A commutative quantale whose multiplication is restricted to map to either the top or bottom element in Q is close being a semiframe.[6] For reference, a pleasingly simple representation result for quantales is given in [BG93b].

2. An **overlap algebra** is a complete Heyting algebra $\mathbf{X}$ with an *overlap relation* $\between \subseteq \mathbf{X} \times \mathbf{X}$ whose intuition is that $x \between y$ when $x \wedge y$ is *inhabited.* The motivation for this comes from constructive logic, in which $\exists p.(p \in x \wedge p \in y)$ is a different and stronger statement than $\neg\forall p.\neg(p \in x \wedge p \in y)$. Accordingly, overlap algebras are described as 'a constructive look at Boolean algebras' [CC20].

   Overlap algebras are not semiframes, but they share an idea with semiframes in making a structural distinction between 'intersect' and 'have a non-empty join'.

23.3 Future work

The list of things that we have *not* done is longer than the list of things that we have. We see this as a feature, not a bug: it suggests that we may have tapped a rich vein of possible future research.

Here are a few comments and ideas. This list is in no particular order and it is not exhaustive!

REMARK 23.3.1 (Theory of Byzantine behaviour). Real networks are subject to Byzantine behaviour — participants that don't follow the rules, e.g. through hostile intent, error, or communication difficulties. Thus, a participant may fall silent due to a communications outage, or a deliberately hostile participant may misreport their view of the network in order to 'invent' or sabotage action and so influence outcomes.

We have asked the question "is p intertwined with p'" but not other important questions like:

- "*How* intertwined are they; what is the minimal number of nodes to corrupt that would split p apart from p'?", or

[6] But not quite! We also need proper reflexivity (Definition 12.2.1(2)), and quantale morphisms do not necessarily map the top element to the top element like semiframe morphisms should (Definitions 12.1.4 and 15.2.1(1)).

- "What conditions can we put on a witness function to guarantee that changing the witness function at one point p will not change $ker(p')$ for any $p' \neq p$?".

Thus at a high level, given a semitopology $(\mathsf{P}, \mathsf{Open})$ we are interested in asking how properties range over an 'ϵ-ball' of perturbed semitopologies — as might be caused by various possible non-standard behaviours from a limited number of Byzantine points — and in particular we are looking for criteria to guarantee that appropriately-chosen good properties be preserved under such perturbation. This exciting and commercially relevant field of research remains to be explored.

REMARK 23.3.2 (Performance). We have considered antiseparation properties such as two points being intertwined, or a space being regular, and we have provided logical specifications of these that could be checked by a SAT solver. It remains to develop optimised algorithms that are quicker on practically relevant use-cases. We speculate on one such algorithm in Remark 21.1.7, but this is just a start. Following up on this and on other algorithms is future work.

REMARK 23.3.3 (Other notions of morphism). In Definition 15.1.1(1) we take a morphism of semitopologies $f : (\mathsf{P}, \mathsf{Open}) \to (\mathsf{P}', \mathsf{Open}')$ to be a continuous function $f : \mathsf{P} \to \mathsf{P}'$.[7]

The reader may be familiar with conditions on maps between topologies other than continuity, such as being *open* (f maps open sets to open sets) and *closed* (f maps closed sets to closed sets).

These conditions also make sense in semitopologies, and furthermore semiframe and graph representations of semitopologies suggests a further design space, that includes conditions on sets intersections and strict inclusions. We briefly list some of the conditions that we could impose on $f : \mathsf{P} \to \mathsf{P}'$:

1. If $O \between O'$ then $f^{-1}(O) \between f^{-1}(O')$. (It is automatic that if $f^{-1}(O) \between f^{-1}(O')$ then $O \between O'$, but the reverse implication is a distinct condition.)
2. If $O \subsetneq O'$ then $f^{-1}(O) \subsetneq f^{-1}(O')$.
3. $O \between O'$ and $f^{-1}(O) \subseteq f^{-1}(O')$ implies $O \subseteq O'$.
   If we write this as a contrapositive — $O \between O'$ and $O \nsubseteq O'$ implies $f^{-1}(O) \nsubseteq f^{-1}(O')$ — then we see the connection to the subintersection relation from Definition 17.2.3.

See also related discussions in Remarks 13.1.3 and 13.1.6, and in Remark 17.2.11.

REMARK 23.3.4 (Exponential spaces). It remains to check whether the category $\mathsf{SemiFrame}$ of semiframes is closed [Mac71, page 180, Section VII.7], or Cartesian.[8] We have checked that the category of semitopologies is Cartesian (it is), but it remains to check whether it is closed.

It also remains to look into the *Vietoris* (also called the *exponential*) semitopologies [Eng89, Exercise 2.7.20, page 120]. In view of our use of **3** to develop a logic for expressing properties of semitopologies, an exponential semitopology based on a many-valued domain might also be relevant. More generally, it remains to consider functors of the from $Hom(\text{-}, B)$ and $Hom(A, \text{-})$, for different values of A and B.

REMARK 23.3.5 (Computational/logical behaviour). Semiframes stand as objects of mathematical interest in their own right (just as frames do) but the original motivation for them comes from semitopologies. It might therefore be useful to think about 'computable' semiframes.

[7] Correspondingly, in Definition 15.2.1(1) we take a morphism of semiframes $g : (\mathsf{X}', \leq', *') \to (\mathsf{X}, \leq, *)$ to be a compatible morphism of complete semilattices.

[8] It would be surprising if it were not Cartesian.

What this would mean is not entirely clear at the moment, but of course this is what would make it research. One possibility is to develop a theory of logic within semiframes. On this topic, we can recall the discussion so far, and note that semiframes support a complementation operation $x^c = \bigvee\{x' \mid \neg(x' * x)\}$, so it is clearly possible to interpret propositional logic in a semiframe (implication would be $x \to y = x^c \mathbf{\vee} y$).

REMARK 23.3.6 (Finiteness and compactness). The relation of semitopologies to finiteness is interesting. On the one hand, our motivating examples are finite because they exist in the real world. On the other hand participants cannot depend on an exhaustive search of the full system being practical (or even permitted — this could be interpreted as a waste of resources or even as hostile or dangerous).

As touched on in Remark 9.4.7, this requires our models and algorithms to at least *make sense* in a world of countably infinitely many points.[9] In fact, arguably even 'countably large' is not quite right. The natural cardinality for semitopologies may be *uncountable*, since network latency means that we cannot even enumerate the network: no matter how carefully we count, we could always in principle discover new participants who have joined in the past (but we just had not heard of them yet).

This motivates us to consider algebraic conditions on a semiframe $(\mathsf{X}, \leq, *)$ that mimic some of the properties of open sets of finite point-set semitopologies, without necessarily insisting on finiteness itself. We saw concrete examples of this in the *chain-completeness* properties on point-set semitopologies from Definition 9.1.2, but for future work we could consider more, such as:

1. We could insist that a $\leq$-descending chain of non-$\bot_{\mathsf{X}}$ elements in X have a non-$\bot_{\mathsf{X}}$ greatest lower bound in X.
2. We could insist that a $\leq$-descending chain of elements strictly $\leq$-greater than some $x \in \mathsf{X}$ have a greatest lower bound that is strictly $\leq$-greater than x.
3. We could insist that if $(x_i \mid i \geq 0)$ and $(y_i \mid i \geq 0)$ are two $\leq$-descending chains of elements, and $x_i * y_i$ for every $i \geq 0$ — in words: x_i is compatible with y_i — then the greatest lower bounds of the two chains are compatible.

The reader may notice how these conditions are reminiscent of compactness conditions from topology: e.g. a metric space is compact if and only if every descending chain of open sets has a nonempty intersection. This is no coincidence, since one of the uses of compactness in topology is precisely to recover some of the characteristics of finite topologies.

Considering semiframes (and indeed semitopologies) with compactness/finiteness flavoured conditions, is future work.

REMARK 23.3.7 (Generalising $*$). In Remark 12.2.2 we mentioned that we can think of semitopologies not as *'topologies without intersections'* so much as *'topologies with a generalised intersection'*. We have studied a relation called $\between$ (for point-set semitopologies) and $*$ (for semiframes), which intuitively measure whether two elements intersect.

But really, this is just a notion of generalised meet. We would take $(\mathsf{X}, \leq)$ and $(\mathsf{X}', \leq')$ to be complete join-semilattices and the generalised meet $* : (\mathsf{X} \times \mathsf{X}) \to \mathsf{X}'$ is any commutative distributive map. Or, we could generalise in a different direction and consider (for example)

[9] This is no different than a programming language including a datatype of arbitrary precision integers: the program must eventually terminate, but because we do not know when, we need the *idea* of an infinity in the language.

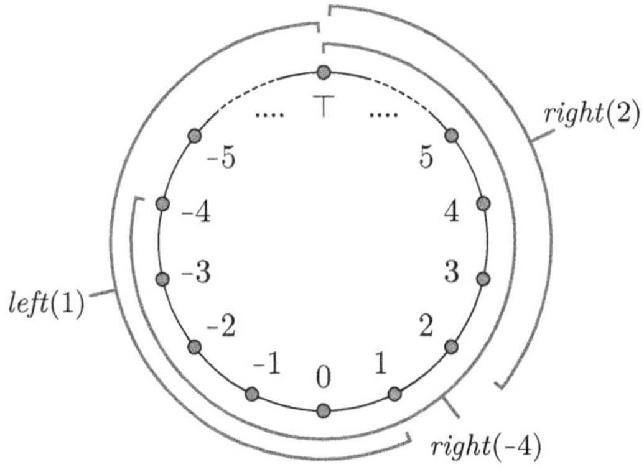

Figure 23.1: A point with two paths to it (Remark 23.3.8)

cocomplete symmetric monoidal categories: $*$ becomes the (symmetric) monoid action. These objects could be studied in their own right, or we could try to translate their structure back to sets, to see what point-set generalisations of semitopologies result.

REMARK 23.3.8 (Homotopy and convergence). We have not looked in any detail at notions of *path* and *convergence* in semitopologies and semiframes. We can give a flavour of why this might be new and different relative to the notions from topologies.

Let $(\mathsf{P}, \mathsf{Open})$ be the semitopology defined as follows, and illustrated in Figure 23.1:

- $\mathsf{P} = \mathbb{Z} \cup \{\top\}$ is thought of intuitively as a circle with 0 at the bottom and $\top$ at the top.
- For each $x \in \mathbb{Z}$ define

  $$left(x) = \{\top\} \cup \{y \in \mathbb{Z} \mid y \leq x\} \quad \text{and} \quad right(x) = \{\top\} \cup \{y \in \mathbb{Z} \mid x \leq y\}$$

  and give P the semitopology Open generated by the sets $left(x)$ and $right(x)$ for all $x \in \mathbb{Z}$.

Intuitively:

- $left(x)$ is a circle segment starting at x (x may be negative) and headed leftwards towards $\top$.
- $right(x)$ is a circle segment starting at x (x may be negative) and headed rightwards towards $\top$.

We can converge on $\top$ from the left (via the negative numbers), and from the right (via the positive numbers) — however, the descending sequences of open neighbourhoods intersect only at $\top$ and do not have a common open intersection. This is not behaviour that would be possible in a topology. This example is really just dressing up one of our earliest observations, from Lemma 2.1.2: in semitopologies a point can have more than one minimal open neighbourhood, and the example illustrates that intuitively each of these minimal open neighbourhoods can be thought of as a distinct direction by which we can converge on the point. Developing this part of the theory is future work.

REMARK 23.3.9 (Constructive mathematics). We have not considered what semiframes would look like in a constructive setting. Much of the interest in frames and locales (versus point-set topologies) comes from working in a constructive setting; e.g. in the topos of sheaves over a base space, locales give a good fibrewise topology of bundles. To what extent similar structures might be built using semiframes, or what other structures might emerge instead, are currently entirely open questions.

It also remains to think about what a suitable Curry-Howard correspondence would be, based on residuated semilattices with a compatibility relation $*$ (instead of a conjunction $\wedge$).

REMARK 23.3.10 (Algebraic semitopology). We mentioned in Subsection 23.2 that our use of semitopology is not directly related to algebraic topology applied to solvability of distributed computing tasks. These are distinct topics: they share a word in their name, but they are no more equal than a Great Dane and a Danish pastry.

But, it is an interesting question what algebraic *semi*topology might look like. Or to put this another way: *What is the geometry of semitopological spaces?* We would very much like to know.

REMARK 23.3.11 (More values). In Section 18 we use a three-valued logic **3** with truth-values corresponding to 'true', 'false', and 'both'. However, in real systems a participant might also wish to return 'neither', or 'don't know', or 'please wait', or even levels confidence of the above. And why should we even restrict ourselves to that? Why not consider an arbitrary lattice, representing all valid combinations of data and/or knowledge? **3** is a good place to start because it is minimal and still very expressive, but when we investigate practical applications it may well turn out to be the case that a richer domain of values is useful. The maths suggests no fundamental obstacles to doing this.

In a related theme, in Definition 2.1.3 we define a *value assignment* $f : \mathsf{P} \to \mathsf{Val}$ to be a function from a semitopology to a codomain Val that is given the discrete semitopology. This is a legitimate starting point, but of course we should consider more general codomains. This could include an arbitrary semitopology on the right (for greatest generality), but even for our intended special case of consensus it would be interesting to try to endow Val with a semilattice structure (or something like it), at least, e.g. to model merging of distinct updates to a ledger.[10] We can easily generate a (semi)topology from a semilattice by taking points to be elements of the lattice and open sets to be up-closed sets, and this would be a natural generalisation of the discrete semitopologies we have used so far.

REMARK 23.3.12. In Proposition 5.3.2 and Remark 5.3.3 we studied how consensus, once achieved on an open set O, propagates to its closure $|O|$. But this is just half of the problem of consensus: it remains to consider (within our semitopological framework) what it is to attain consensus on some open set in the first place.

That is: suppose $(\mathsf{P}, \mathsf{Open})$ is a semitopology and $f : \mathsf{P} \to \mathsf{Var}$ is a value assignment. Then what does it mean, in maths and algorithms, to find a value assignment $f' : \mathsf{P} \to \mathsf{Var}$ that is 'close' to f but is continuous on some open set O? We have constructed a detailed theory of what it would then be to extend f' to an f'' that continuously extends f' to regular points; but we have not yet looked at how to build the f'. We speculate that unauthenticated Byzantine consensus algorithms (like Information-Theoretic HotStuff [AS20]) can be understood in our setting; unlike

[10]We write 'something like it' because we might also consider, or consider excluding, possibly conflicting updates.

authenticated algorithms, unauthenticated algorithms do not rely on one participant being able to prove to another, by exhibiting signed messages, that a quorum has acted in a certain way.

23.4 Open problems

In addition to the future work mentioned above, we note some technical questions that have arisen in this text which we have not yet had time to answer:

1. In view of Theorem 6.5.8, does there exist a space such that every point is quasiregular and unconflicted, but no point is hypertransitive (Definition 6.5.4)?

2. As per Remark 6.5.9, it remains an open problem to check whether there is some natural property X' such that regular = indirectly regular + X' (Definition 9.3.2).

3. In Remark 9.1.3 we draw an analogy between the chain-completeness condition on semitopologies and the Alexandrov condition (closure under arbitrary intersections) on topologies. It is known that Alexandrov spaces are uniquely characterised by their specialisation preorder; it would be interesting to check whether the analogy extends and some characterisation in a similar spirit can be found for chain-complete semitopologies.

4. As per Remark 9.5.13, we have no topological characterisation of witness semitopologies. That is, it is an open problem to abstractly characterise the class of semitopologies that can be generated from a witness function.

5. Also as per Remark 9.5.13, it remains to investigate conditions on witness functions to guarantee good behaviour, such as quasiregularity, weak regularity, or regularity, of points — or even the *existence* of some (quasi/weakly) regular point.

6. In Example 14.3.4(4) we prove that $\mathbb{R}$ is sober as a semitopological space. This seems innocuous enough, since $\mathbb{R}$ is Hausdorff and it is known that Hausdorff topologies are sober — however, Lemma 14.3.10(1) shows that Hausdorff *semi*topologies are *not* necessarily sober. Thus $\mathbb{R}$ is not a sober semitopology just because it is Hausdorff; it has other properties that make this work. Subsection 14.3 contains some results, but developing a more rounded understanding of what makes a semitopology sober, is future work.

7. As per the discussion in Remark 22.6.18, how does the property that $K(p)$ is a minimal closed neighbourhood (though not necessarily of p) fit in to the other well-behavedness properties we have considered, such as (quasi/weak/indirect) regularity, being unconflicted, and being hypertransitive?

23.5 Final comments

Distributed systems are an old idea; think: telephone exchanges, satellite systems — and of course the generals in an army, as per the classic paper [LSP82]. However, it is not hyperbole to note that the use and importance of distribution and decentralisation has expanded exponentially in the last twenty years. This has provoked an explosion of new algorithms and new mathematics with which to understand them. This includes looking into generalisations of the notion of consensus and quorum [ACTZ24, SWRM21, CLZ23, LCL23, BKK22, GPG18, LLM+19, LGM19, FHNS22, LL23], and new systems [Mac18, Maz15, SWRM21, SYB14].

We have combined the research on consensus with a long mathematical tradition of studying topologies, algebras, and the dualities between them (references in Remark 15.4.3 and at the start of Subsection 23.2). We do this by applying a classic technique: *topologise, then dualise*. And, we think is fair to say that it works: we get new and interesting structures, a duality result, and a new logic. As noted in Subsections 23.3 and 23.4, there is no shortage of scope for future research.

List of Figures

2.1 An example of a point with two minimal open neighbourhoods (Lemma 2.1.2) . . 16
2.2 Two nonidentical semitopologies (Remark 2.1.7) 19

3.1 Examples of topens (Example 3.3.3) 30
3.2 Two counterexamples for (strong) transitivity 37

4.1 Illustration of Example 4.4.1(3&4) 52

5.1 Examples of open neighbourhoods (Remarks 5.4.6 and 13.1.3) 62
5.2 An unconflicted, irregular space (Proposition 6.2.1) in which every minimal closed neighbourhood has a non-transitive open interior (Example 5.6.8) 68
5.3 The Sierpiński space Sk (Example 5.8.4) 74

6.1 Examples of boundary points (Example 6.3.3). 79
6.2 A weakly regular, conflicted space (Proposition 6.4.11); the opens are the down-closed sets 83
6.3 Two counterexamples 85

7.1 Example 7.3.5: $|*| \subsetneq *_{\between} \subsetneq \{0,1,*\}$ 96

9.1 Lemma 9.3.9(2): a point $*$ that is quasiregular but not indirectly regular 119

10.1 Illustration of Example 10.1.3(3&4) 129
10.2 The semitopologies in Example 10.3.3 135

11.1 A weakly dense subset is not enough for uniqueness (Remark 11.5.6(3)) 149

13.1 A semiframe with no abstract points (Lemma 13.2.7) 168

14.1 Two counterexamples for sobriety 179
14.2 Example soberifications (Remark 14.3.12) 180

16.1 Strongly compatible filter that contains no transitive element 196

17.1 $O \leftrightarrow O'$ but $O \not\leq O'$ (Lemma 17.1.7(2)) 209
17.2 Semitopologies with isomorphic intersection graphs (Remark 17.1.16) 212

18.1 The semitopology **3** / the set of truth values of three-valued logic 220
18.2 Truth-tables for three-valued logic (Definition 18.2.1) 223
18.3 Some truth-table equivalences (Lemma 18.2.2) 223
18.4 Predicate syntax 226
18.5 Denotation of a predicate ϕ with respect to a valuation f 227
18.6 Derivable tag-sequents 233

22.1 Correspondence between extremal valuations, maximal disjoint pairs of open sets, and regular open sets 265
22.2 Summary of antiseparation properties 270
22.3 Summary of Lemmas 22.6.12 & 22.6.13 270
22.4 1 and 2 are hypertwined and hyperdefinite, but they are not even quasiregular (Example 22.6.8) 272

23.1 A point with two paths to it (Remark 23.3.8) 287

A word on the figures: As per a comment in Figure 3.1, we may omit open sets that are unions of open sets that are already illustrated. In particular, we usually do not include the universal open set, where this is a union of open sets that are already illustrated. Thus in Figure 3.1 we *do* draw the universal open set in the top left diagram, which is not a union of $\{0\}$ and $\{2\}$; but we *do not* draw it in the top right diagram, since $\{0,1,2\} = \{0,1\} \cup \{1,2\}$.

Bibliography

[AAZ11] Ofer Arieli, Arnon Avron, and Anna Zamansky, *Ideal paraconsistent logics*, Studia Logica **99** (2011), no. 1-3, 31–60.

[ACTZ24] Orestis Alpos, Christian Cachin, Björn Tackmann, and Luca Zanolini, *Asymmetric distributed trust*, Distributed Computing (2024), Available online at https://doi.org/10.1007/s00446-024-00469-1.

[AS20] Ittai Abraham and Gilad Stern, *Information Theoretic HotStuff*, 24th International Conference on Principles of Distributed Systems, OPODIS 2020, December 14-16, 2020, Strasbourg, France (Virtual Conference) (Quentin Bramas, Rotem Oshman, and Paolo Romano, eds.), LIPIcs, vol. 184, Schloss Dagstuhl - Leibniz-Zentrum für Informatik, 2020, pp. 11:1–11:16.

[BG93a] Elizabeth Borowsky and Eli Gafni, *Generalized FLP Impossibility Result for T-resilient Asynchronous Computations*, Proceedings of the Twenty-fifth Annual ACM Symposium on Theory of Computing (New York, NY, USA), STOC '93, ACM, 1993, pp. 91–100.

[BG93b] Carolyn Brown and Doug Gurr, *A representation theorem for quantales*, Journal of Pure and Applied Algebra **85** (1993), no. 1, 27–42.

[BKK22] João Paulo Bezerra, Petr Kuznetsov, and Alice Koroleva, *Relaxed reliable broadcast for decentralized trust*, Networked Systems (Mohammed-Amine Koulali and Mira Mezini, eds.), Lecture Notes in Computer Science, Springer International Publishing, 2022, pp. 104–118.

[Bou98] Nicolas Bourbaki, *General topology: Chapters 1-4*, Elements of mathematics, Springer, 1998.

[Bra87] Gabriel Bracha, *Asynchronous Byzantine agreement protocols*, Information and Computation **75** (1987), no. 2, 130–143.

[Bre84] D. A. Bredhikin, *A representation theorem for semilattices*, Proceedings of the American Mathematical Society **90** (1984), no. 2, 219–220.

[BRV01] Patrick Blackburn, Maarten de Rijke, and Yde Venema, *Modal logic*, Cambridge Tracts in Theoretical Computer Science, Cambridge University Press, 2001.

[Cam78] Paul J Campbell, *The origin of "Zorn's Lemma"*, Historia Mathematica **5** (1978), no. 1, 77–89.

[Car11] Olivia Caramello, *A topos-theoretic approach to Stone-type dualities*, 2011, https://doi.org/10.48550/arXiv.1103.3493.

[CC20] Francesco Ciraulo and Michele Contente, *Overlap Algebras: a Constructive Look at Complete Boolean Algebras*, Logical Methods in Computer Science **16** (2020).

[Chu36] Alonzo Church, *An unsolvable problem of elementary number theory*, American Journal of Mathematics **58** (1936), no. 2, 345–363.

[CL02] Miguel Castro and Barbara Liskov, *Practical Byzantine fault tolerance and proactive recovery*, ACM Transactions on Computer Systems (TOCS) **20** (2002), no. 4, 398–461.

[CLZ23] Christian Cachin, Giuliano Losa, and Luca Zanolini, *Quorum systems in permissionless networks*, 26th International Conference on Principles of Distributed Systems (OPODIS 2022) (Eshcar Hillel, Roberto Palmieri, and Etienne Rivière, eds.), Leibniz International Proceedings in Informatics (LIPIcs), vol. 253, Schloss Dagstuhl – Leibniz-Zentrum für Informatik, 2023, ISSN: 1868-8969, pp. 17:1–17:22.

[DG84] William F. Dowling and Jean H. Gallier, *Linear-time algorithms for testing the satisfiability of propositional horn formulae*, The Journal of Logic Programming **1** (1984), no. 3, 267–284.

[DP02] B. A. Davey and Hilary A. Priestley, *Introduction to lattices and order*, 2 ed., Cambridge University Press, 2002.

[Edg20] Dorothy Edgington, *Indicative conditionals*, The Stanford Encyclopedia of Philosophy (Edward N. Zalta, ed.), Metaphysics research lab, CSLI, Stanford University, fall 2020 ed., 2020, Available online at https://plato.stanford.edu/archives/fall2020/entries/conditionals/ (permalink: https://archive.ph/FzVOF).

[ELW] Jonathan Emerson, Mark Lezama, and Eric W. Weisstein, *Transfinite induction*, Available online at https://mathworld.wolfram.com/TransfiniteInduction.html.

[Eng89] Ryszard Engelking, *General topology*, Sigma Series in Pure Mathematics, Heldermann Verlag, Berlin, 1989.

[Erd18] John M. Erdman, *A problems based course in advanced calculus*, Pure and Applied Undergraduate Texts, no. 32, American Mathematical Society, 2018, Available online at https://web.archive.org/web/20221128144749/https://web.pdx.edu/~erdman/PTAC/problemtext_pdf.pdf.

[Ern09] Marcel Erné, *Closure*, Beyond topology, vol. 486, American Mathematical Society, 2009, pp. 163–238.

[Fey14] Mark Fey, *A straightforward proof of Arrow's theorem*, Economics Bulletin **34** (2014), no. 3, 1792–1797.

[FHNS22] Martin Florian, Sebastian Henningsen, Charmaine Ndolo, and Björn Scheuermann, *The Sum of Its Parts: Analysis of Federated Byzantine Agreement Systems*, Distributed Computing **35** (2022), no. 5, 399–417 (en).

[FM88] Paul Feldman and Silvio Micali, *Optimal algorithms for Byzantine agreement*, Proceedings of the twentieth annual ACM symposium on Theory of computing, STOC '88, Association for Computing Machinery, 1 1988, pp. 148–161.

[Gab26] Murdoch J. Gabbay, *Decentralised collaborative action: cryptoeconomics in space*, Cryptoeconomic Theory: Blockchain-AI Integration (M. Swan, S. Takagi, and F. Witte, eds.), World Scientific, London, 2026, to appear; available at arXiv: 2504.12493 (permalink).

[Goo14] L. M. Goodman, *Tezos – a self-amending crypto-ledger (white paper)*, September 2014.

[GPG18] Álvaro García-Pérez and Alexey Gotsman, *Federated byzantine quorum systems*, 22nd International Conference on Principles of Distributed Systems (OPODIS 2018), Schloss Dagstuhl-Leibniz-Zentrum fuer Informatik, 2018.

[Gut22] Valentin Gutev, *Simultaneous extension of continuous and uniformly continuous functions*, Studia Mathematica **265** (2022), no. 2, 121–139.

[GvG24] Mai Gehrke and Sam van Gool, *Topological duality for distributive lattices: Theory and applications*, Cambridge Tracts in Theoretical Computer Science, Cambridge University Press, 2024, Available online at https://arxiv.org/abs/2203.03286.

[HÖ3] Reiner Hähnle, *Complexity of many-valued logics*, Beyond two: theory and applications of multiple-valued logic (Melvin Fitting, Ewa Orłowska, and Janusz Kacprzyk, eds.), Studies in fuzziness and soft computing, Springer, January 2003.

[Han11] Zuzana Haniková, *Computational complexity of propositional fuzzy logics*, vol. 2, pp. 793–851, College Publications, 01 2011.

[HKR13] Maurice Herlihy, Dmitry Kozlov, and Sergio Rajsbaum, *Distributed computing through combinatorial topology*, Morgan Kaufmann, 2013.

[HM00] Martin Hirt and Ueli Maurer, *Player Simulation and General Adversary Structures in Perfect Multiparty Computation*, Journal of Cryptology **13** (2000), no. 1, 31–60.

[HS93] Maurice Herlihy and Nir Shavit, *The asynchronous computability theorem for t-resilient tasks*, Proceedings of the twenty-fifth annual ACM symposium on Theory of computing, 1993, pp. 111–120.

[Hul03] Klaus Hulek, *Elementary algebraic geometry*, Student Mathematical Library, vol. 20, American Mathematical Society, 2003.

[Jec73] Thomas Jech, *The axiom of choice*, North-Holland, 1973, ISBN 0444104844.

[Joh86] Peter T. Johnstone, *Stone spaces*, vol. 3, Cambridge University Press, 1986.

[Joh87] ______, *Notes on logic and set theory*, Cambridge University Press, 1987.

[JT04] Tyler Jarvis and James Tanton, *The hairy ball theorem via Sperner's lemma*, The American Mathematical Monthly **111** (2004), no. 7, 599–603.

[Kop89] Sabine Koppelberg, *Handbook of boolean algebras, volume 1*, North-Holland, 1989, Series editors Robert Bonnet and James Donald Monk.

[Lam98] Leslie Lamport, *The part-time parliament*, ACM Transactions on Computer Systems **16** (1998), no. 2, 133–169.

[LCL23] Xiao Li, Eric Chan, and Mohsen Lesani, *Quorum Subsumption for Heterogeneous Quorum Systems*, August 2023, arXiv:2304.04979 [cs].

[LGM19] Giuliano Losa, Eli Gafni, and David Mazières, *Stellar consensus by instantiation*, 33rd International Symposium on Distributed Computing (DISC 2019) (Dagstuhl, Germany) (Jukka Suomela, ed.), Leibniz International Proceedings in Informatics (LIPIcs), vol. 146, Schloss Dagstuhl–Leibniz-Zentrum fuer Informatik, 2019, pp. 27:1–27:15.

[Lif08] Vladimir Lifschitz, *What is answer set programming?*, Proceedings of the 23rd National Conference on Artificial Intelligence - Volume 3, AAAI'08, AAAI Press, 2008, p. 1594–1597.

[Lif19] Vladimir Lifschitz, *Answer set programming*, Springer, 2019.

[LL23] Xiao Li and Mohsen Lesani, *Open Heterogeneous Quorum Systems*, April 2023, arXiv:2304.02156 [cs].

[LLM+19] Marta Lokhava, Giuliano Losa, David Mazières, Graydon Hoare, Nicolas Barry, Eli Gafni, Jonathan Jove, Rafał Malinowsky, and Jed McCaleb, *Fast and secure global payments with Stellar*, Proceedings of the 27th ACM Symposium on Operating Systems Principles (New York, NY, USA), SOSP '19, Association for Computing Machinery, 2019, p. 80–96.

[LSP82] Leslie Lamport, Robert Shostak, and Marshall Pease, *The Byzantine Generals Problem*, ACM Trans. Program. Lang. Syst. **4** (1982), no. 3, 382–401.

[Mac71] Saunders Mac Lane, *Categories for the working mathematician*, Graduate Texts in Mathematics, vol. 5, Springer, 1971.

[Mac18] Ethan MacBrough, *Cobalt: BFT Governance in Open Networks* (en), Available online at https://doi.org/10.48550/arXiv.1802.07240.

[Maz15] David Mazières, *The Stellar consensus protocol: a federated model for Internet-level consensus*, Tech. report, Stellar Development Foundation, 2015, https://www.stellar.org/papers/stellar-consensus-protocol.pdf (permalink: https://web.archive.org/web/20240629063518/https://stellar.org/learn/stellar-consensus-protocol).

[MM92] Saunders Mac Lane and Ieke Moerdijk, *Sheaves in geometry and logic: A first introduction to topos theory*, Universitext, Springer, 1992.

[MNPS91] Dale Miller, Gopalan Nadathur, Frank Pfenning, and Andre Scedrov, *Uniform proofs as a foundation for logic programming*, Annals of Pure and Applied Logic **51** (1991), no. 1, 125–157.

[MR98] Dahlia Malkhi and Michael Reiter, *Byzantine quorum systems*, Distributed computing **11** (1998), no. 4, 203–213.

[NW94] Moni Naor and Avishai Wool, *The load, capacity and availability of quorum systems*, Proceedings 35th Annual Symposium on Foundations of Computer Science, vol. 27, IEEE, 1994, pp. 214–225.

[PN01] Valeriu Popa and Takashi Noiri, *On the definitions of some generalized forms of continuity under minimal conditions*, Memoirs of the Faculty of Science. Series A. Mathematics **22** (2001), 9–18.

[Poo92] Bjorn Poonen, *Union-closed families*, Journal of Combinatorial Theory, Series A **59** (1992), no. 2, 253–268.

[PP12] Jorge Picado and Aleš Pultr, *Frames and locales: Topology without points*, 1 ed., Frontiers in Mathematics, Birkhäuser, 2012.

[PP21] ———, *Separation in point-free topology*, 1 ed., Birkhäuser, 2021.

[Rik62] William H. Riker, *The theory of political coalitions*, Yale University Press, 1962.

[Riv85] Ivan Rival (ed.), *Graphs and order. the role of graphs in the theory of ordered sets and its applications*, Proceedings of NATO ASI, Series C (ASIC), vol. 147, Banff/Canada, 1985.

[Ros90] Kimmo I. Rosenthal, *Quantales and their applications*, Pitman Research Notes in Mathematics, no. 234, Longman Scientific & Technical, UK, 1990.

[Rus03] Bertrand Russell, *The principles of mathematics*, Cambridge University Press, 1903, http://fair-use.org/bertrand-russell/the-principles-of-mathematics/s37 (permalink: https://web.archive.org/web/20231015061016/http://fair-use.org/bertrand-russell/the-principles-of-mathematics/s37).

[Rya10] Johnny Ryan, *A history of the internet and the digital future*, Reaktion Books, 2010.

[Sto36] Marshall H. Stone, *The theory of representation for boolean algebras*, Transactions of the American Mathematical Society **40** (1936), no. 1, 37–111.

[SWRM21] Isaac Sheff, Xinwen Wang, Robbert van Renesse, and Andrew C. Myers, *Heterogeneous Paxos*, 24th International Conference on Principles of Distributed Systems (OPODIS 2020) (Dagstuhl, Germany) (Quentin Bramas, Rotem Oshman, and Paolo Romano, eds.), Leibniz International Proceedings in Informatics (LIPIcs), vol. 184, Schloss Dagstuhl–Leibniz-Zentrum für Informatik, 2021, ISSN: 1868-8969, pp. 5:1–5:17.

[SYB14] David Schwartz, Noah Youngs, and Arthur Britto, *The Ripple Protocol Consensus Algorithm*, Ripple Labs Inc White Paper **5** (2014), no. 8, 151.

[SZ93] Michael Saks and Fotios Zaharoglou, *Wait-free k-set agreement is impossible: The topology of public knowledge*, Proceedings of the twenty-fifth annual ACM symposium on Theory of computing, 1993, pp. 101–110.

[Szá07] Árpád Száz, *Minimal structures, generalized topologies, and ascending systems should not be studied without generalized uniformities*, Filomat (Nis) **21** (2007), 87–97.

[Tec11] Ars Technica, *How the atom bomb helped give birth to the internet*, https://arstechnica.com/tech-policy/2011/02/how-the-atom-bomb-gave-birth-to-the-internet/, 2 2011, Permalink: http://web.archive.org/web/20240622221756/https://arstechnica.com/tech-policy/2011/02/how-the-atom-bomb-gave-birth-to-the-internet/.

[Vic89] Steven Vickers, *Topology via logic*, Cambridge University Press, USA, 1989.

[Vid21] Amanda Vidal, *On transitive modal many-valued logics*, Fuzzy Sets and Systems **407** (2021), 97–114, Knowledge Representation and Logics.

[Wes11] Dag Westerståhl, *Generalized quantifiers*, The Stanford Encyclopedia of Philosophy (Edward N. Zalta, ed.), Metaphysics research lab, CSLI, Stanford University, summer 2011 ed., 2011, Available online at https://plato.stanford.edu/archives/sum2011/entries/generalized-quantifiers/ (permalink: https://archive.ph/YlP09).

[Wik24a] Wikipedia, *Equivalence of categories*, http://en.wikipedia.org/w/index.php?title=Equivalence%20of%20categories&oldid=1227082771, 2024, Permalink: https://web.archive.org/web/20230316075107/https://en.wikipedia.org/wiki/Equivalence_of_categories.

[Wik24b] ———, *Ideal (order theory); maximal ideals*, http://en.wikipedia.org/w/index.php?title=Ideal%20(order%20theory)&oldid=1200832357, 2024, Permalink: https://web.archive.org/web/20230724184908/https://en.wikipedia.org/wiki/Ideal_(order_theory)#Maximal_ideals.

[Wik24c] ———, *Intersection graph*, http://en.wikipedia.org/w/index.php?title=Intersection%20graph&oldid=1205563104, 2024, Permalink: https://web.archive.org/web/20230324152732/https://en.wikipedia.org/wiki/Intersection_graph.

[Wik24d] ———, *Paraconsistent logic; an ideal three-valued paraconsistent logic*, https://en.wikipedia.org/w/index.php?title=Paraconsistent%20logic&oldid=1228606179#An_ideal_three-valued_paraconsistent_logic, 2024, Permalink: http://web.archive.org/web/20231004082704/https://en.wikipedia.org/wiki/Paraconsistent_logic#An_ideal_three-valued_paraconsistent_logic.

[Wik24e] ———, *Separation axiom; main definitions*, https://en.wikipedia.org/w/index.php?title=Separation%20axiom&oldid=1230514922#Main_definitions, 2024, Permalink: https://web.archive.org/web/20221103233631/https://en.wikipedia.org/wiki/Separation_axiom#Main_definitions.

[Wik24f] ———, *Sierpiński space; categorical description*, http://en.wikipedia.org/w/index.php?title=Sierpi%C5%84ski%20space&oldid=1228037897, 2024, Permalink: https://web.archive.org/web/20231015150722/https://en.wikipedia.org/wiki/Sierpi%C5%84ski_space#Categorical_description.

[Wil70] Stephen Willard, *General topology*, Addison-Wesley, 1970, Reprinted by Dover Publications.

[WM16] Stefan Wintein and Reinhard Muskens, *A Gentzen Calculus for Nothing but the Truth*, Journal of Philosophical Logic **45** (2016), 451–465.

[WR10] Alfred North Whitehead and Bertrand Russell, *Principia mathematica*, vol. 1, Cambridge University Press, 1910.

Comments on citations and typesetting

I have typeset URL pointers in full (even though this can be a bit long) so that readers of the print version can see them. Of course in the pdf, these links are clickable. Where online sources are relevant, I provide links, and also permalinks to archived versions of the webpage.

Where I need to reference standard definitions or lemmas, I cite standard textbooks where possible; but Wikipedia and the Stanford Encyclopedia of Philosophy are also very good, and being online they are easily accessible, so for the reader's convenience I cite them too (again, with permalinks).

Acknowledgements

I would like to thank three anonymous referees for generously donating their time and careful attention to this material. Thanks to Giuliano Losa and the Stellar Development Foundation for their generous support and funding. Giuliano was instrumental in explaining the issues, reviewing material, and guiding me to interesting problems; thank you for the many interesting discussions and insightful comments. Thanks to Luca Zanolini and to Michael Gabbay for their editorial input. Thanks to the mathematics StackExchange community for helping to find citations for a simple but obscure lemma (see Remark 5.7.9), and to Jobst Heitzig and Steve Vickers for their interest in these ideas and for asking good questions. Thanks to Jane Spurr for editorial work.

This work would not have been possible in its current form without the support of these people and others. All remaining errors are mine.

This text is a revised version dated July 2025, which corrects some minor errors and omissions in the original version published August 2024.

Index

$(\mathsf{P}, \mathsf{Open})$-extremal valuation, 260
$O \leq O'$ (node preorder on a graph), 208
$P_1 \times P_2$ (square), 91
T_0 space, 171
T_2 space (Hausdorff condition), 35
$X \between Y$ (intersection of sets), 25
$X \between_Y Z$ (X and Z intersect in Y), 37
$X \ltimes Y$ (subintersection of sets), 212
$\Phi \vDash \Psi$ (sequent), 108
$\Phi \vdash \Psi$ (sequent), 228
Σ (tag-sequent), 233
$\mathbf{3} = \{\mathbf{t}, \mathbf{b}, \mathbf{f}\}$ (semitopology), 220
F^* (compatibility system of a set), 194
$\langle X \rangle_P$ (atoms in P intersecting X), 143
$f \leq f'$ (partial order on valuations), 260
$f \vDash \phi$ (validity of a predicate in a valuation), 228
$f^{\vDash}$ (designated set of a valuation), 229
$f^{\vDash\neg}$ (neg-designated set of a valuation), 229
$K(p)$ (community of a point), 43
$k(F)$ (abstract community of a set), 198
$interior(P)$ (open interior), 43
$p_{\between}$ (points intertwined with a point p), 33
$\between$-complete semitopology, 80
$\mathsf{Closed}(wf)$, 101
$\overset{\mathsf{x}}{\vDash}$, 260
$nbhd^c(p)$ (closed neighbourhood system of a point), 61
$\mathsf{P} \vDash (\Phi \vdash \Psi)$ (sequent validity), 228
$\mathsf{Open}(wf)$, 101
π_1 (first projection), 91
π_2 (second projection), 91
$\prec$-fixedpoint, 104
One (singleton semitopology), 248
$\mathsf{St}(\mathbf{X}, \leq, *)$ (semitopology of abstract points), 170
$\mathsf{St}\,\mathsf{Fr}(\mathsf{P}, \mathsf{Open})$ (soberification), 184
$\supset$, $\equiv$, $\wedge$, $\vee$, $\Rightarrow$, $\Leftrightarrow$, T, TB, B (connectives and modalities in $\mathbf{3}$), 223
$\{tb, ff, fb, tt\}$ (tags), 233
$p \between p'$ (two intertwined points), 33
$p \between^{=} p'$ (consensus equivalent points), 271
$p \between^{\mathsf{T}} p'$ (hypertwined points), 271
$p \between^{*} p'$ (transitively intertwined points), 276
$p \between_{wf} p'$ (logical predicate), 245
$p \leq_K p'$ (community preorder on points), 81
$p \leq_{\between} p'$ (intertwined preorder on points), 80
$p \mathrel{\mathring{=}} p'$ (topologically indistinguishable points), 171
$x * x'$ (compatibility relation on elements), 157
$x \ltimes y$ (subintersection of semiframe elements), 212
x^* (compatibility system of a semiframe element), 193
(wf-)closed set, 101
(wf-)open set, 101
(open) atom, 124
(open) cover of p, 122
(open) interior of P, 43

(open) neighbourhood of p, 122
(partial) continuous extension of f, 149
(semitopological) subspace, 36
2-satisfiable (propositional theory), 257
3-satisfiable (propositional theory), 257
3Horn clause, 257
3Horn clause theory, 258

abstract community (of a set: $k(F)$), 198
abstract point (completely prime semifilter), 163
actionable coalitions, 22
Alexandrov topology, 113
all-but-one semitopology, 17
ambivalent (valuation at p), 259
ambivalent element ($\mathbf{b} \in \mathbf{3}$), 259
answer set (for $\mathsf{Ax}(\mathbf{P}, wf)$), 109
ascending chain of sets, 31
atomic proposition symbols $\bar{\mathbf{P}}$, 108
axiom $\bar{w}f(p)$, 108
axioms arising from wf, 237

blocking set for p, 101
blocks (P blocks p), 101
boundary of P, 78
boxed negative literal $\mathsf{T}\neg p$, 257
boxed positive literal $\mathsf{T}p$, 257

category of semiframes $\mathsf{SemiFrame}$, 185
category of semitopologies $\mathsf{SemiTop}$, 183
category of sober semitopologies Sober, 183
category of spatial semiframes $\mathsf{Spatial}$, 185
chain of sets, 31
chain-complete semitopology, 113
characteristic sets of a valuation $char(f)$, 221
clique of sets, 32
clopen set, 56
closed neighbourhood, 60
closed neighbourhood of $p \in \mathbf{P}$, 60
closed neighbourhood system $nbhd^c(p)$, 61
closed predicate, 226
closed set, 56
closed sets Closed, 56
closure of P, 55
cluster point, 62
community of p ($K(p)$), 43
community preorder $p \leq_K p'$, 81
compatibility (of x with Y: $x * Y$), 192
compatibility (of morphism of semiframes), 185
compatibility relation $* \subseteq \mathbf{X} \times \mathbf{X}$, 157
compatibility system (of a set; F^*), 194
compatibility system (of an element: x^*), 193
compatible elements, 157
compatible subset of a semiframe, 163
complement of X (X^c), 212
complete join-semilattice, 156
completely prime subset of a semiframe, 163
componentwise inclusion ordering on pairs of sets, 263
confident at $p \in \mathbf{P}$, 149
conflicted point, 75
conflicted/unconflicted set, 75
conjunctive normal form, 253
connected (semi)topology, 28
consensus equivalent points $p \between^{=} p'$, 271
continuous extension (in semitopology), 147
continuous extension (in topology), 145
continuous function, 21
continuous function at a point, 21
continuous function on a set, 21
continuous valuation $f : \mathbf{P} \to \mathbf{3}$, 220
continuously extends, 147
counting semitopologies, 20
covers (O covers p; $O \gtrdot \{p\}$), 122

deductively closed (set of predicates), 18
definite (valuation at p), 259
definite elements ($\mathbf{t}, \mathbf{b} \in \mathbf{3}$), 259
dense set, 139
strongly dense, 139
weakly dense, 139
derivable sequent, 108
derivable tag-sequents, 233
derived logic $\mathsf{Prop}(\mathbf{P}, wf)$, 108
descending chain of sets, 31
designated set $f^{\models} \subseteq \mathbf{P}$, 229
designated values $\{\mathbf{t}, \mathbf{b}\}$, 228
deterministic Horn clause theory, 110

deterministic semitopology, 110
deterministic witnessed set, 110
disconnected (semi)topology, 28
discrete semitopology on **P**, 16
distributive law (for compatibility relation), 157

elements (of a poset), 156
empty Horn clause, 257
empty semiframe, 159
enables (P enables p), 101
enabling set for p, 101
extensionally equivalent (nodes in a graph: $O \approx O'$), 210
extremal axioms ClosedPred3(**P**), 265
extremal theory AxEx(wf), 266
extremal valuation, 260

final segment semitopology, 39, 165
final semitopology, 16
first projection $\pi_1 : \mathsf{P}_1{\times}\mathsf{P}_2{\to}\mathsf{P}_1$, 91
free variables of a predicate, 226
functorial map, 187

greatest element (in poset), 29

Hausdorff space, 35
head (of a 3Horn clause), 258
Horn clause, 256
Horn clause theory, 257
hyperconnected set, 40
hyperdefinite point, 271
hypertransitive point, 86
hypertransitive semitopology, 86
hypertwined points $p \between^{\mathsf{T}} p'$, 271

indicator function δ, 221
indirectly regular point, 116
initial semitopology, 16
intersect in Y ($X \between_Y Z$), 37
intersecting sets $X \between Y$, 25
intersection graph IntGr(**P**, Open), 207
intertwined (a set **P**), 34
intertwined (two points $p \between p'$), 33
intertwined of p ($p_{\between}$), 33
intertwined preorder $p \leq_{\between} p'$, 80
intertwined-complete, 80
intertwined-complete semitopology, 80
inverse image $f^{-1}(O')$, 21
irreducible closed set, 179
irregular point, 44
isomorphism between semiframes, 174

join (of elements in a poset), 156
join-irreducible element, 38

kernel atom of p, 127
kernel of p ($ker(p)$), 127
kissing set of P and P', 84

least element (in poset), 29
limit of f at p ($\lim_p f$), 151
limit points of P ($lim(P)$), 106
literal, 256
literal (arity + propositional atom), 254
literal (three-valued logic), 257
locally constant at a point, 22
locally constant on a set, 22

many semitopology, 17
material equivalence $p{\equiv}q$, 224
material implication $p{\supset}q$, 224
maximal disjoint sets, 263
maximal element (in poset), 29
maximal semifilter, 163
maximal topen set, 26
meet in Y ($X \between_Y Z$), 37
meet-irreducible element, 38
minimal element (in poset), 29
minimally strongly dense open subset of P, 142
minimally weakly dense in P, 142
model (for Ax(**P**, wf)), 109
monotone morphism, 156
more-than-one semitopology, 18
morphism, 156
morphism of semiframes, 185
morphism of semitopologies, 183

neg-designated set $f^{\models\neg} \subseteq \mathsf{P}$, 229
negative arity, 253
negative literal, 254, 256
negative literal (boxed $\mathsf{T}\neg p$; unboxed $\neg p$), 257

neighbourhood, 60
neighbourhood of p, 60
node preorder $\leq$ (on a graph), 208
nonempty semiframe, 159
number of semitopologies, 20

open covers of p (set of; $Covers(p)$), 123
open neighbourhood, 23
open neighbourhood system $nbhd(p)$, 61
open sets Open, 5
overlap algebra, 284

pairwise compatible subset of a semiframe, 163
points, 5
points (of NP-completeness construction), 254
pointwise application $f(X)$, 91
poset, 156
positive arity, 253
positive literal, 254, 256
positive literal (boxed $\mathsf{T}p$; unboxed p), 257
predicates over $\mathbf{P}$, 226
preimage $f^{-1}(O')$, 21
prime subset of a semiframe, 163
product semitopology $\mathbf{P}_1 \times \mathbf{P}_2$, 91
proper subintersection (in semiframes: $x \ltimes y$), 212
proper subintersection (in sets: $X \ltimes Y$), 212
properly reflexive relation, 157
propositional atom, 226
propositional atoms, 253
propositional constant, 226
pseudocomplement, 161

quantale, 284
quasiregular point, 44
quasiregular semifilter, 199

rational lines, 69
regular closed set, 70
regular open set, 70
regular point, 44
regular semifilter, 199
regular/weakly regular/quasiregular/irregular set, 44
restriction (of a function to a set), 152

satisfiable propositional theory (three-valued), 257
satisfiable propositional theory (two-valued), 257
second projection $\pi_2 : \mathbf{P}_1 \times \mathbf{P}_2 \rightarrow \mathbf{P}_1$, 91
semifilter, 163
semiframe, 159
semiframe of open sets $\mathsf{Fr}(\mathbf{P}, \mathsf{Open})$, 159
semitopological space, 5
semitopology, 5
semitopology induced by $(\mathbf{P}, \mathsf{Open})$, 36
semitopology of abstract points $\mathsf{St}(\mathbf{X}, \leq, *)$, 170
semitopology of values, 16
sequent $\Phi \vDash \Psi$, 108
sequent $\Phi \vdash \Psi$, 228
sequent validity $\mathbf{P} \models (\Phi \vdash \Psi)$, 228
sides of the square, 91
Sierpiński space, 74
singleton semiframe, 159
singleton semitopology $\mathsf{One} = \{*\}$, 248
sober semitopology, 177
soberification $\mathsf{St}\,\mathsf{Fr}(\mathbf{P}, \mathsf{Open})$, 184
spatial semiframe, 173
splits (value assignment splits a set), 28
square $P_1 \times P_2$, 91
strong compatibility (of a set), 195
strong topen, 37
strongly chain-complete semitopology, 113
strongly compatible semiframe, 195
strongly compatible semitopology, 205
strongly dense for P, 144
strongly dense in P, 139
strongly dense neighbourhood in P, 140
strongly topen set, 37
strongly transitive set, 37
subintersection (in sets: $X \ltimes Y \vee X = Y$), 212
subspace, 36
supermajority semitopology, 17

tag-sequent Σ, 233
tags *tb*, *ff*, *fb*, and *tt*, 233
theory arising from wf, 237

THREE **3** (semitopology), 220
topen set, 26
topologically distinguishable, 171
topologically indistinguishable points, 171
transitive element (in a semiframe), 192
transitive node in a graph, 209
transitive set, 26
transitively intertwined, 276
trivial semitopology, 17

unanimous at $p \in \mathsf{P}$, 149
unboxed negative literal $\neg p$, 257
unboxed positive literal p, 257
uncomputable semitopology, 99
unconflicted point, 75
unique continuous extension, 147
unit 3Horn clause, 258
unit clause, 257
unsatisfiable 3Horn clause, 258
up-closed subset of a semiframe, 163

valid (Φ in f), 228
valid (ϕ in f), 228
valid tag-sequent, 234
valuation $f : \mathsf{P} \to \mathbf{3}$, 220
value assignment $f : \mathsf{P} \to \mathsf{Val}$, 16
variable symbols Var, 226

weakly dense in P, 139
weakly regular point, 44
weakly regular semifilter, 199
witness function $wf : \mathsf{P} \to \mathcal{W}(\mathsf{P})$, 100
witness ordering $X \prec X'$, 104
witness semitopology $\mathsf{Open}(wf)$, 101
witness-set, 100
witness-set $w \in wf(p)$, 100
witnessed set (P, wf), 100
witnessing universe of P, 100

www.ingramcontent.com/pod-product-compliance
Lightning Source LLC
Chambersburg PA
CBHW060326200326
41519CB00011BA/1852
* 9 7 8 1 8 4 8 9 0 4 6 5 1 *